Mancelona Township Library
202 W State Street
P.O. Box 499
Mancelona, MI 49659

001.942 Mack, John E.,
M 1929-

Abduction.

$22.00

DATE			
APR 23 2003	5364		
APR 26 2005			
AUG 05 2005			
JUL 27 2007			

Mancelona Township Library
202 W State Street
P.O. Box 499
Mancelona, MI 49659

BAKER & TAYLOR BOOKS

ABDUCTION

ALSO BY JOHN E. MACK, M.D.

Nightmares and Human Conflict

Borderline States in Psychiatry (edited)

A Prince of Our Disorder

Vivienne: The Life and Suicide of an Adolescent Girl
(with Holly Hickler)

Last Aid
(edited with E. Chivian, S. Chivian, and R. J. Lifton)

The Development and Sustaining of Self-Esteem in Childhood
(edited with Steven Ablon)

The Alchemy of Survival: One Woman's Journey
(with Rita S. Rogers)

Human Feelings: Explorations in Affect Development and Meaning
(edited with Steven Ablon, Daniel Brown, and Edward J. Khantzian)

ABDUCTION

Human Encounters with Aliens

John E. Mack, M.D.

Charles Scribner's Sons
New York

Maxwell Macmillan Canada
Toronto

Maxwell Macmillan International
New York Oxford Singapore Sydney

In accordance with the wishes of some of the people whose experiences are described in this book, certain names and identifying details have been changed to protect their privacy.

Copyright © 1994 by John E. Mack, M.D.

All rights reserved. No part of this book may be reproduced or transmitted in any form or by any means, electronic or mechanical, including photocopying, recording, or by any information storage and retrieval system, without permission in writing from the Publisher.

Charles Scribner's Sons
Macmillan Publishing Company
866 Third Avenue
New York, NY 10022

Maxwell Macmillan Canada, Inc.
1200 Eglinton Avenue East
Suite 200
Don Mills, Ontario M3C 3N1

Macmillan Publishing Company is part of the Maxwell Communication Group of Companies.

Library of Congress Cataloging-in-Publication Data
Mack, John E., date.
Abduction: human encounters with aliens/John E. Mack.
p. cm.
Includes bibliographical references and index.
ISBN 0-684-19539-9
1. Unidentified flying objects—Sightings and encounters.
I. Title.
TL789.3.M33 1994
001.9'42—dc20 93-38116
CIP

Macmillan books are available at special discounts for bulk purchases for sales promotions, premiums, fund-raising, or educational use. For details, contact:

Special Sales Director
Macmillan Publishing Company
866 Third Avenue
New York, NY 10022

10 9 8 7 6 5 4 3 2 1

Printed in the United States of America

To Budd Hopkins,
who led the way

Contents

ACKNOWLEDGMENTS

There are many people to whom I am grateful for the help they gave me in this work. There are those who contributed to the preparation of the manuscript and others who helped with information and ideas. Some people offered direct personal support. A few contributed in more than one of these ways. Some did me the kindness of not speaking critically behind my back when confronted with a subject that may have radically challenged their view of reality. Obviously I cannot name these individuals, but you know who you are.

I wish to acknowledge Miriam Altshuler, Walt Andrus, Michael and Isabel Blumenthal, Thomas E. Bullard, John Carpenter, Blanche Chavonstie, David Cherniack, Jerome Clark, Barbara Corbisier, Joan Erikson, C. Richard Farley, Penelope Franklin, Mary Westbrook Geha, Bill Goldstein, David Gotlib, Stanislav Grof, Hugh Gusterson, Joanne Hager, Richard Haines, Judith Herman, Robert and Joan Holt, Budd Hopkins, Linda Moulton Howe, Barbara Marx Hubbard, David Jacobs, Douglas Jacobs, Eric Jacobson, C. B. Scott Jones, Honey Black Kay, Gurucharan Singh Khalsa, Thomas and Jehane Kuhn, Roberto Lewis-Fernandez, Robert Jay Lifton, Caroline McLeod, John Miller, Malkah Notman, Joseph Nyman, David and Andrea Pritchard, Joseph Regal, Kenneth Ring, Laurance Rockefeller, Mark Rodeghier, Rudolf Schild, Timothy Seldes, Vivienne Simon, Karen Speerstra, Joel Speerstra, Ervin Staub, Myron Stocking, Richard Tarnas, Keith Thompson, George Vaillant, Jacques Vallee, Roger Walsh, Kenneth Warren, Karen Wesolowski and Jennie Zeidman.

I wish especially to thank Pam Kasey, whose partnership throughout this project has made it possible; Dominique Callimanopulos, for her caring and support in many aspects of my work; Pat Carr and Leslie Hansen, for their indispensible care in bringing this book into being; and my wife, Sally, for the gift of her love and support throughout the time that this work was being created.

Finally, and perhaps most important, I want to thank the experiencers themselves, both those whom I have written about in these pages and the many others who have taught me about this phenomenon, for the remarkable courage they have shown in sharing their stories.

Preface

An author embarking on a venture as manifestly novel as this one must inevitably ask if some link may be found with his previous work. For me, the connection resides in the matter of identity—who we are in the deepest and broadest sense. In retrospect, this focus has been with me from the beginning, driving my clinical explorations of dreams, nightmares, and adolescent suicide, my biographical researches, as well as the studies of the nuclear arms race and ethnonational conflict and, more recently, transpersonal psychology, with which I have been involved. The abduction phenomenon, I have come to realize, forces us, if we permit ourselves to take it seriously, to reexamine our perception of human identity—to look at who we are from a cosmic perspective.

This book is not simply about UFOs or even alien abductions. It is about how this phenomenon, both traumatic and transformative, can expand our sense of ourselves and our understanding of reality, and awaken our muted potential as explorers of a universe rich in mystery, meaning, and intelligence.

When we explore phenomena that exist at the margins of accepted reality, old words become imprecise or must be given new meanings. Terms like "abduction," "alien," "happening," and even "reality" itself, need redefinition lest subtle distinctions be lost. In this context, thinking of memory too literally as "true" or "false" may restrict what we can learn about human consciousness from the abduction experiences I recount in the pages that follow.

ABDUCTION

CHAPTER ONE

UFO ABDUCTIONS: AN INTRODUCTION

In the fall of 1989, when a colleague asked me if I wished to meet Budd Hopkins, I replied, "Who's he?" She told me that he was an artist in New York who worked with people who reported being taken by alien beings into spaceships. I then said something to the effect that he must be crazy and so must they. No, no, she insisted, it was a very serious and real matter. A day came soon when I would be in New York for another purpose—it was January 10, 1990, one of those dates you remember that mark a time when everything in your life changes—and she took me to see Budd.

Nothing in my then nearly forty years of familiarity with the field of psychiatry prepared me for what Hopkins had to say. I was impressed with his warmth, sincerity, intelligence, and caring for the people with whom he had been working. But more important than that were the stories he told me from people all over the United States who had come forth to tell him about their experiences after reading one of his books or articles or hearing him on television. These corresponded, sometimes in minute detail, to those of other "abductees" or "experiencers," as they are called.

Most of the specific information that the abductees provided about the means of transport to and from the spaceships, the descriptions of the insides of the ships themselves, and the procedures carried out by the aliens during the abductions had never been written about or shown in the media. Furthermore, these individuals were from many parts of the country and had not communicated with each other. They seemed in other respects quite sane, had come forth reluctantly, fearing the discrediting of their stories or outright ridicule they had encountered in the past. They had come to see Hopkins at considerable expense, and, with rare exceptions, had nothing to gain materially from telling their stories. In one example a woman was startled when Hopkins showed her a drawing of an alien being. She asked how he had been able to depict what she had seen when they had only just begun talking. When he explained that the drawing had been made by another person from a different part of the country she became intensely upset, for an experience that she had wanted to believe was a dream, now, she felt, must be in some way real.

My reaction was in some respects like this woman's. What Hopkins had encountered in the more than two hundred abduction cases he had seen over a fourteen-year period were reports of experiences that had the characteristics of real events: highly detailed narratives that seemed to have no obvious symbolic pattern; intense emotional and physical traumatic impact, sometimes leaving small lesions on the experiencers' bodies; and consistency of stories down to the most minute details. But if these experiences were in some sense "real," then all sorts of new questions opened up. How often was this occurring? If there were large numbers of these cases, who was helping these individuals deal with their experiences and what sort of support or treatment was called for? What was the response of the mental health profession? And, most basic of all, what was the source of these encounters? These and many other questions will be addressed in this book.

In response to my obvious but somewhat confused interest, Hopkins asked if I wished to see some of these experiencers myself. I agreed, with curiosity tinged by slight anxiety. At his home a month later Hopkins arranged for me to see four abductees, one man and three women. Each told similar stories of their encounters with alien beings and abduction experiences. None of them seemed psychiatrically disturbed except in a secondary sense, that is they were troubled as a *consequence* of something that had apparently happened to them. There was nothing to suggest that their stories were delusional, a misinterpretation of dreams, or the product of fantasy. None of them seemed like people who would concoct a strange story for some personal purpose. Sensing my now obvious interest, Hopkins asked if I wanted him to refer cases to me in the Boston area, of which he already knew quite a few. Again I agreed, and in the spring of 1990 I began to see abductees in my home and hospital offices.

In the more than three and a half years I have been working with abductees I have seen more than a hundred individuals referred for evaluation of abductions or other "anomalous" experiences. Of these, seventy-six (ranging in age from two to fifty-seven; forty-seven females and twenty-nine males, including three boys eight and under) fulfill my quite strict criteria for an abduction case: conscious recall or recall with the help of hypnosis, of being taken by alien beings into a strange craft, reported with emotion appropriate to the experience being described and no apparent mental condition that could account for the story. I have done between one and eight several-hour modified hypnosis sessions with forty-nine of these individuals, and have evolved a therapeutic approach I will describe shortly.

Although I have a great debt and profound respect for the pioneers in this field, like Budd Hopkins, who have had the courage to investigate and report information that runs in the face of our culture's consensus reality, this book is based largely on my own clinical experience. For this is a subject that is so controversial that virtually no accepted scientific authority has evolved that I might use to bolster my arguments or conclusions. I will report, therefore, what I have learned primarily from my own cases and will make interpretations and draw conclusions on the basis of this information.

The experience of working with abductees has affected me profoundly. The intensity of the energies and emotions involved as abductees relive their experiences is unlike anything I have encountered in other clinical work. The immediacy of presence, support, and understanding that is required has influenced the way I regard the psychotherapeutic task in general. Furthermore, I have come to see that the abduction phenomenon has important philosophical, spiritual, and social implications. Above all, more than any other research I have undertaken, this work has led me to challenge the prevailing worldview or consensus reality which I had grown up believing and had always applied in my clinical/scientific endeavors. According to this view—called variously the Western, Newtonian/Cartesian, or materialist/dualist scientific paradigm—reality is fundamentally grounded in the material world or in what can be perceived by the physical senses. In this view intelligence is largely a phenomenon of the brain of human beings or other advanced species. If, on the contrary, intelligence *is* experienced as residing in the larger cosmos, this perception is an example of "subjectivity" or a projection of our mental processes.

What the abduction phenomenon has led me (I would now say inevitably) to see is that we participate in a universe or universes that are filled with intelligences from which we have cut ourselves off, having lost the senses by which we might know them. It has become clear to me also that our restricted worldview or paradigm lies behind most of the major destructive patterns that threaten the human future—mindless corporate acquisitiveness that perpetuates vast differences between rich and poor and contributes to hunger and disease; ethnonational violence resulting in mass killing which could grow into a nuclear holocaust; and ecological destruction on a scale that threatens the survival of the earth's living systems.

There are, of course, other phenomena that have led to the challenging of the prevailing materialist/dualistic worldview. These include near death experiences, meditation practices, the use of psychedelic substances, shamanic journeys, ecstatic dancing, religious rituals, and other practices that open our being to what we call in the West nonordinary states of consciousness. But none of these, I believe, speaks to us so powerfully in the language that we know best, the language of the physical world. For the abduction phenomenon reaches us, so to speak, where we live. It enters harshly into the physical world, whether or not it is *of* this world. Its power, therefore, to reach and alter our consciousness is potentially immense. All of these matters will be discussed more fully in the clinical case examples that constitute the bulk of this book, and, especially, in the concluding chapter.

One of the important questions in abduction research has been whether the phenomenon is fundamentally new—related to the sightings of "flying saucers" and other unidentified flying objects (UFOs) in the 1940s and the discovery in the 1960s that these craft had "occupants"—or is but a modern chapter in a long story of humankind's relationship to vehicles and creatures appearing from the heavens that goes back to antiquity.

BEINGS FROM THE SKY OR OTHER DOMAINS THROUGH HISTORY*

The connection between humans and beings from other dimensions has been illustrated in myths and stories from various cultures for millennia. In contradiction to the post-Renaissance metaphysic, predominant in Western societies, that places man at the center of creation, above and separate from other forms of life, there are peoples around the world who customarily communicate with nonhuman intelligences and spirits through a variety of means. These communications and the myths they generate are an integral part of the cosmologies of many non-Western cultures, constituting for each a kind of ontological skeleton upon which hangs the balance of culture, customs, and lifestyle.

Throughout history, many societies have acknowledged conscious-

* Dominique Callimanopulos researched and contributed much of the writing of this and the following section.

ness as something more potent than we have in the West—as a sieve or receiver and transmitter of communication with forces, not always visible, other than ourselves. The contemporary Western tenet that we are alone in the universe, conversant only with ourselves, is, in fact, a minority perspective, an anomaly.

Across many epochs, humans have reported making contact with a multitude of gods, spirits, angels, fairies, demons, ghouls, vampires, and sea monsters. All have been said to instruct, direct, harass, or befriend humans with varying dispositions, motives, and purposes. While many of these beings have seemed quite at home on Earth, the majority made their visits from other habitats or dimensions. The sky, in particular, has always been a popular haven for nonhumans and has come to represent extraterrestrial dimensionality rather opulently, especially as Earth's frontiers seemed, in recent times, to have shrunk. As Ralph Noyes has noted, "we used to populate the Earth with spirits and Gods. Now they have been chased away and the sky is their haven" (Noyes 1990).

In Truk, located in the Marshall Islands, people have traditionally believed in an outer world that corresponds in some ways to our modern conception of outer space. It is a world of mystery and power, a world from which people in this world derived their being. There was, moreover, a continual dialogue between the people of this world and the inhabitants of the outer spirit world (Goodenough 1986, p. 558). Likewise, Native American Hopi were traditionally taught by the Kachinas, spiritlike beings from other planets, who instructed them in agricultural techniques and gave them philosophical and moral guidelines that have shaped Hopi culture (Clark and Coleman 1975, p. 215). People in Ireland believed that fairies or the gentle folk were not earthly, having originated on other planets. Fairies often travel about the skies in cloudlike aerial boats called "fairy boats" or "spectre ships" (Rojcewicz 1991, p. 481).

Mircea Eliade, the renowned mythologist, has amply documented the symbolic significance of the differentiation of sky and Earth as illustrating both the separation and connection between the human and spirit worlds. According to Eliade, "archaic myths worldwide speak of an extremely close proximity that existed primordially between Heaven and Earth. In illo tempore, the gods came down to Earth and mingled with men and men, for their part, could go up to Heaven by climbing the mountain, tree creeper or ladder, or might even be taken up by birds" (Eliade 1957, p. 59).

These ascension myths, Eliade says, these images of the earth and the heavens being somehow joined, are found in many tribes (including Australian, Pygmy, and Arctic) and have been elaborated by pastoral and sedentary cultures and transmitted right down to the great urban cultures of Oriental antiquity. When Heaven was abruptly separated from Earth, when the tree of the Liana connecting Earth to Heaven was cut, or the mountain which used to touch the sky was flattened out—then the paradisiac stage was over and humans entered into their present condition (Eliade 1957, p. 59).

"In effect, all these myths show us primordial man enjoying a beatitude, a spontaneity and freedom, which he has unfortunately lost in consequence of the fall—that is, of what followed upon the mythical that caused the rupture between Heaven and Earth . . . Immorality, spontaneity and freedom; the possibility of ascension into Heaven and easily meeting with the gods, friendship with the animals and knowledge of their language. These freedoms and abilities have been lost, as the result of a primordial event—the fall of man, expressed as an on tological mutation of his own condition, as well as a cosmic schism" (Eliade 1957, p. 61). Only special members of every culture, like shamans, could continue to move between Heaven and Earth, between humans and the spirit world.

The Koryaks of Siberia remember the mythical era of their Great Raven, when humans could go up to Heaven without difficulty: in our days, they add, it is only shamans who are capable of this. The Bakairi of Brazil think that, for the shaman, Heaven is no higher than a house, so that he reaches it in the twinkling of an eye (Eliade 1957, p. 65).

There are countless myths, tales, and legends concerning human or superhuman beings who fly away into Heaven and travel freely between Heaven and Earth. Again, according to Eliade, "the motifs of flight and ascension are attested to at every level of the archaic cultures, as much in the rituals and mythologies of the shamans and the ecstatics as in the myths and folklore of other members of the society who make no pretense to be distinguished by the intensity of their religious experience. A great many symbols and significations to do with spiritual life and above all, with the power of intelligence, are connected with the images of 'flight' and 'wings'; they all express a break with the universe of everyday experience . . . both transcendence and freedom are to be obtained through the flight" (Eliade 1957, p. 105).

It would seem that today's UFO abductees are continuing an amply documented tradition of ascent and extraterrestrial communication. But alien abductions and their effects on abductees possess their own uniqueness. Peter Rojcewicz, a folklorist, has compared the experience of today's abductees or experiencers with other aerial and abduction phenomena and alludes to the possibility of the existence of an intelligence, a spirit, an energy, a consciousness behind UFO experiences and extraordinary encounters of all types, that adapts its form and appearance to fit the environment of the times (Rojcewicz 1991).

Rojcewicz cites the long history of sightings of unusual aerial phenomena, and beings or objects of light. In ancient times there were sightings of "celestial cars, chariots that flew in the sky, flying palaces that shined and moved about in the sky . . . There are also many different descriptions of fiery shields in the sky, like triangles. Fiery crosses were also seen over western Europe." He also notes the presence of clouds or cloudy light surrounding unusual objects, including UFOs, as well as the spontaneous appearance of luminous religious images in the sky, frequently witnessed by thousands of people. In the United States, as recently as the last century, Americans witnessed ships—schooners and boats—sailing in the sky (Rojcewicz 1992). Jerome Clark, after a careful investigation of the airship sightings of the late 1890s, concluded that the vehicles in the sky observed frequently in the United States may have been related to contemporary UFOs but interpreted according to the technology and mythology of the time (Clark 1991).

According to Mario Pazzaglini, a psychologist who has been interested in the abduction experience for a number of years, manifestations of a "UFO-associated" nature have been recorded for the last ten thousand years, starting with an engraving of Ezekiel in the Old Testament in which he depicts a vision containing wheels, angels, light, and clouds (Pazzaglini 1992).

Unusual sky phenomena were also recorded by the Romans, the Greeks in the fourth century, and in the Middle Ages. Sometimes manifesting as stars, fires in the sky, crosses, lights, or beams, apparitions would often simply disappear or sometimes leave their mark. Many of these sightings were viewed by thousands of people and interpreted as religious miracles. Often these phenomena fit nicely into already existing spiritual beliefs held by the viewers.

The phenomenon of humans being transported into other dimensions also has a long history in most cultures. Tibetans have long believed that humans could separate from the "etheric" or "subtle" body and go traveling in an "out of body" capacity for hours or days at a time. "They have experiences in different places and then return." Tibetans distinguish between different gradations of subtlety and grossness, or density of beings. "The mind or consciousness produced by grosser matter cannot communicate with these subtle things. In some, you witness the grosser level of mind subdued and the more subtle mind become active. Then there's an opportunity, a chance to communicate with or sometimes see another being who is more subtle than our mind or body." (The Dalai Lama 1992). Contemporary examples of such entities in the West might be the "spirit guides" that are reported by many individuals. Descriptions of these spirits who appear to individuals or to intermediaries vary widely.

Rojcewicz includes UFO abductions within a wide range of paranormal experiences, including near death experiences; powerful psychic, spiritual, mystical, and out-of-body experiences; and encounters with a range of beings—such as witches, fairies, werewolves—that often result, for an individual, in a substantive transformation of values and orientation. The question of whether and why these events occur, of course, remains unanswered. There is much debate even about how to frame such questions.

The most commonly debated issue, whether abductions are really taking place, leads us to the center of questions about perception and levels of consciousness. The most glaring question is whether there is any reality independent of consciousness. At the level of personal consciousness, can we apprehend reality directly, or are we by necessity bound by the restrictions of our five senses and the mind that organizes our worldview? Is there a shared, collective consciousness that operates beyond our individual consciousness? If there is a collective consciousness, how is it influenced, and what determines its content? Is UFO abduction a product of this shared consciousness? If, as in some cultures, consciousness pervades all elements of the universe, then what function do events like UFO abductions and various mystical experiences play in our psyches and in the rest of the cosmos?

These are questions that are not easily answered. Perhaps all we can do at this point in time is to acknowledge the questions as we listen to the experiences of those who have moved beyond our culturally shared ideas of "reality." The UFO abduction experience, while unique in many respects, bears resemblance to other dramatic, transformative experiences undergone by shamans, mystics, and ordinary citizens who have had encounters with the paranormal. In all of these experiential realms, the individual's ordinary consciousness is radically transformed. He or she is initiated into a non-ordinary state of being which results, ultimately, in a reintegration of the self, an immersion or entrenchment into states and/or knowledge not previously accessible.

Sometimes the process is brought on by illness or a traumatic event of some kind, and sometimes the individual is simply pulled into a sequence of states of being from which he or she emerges with new powers and sensitivities. "During his initiation, the Shaman learns how to penetrate into other dimensions of reality and maintain himself there; his trials, whatever the nature of them, endow him with a sensitivity that can perceive and integrate these new experiences . . . through the strangely sharpened senses of the Shaman the sacred manifests itself" (Eliade 1957, p. 66). Like many abductees, the initiate hones his new sensibilities in the service of wisdom that can be used by his people.

Revelation is not just accessible to those in pursuit of enlightenment, but it can knock on any door at any time. Earlier in this century, a Dr. Buche described what seems to be a certain kind of archetypal experience: "He and two friends had spent the evening reading Wordsworth, Keats, Browning and especially Whitman. He was in a state of almost passive enjoyment. All at once, without warning of any kind, he found himself wrapped round, as it were, by a flame-colored cloud . . . The next he knew, the light was within himself. Directly afterwards came upon him a sense of exaltation, of immense joyousness accompanied or followed by an intellectual illumination impossible to describe. Into his brain streamed one momentary lightning flash of Brahmic splendour, leaving him thenceforth for always an aftertaste of Heaven" (Eliade 1965, p. 69).

The experience of internalizing what is first perceived as external light happens frequently during mystical flashes or transcendental journeys that result in spiritual rebirth. Perhaps an analogy might be drawn to UFO encounters where the abductee is initially "struck by a beam of light," spies a bright ship, and is then taken inside. Brazilian abductees in particular seem to have perceived illuminated clouds, frequently red in color, in association with spaceships (Story 1980).

The mystic or the shaman, like the abductee, makes a pilgrimage, usually with ardor, to receive a new dimension of experience or knowledge. This involves a rebirth which is sometimes very distressing, a retracing of one's steps to a preternatural, primordial arena to recondition the consciousness of the experiencer. The resulting psychic chaos is a metaphor for the precosmogenic chaos, amorphic yet penetrating, that the individual has been exposed to. The abductee is a modern Dante, whose ontological underpinnings are unraveled. Returned to his bed or his car after his time with aliens, he struggles to reassemble his worldview. Most often, he undertakes his journey alone, and many times his absence is not even noticed by those to whom he might turn to corroborate his coordinates.

Jacques Vallee, perhaps the most comprehensive cross-cultural ufology investigator, discusses the international history of UFO encounters in his two books *Dimensions* and *Passport to Magonia*. Describing hundreds of sightings of strange sky-born objects and their occupants across time, continents, and societies, he cites the seemingly unexplainable presence of disks in the symbology of various civilizations—the Phoenicians and early Christians, for instance, associated them with communications between angels and God. He compares some of the phenomenology of a UFO encounter with historical records of experiences of a mystical nature. Beams of light commonly play a role in both UFO and out-of-body encounters (Vallee 1988, p. 34). As for the beings themselves, Vallee draws many analogies to the worldwide sightings of nonhuman, shape-shifting, aerially adept beings throughout history. These beings appear to mankind in thousands of different guises; they possess extraordinary powers and frequently aim to partake of and/or take away something belonging to humans, desiring to communicate with or simply play with tricks on them. He concludes, "The UFO occupants, like the elves of old, are not extraterrestrials. They are denizens of another reality" (1988, p. 96). Vallee believes that abductees' interaction with aliens is a part of "an age-old and worldwide myth that has shaped our belief structures, our scientific expectations, and our view of ourselves" (1988, p. 99). He writes, "The same power attributed to saucer people was once the exclusive property of fairies" (1988, p. 134).

Vallee draws parallels between religious apparitions, the fairy-faith, the reports of dwarf-like beings with supernatural powers, the airship tales in the United States in the last century, and the present stories of UFO landings (1988, p. 140). He speculates broadly:

> Or should we hypothesize that an advanced race somewhere in the universe and sometime in the future has been showing us three-dimensional space operas for the last two thousand years, in an attempt to guide our civilization? If so, do they deserve our congratulations? . . . Are we dealing instead with a parallel universe, another dimension, where there are human races living, and where we may go at our expense, never to return to the present? Are these races only semihuman, so that in order to maintain contact with us, they need cross-breeding with men and women of our planet? Is this the origin of the many tales and legends where genetics plays a great role: the symbolism of the Virgin in occultism and religion, the fairy tales involving human midwives and changelings, the sexual overtones of

> the flying saucer reports, the biblical stories of intermarriage between the Lord's angels and terrestrial women, whose offspring were giants? From that mysterious universe, are higher beings projecting objects that can materialize and dematerialize at will? Are the UFOs "windows" rather than "objects"? There is nothing to support these assumptions, and yet, in view of the historical continuity of the phenomenon, alternatives are hard to find, unless we deny the reality of all the facts, as our peace of mind would indeed prefer. (1988, p. 143–44)

Where facts are lean, inconsistent, or disparate, human beings, Vallee assures us, are quick to fill in the gaps. "Because many observations of UFO phenomena appear self-consistent and at the same time irreconcilable with scientific knowledge, a logical vacuum has been created that human imagination tries to fill with fantasy" (1988, p. 145).

Ultimately, Vallee prescribes our remaining open to learning from phenomena that we do not yet understand. "These unexplained observations need not represent a visitation from space visitors, but something even more interesting: a window toward undiscovered dimensions of our environment" (1988, p. 203). "I believe that the UFO phenomenon represents evidence for other dimensions beyond spacetime; the UFOs may not come from ordinary space, but from a multiverse which is all around us, and of which we have stubbornly refused to consider the disturbing reality in spite of the evidence available to us for centuries" (1988, p. 253).

THE UFO ABDUCTION PHENOMENON WORLDWIDE

Another question concerns the worldwide distribution of abductions, or reports of the phenomenon, which may be quite a different matter. UFO abductions have been reported and collected most frequently in Western countries or countries dominated by Western culture and values. Insofar as the abduction phenomenon may be seen as occurring in the context of the global ecological crisis, which is an outcome of the Western materialist/dualistic worldview, it may be that its "medicine" is being administered primarily where it is most needed—in the United States and the other Western industrial countries. Related to this would be the fact that in many cultures the entry into the physical world of vehicles, and even contact with creatures, seemingly from space or another dimension, would not be as noteworthy as in societies where traffic from the spirit world or the "world beyond" into our physical existence would be considered remarkable.

The first publication of an abduction case took place in Brazil, involving the reported abduction of the son of a rancher, Antonio Villas-Boas, in 1957. Reports of UFO sightings worldwide, however, far outnumber accounts of actual abductions. The most comprehensive guide to abductions overseas was put together in 1987 by Thomas Bullard, a folklorist at the University of Indiana (Bullard 1987). Bullard lists reported abductions from seventeen countries, including Argentina, Australia, Bolivia, Brazil, Canada, Chile, England, Finland, France, Poland, South Africa, the Soviet Union, Spain, Uruguay, and West Germany.

The United States leads the way in sheer numbers of abductions, with England and Brazil following behind, largely because of the availability of practicing hypnotists and therapists working with abductees in these countries. In illustration of this point, China boasts the largest number of witnesses of a single UFO sighting—on August 24, 1981, one million Chinese saw a spiral-shaped UFO simultaneously (Chiang 1993)—but there is no record of any follow-up questioning of individual witnesses.

The therapeutic exploration of abduction experiences is, however, gradually catching hold. In May 1993, Germany's second largest television station presented a forty-five-minute documentary about the abduction phenomenon which won Germany's highest television award. While two therapists offered their services free of charge to abductees subsequent to the broadcast, only twenty people have responded. As elsewhere, abduction remains a frightening experience that many would rather not confront unless symptoms resulting from the encounter require them to do so.

Even accounts of UFO sightings are, throughout the world, shrouded in secrecy. Spain's Ministry of Defense UFO files were released in 1992. These contain, mostly, reports of sightings by Air Force personnel. Much work remains to be done in persuading other nations similarly to open up classified files on the subject.

In some countries, where people hold all sorts of beliefs in supernatural beings, abduction experiences are confused or simply connected with other visitations. Cynthia Hind, a researcher from South Africa, reports, "Their reactions are as perhaps Westerners would react to ghosts; not necessarily terrified (or not always so) but certainly wary of what they see" (Hind 1993, p. 17).

Abductees overseas seem to have contact with a greater variety of entities than Americans. These range from tiny men to tall, hooded beings, and include naked individuals of both sexes and humanoid beings with every manner or shape of head, feet, and hands. A Dutch couple recently described their UFO visitors as being tiny and appearing in rainbow hues—green, orange, and purple (personal communication, September 1992).

But universal properties of the abduction experience remain. Most often, abductees everywhere are compellingly drawn toward a powerful light, often while they are driving or asleep in their beds. Invariably, they are later unable to account for a "lost" period of time, and they frequently bear physical and psychological scars of their experience. These range from nightmares and anxiety to chronic nervous agitation, depression, and even psychosis, to actual physical scars—puncture and incision marks, scrapes, burns, and sores.

Some encounters are more sinister, traumatizing, and mysterious. Others seem to bear a healing and educational intent. Most often, say abductees, they are told or warned by the beings or people not to tell about their experiences. In Puerto Rico, Miguel Figueroa, for example, reported receiving threatening phone calls the day after he saw five little, gray men in the middle of the road (Martin 1993).

Even less well documented than the actual abductions are the consequences of the experience. In working with abductees, Gilda Moura, a Brazilian psychologist, reports on the paranormal abilities many Brazilian abductees experience after an encounter. These include increased telepathic abilities, clairvoyance, visions, and the receiving of spiritual messages which are often concerned with world ecology, the future of humankind, and social justice. Many abductees decide to change their profession after their experience (Moura, in press).

It is likely that with the publicizing of therapeutic and hypnosis techniques currently being pioneered in the United States, much more information about abduction experiences overseas will be available in coming years; for the rest of the world certainly does not lack awareness of the UFO phenomenon, as is evidenced by the proliferation of UFO bureaus, offices, and research organizations abroad.

MODERN-DAY ABDUCTIONS

The modern history of abductions begins with the experience of Barney and Betty Hill in September 1961 (Fuller 1966). The Hills, a stable, respectable interracial couple living in New Hampshire, had suffered from disturbing symptoms for more than two years when they reluctantly consulted Boston psychiatrist Benjamin Simon. Barney was an insomniac and Betty had frequent nightmares. Both were so persistently anxious that it became intolerable for them to continue their lives without looking into disturbing repercussions of the September night in which they could not account for two hours during the return journey from a holiday in Montreal. Except for the distresses related to the incident they described, Dr. Simon reported no psychiatric illness.

On the night of September 19, 1961, the Hills reported that their car was "flagged down" by small, gray humanoid beings with unusual eyes. Before this they had noticed an erratically moving light and then a strange craft. With binoculars Barney had been able to see the creatures inside the craft. The Hills were amnesic about what happened to them during the missing hours until undergoing repeated hypnosis sessions with Dr. Simon. In their meetings with him, Dr. Simon instructed the Hills not to tell each other details of the memories that were emerging. After being taken from their car the Hills said they were led by the beings against their wills onto a craft. Each reported that on the craft they were placed on a table and subjected to detailed medical-like examinations with taking of skin and hair "samples." A needle was inserted into Betty's abdomen and a "pregnancy test" performed. Researchers have discovered recently that a sperm sample was taken from Barney, a fact that was withheld by him and John Fuller, who later wrote about their case, because it was too humiliating at the time for Barney to admit (Jacobs 1992). The beings communicated with the Hills telepathically, nonverbally, "as if it were in English." The Hills were "told to" forget what had happened.

Despite Dr. Simon's belief that the Hills had experienced some sort of shared dream or fantasy, a kind of folie à deux, they persisted in their conviction that these events really happened, and that they had not communicated the corroborating details to each other during the investigation of their symptoms. Barney, who died in 1969 at the age of forty-six, had been particularly reluctant to believe in the reality of the experience lest he appear irrational. "I wish I could think it was an hallucination," he told Dr. Simon when the doctor pressed him. But in the end Barney concluded, "we had seen and been a part of something different than anything I had seen before," and "these things did happen to me." Betty, who continues to speak publicly about her experience, also believes in the reality of these events. In 1975 a film about the Hill case, *The UFO Incident,* starring James Earl Jones as Barney, was shown on television in the United States.

A number of books and articles documented abduction experiences by other individuals in the years following the Hills's testimony

(Lorenzen and Lorenzen 1976; Lorenzen and Lorenzen 1977; Haisell 1978; Fowler 1979; Rogo 1980; Druffel and Rogo 1980; Bullard 1987, pp. 1-15; Clark 1990, p. 2). It has been the pioneering research of New York artist and sculptor Budd Hopkins, however, over almost two decades with hundreds of abductees, that has established the essential consistency of the abduction phenomenon. Hopkins's first book, *Missing Time*, published in 1981, documented the unaccounted-for time periods and associated symptoms that indicate that abduction experiences have taken place, as well as the characteristic details of such experiences (Hopkins 1981). Hopkins also found that abduction experiences were possibly associated with previously unexplainable small cuts, body scars, and scoop marks; the narratives even suggested that small objects or "implants" may have been inserted in victims' noses, legs, and other body parts. In his second book, *Intruders*, published in 1987, Hopkins defined the sexual and reproductive episodes that have come to be associated with the abduction phenomenon (Hopkins 1987). Temple University historian David Jacobs has further refined the basic reported pattern of an abduction experience (Jacobs 1992). Jacobs identifies primary phenomena such as manual or instrument examination, staring, and urological-gynecological procedures; secondary events, including machine examination, visualization, and child presentation; and ancillary events, among them miscellaneous additional physical, mental, and sexual activities and procedures.

None of this work, in my view, has come to terms with the profound implications of the abduction phenomenon for the expansion of human consciousness, the opening of perception to realities beyond the manifest physical world and the necessity of changing our place in the cosmic order if the earth's living systems are to survive the human onslaught.

Polls of the prevalence of the UFO abduction phenomenon in the United States, including a survey of nearly six thousand Americans conducted by the Roper organization between July and September 1991 (Hopkins, Jacobs, and Westrum 1991) indicate that from several hundred thousand to several million Americans may have had abduction or abduction-related experiences. The Roper poll has been criticized on the grounds that the indicators of possible abduction used—such as seeing unusual lights, missing time, or a feeling of flying—may not in fact actually mean that an abduction has occurred. But a more serious difficulty in estimating the prevalence of abductions lies in the fact that we do not know what an abduction really is—the extent, for example, to which it represents an event in the physical world or to which it is an unusual subjective experience with physical manifestations. A still greater problem resides in the fact that memory in relation to abduction experiences behaves rather strangely. As in the cases, for example, of Ed (chapter 3) or Arthur (chapter 15) the memory of an abduction may be outside of consciousness until triggered many years later by another experience or situation that becomes associated with the original event. The experiencer in a situation such as this could be counted on the negative side of the ledger *before* the triggering experience and on the positive side *after* it.

WHO ARE THE ABDUCTEES?

None of the efforts to characterize abductees as a group have been successful. They seem to come, as if at random, from all parts of society (Bullard 1987; Hopkins 1981, 1987; Jacobs 1992, pp. 327–28). My own sample includes students, housewives, secretaries, writers, business people, computer industry professionals, musicians, psychologists, a nightclub receptionist, a prison guard, an acupuncturist, a social worker, and a gas station attendant. At first I thought that working class people predominated, but that appears to be an artifact related to the fact that those with less of an economic and social stake in the society seem less reluctant to come forward. Conversely, more professionally and politically prominent abductees fear the humiliation, rejection, and threat to their position that public revelation of their experiences might bring. One of the men with whom I have worked left me a note with a telephone number and a post office box in a town in which he did not live. He did not tell me his real name until some trust had been established between us. A highly renowned political figure who is well known in UFO circles to be an abduction witness has applied the skills of his profession to the fullest to avoid public identification and embarrassment (Hopkins 1992).

Efforts to establish a pattern of psychopathology other than disturbances associated with a traumatic event have been unsuccessful. Psychological testing of abductees has not revealed evidence of mental or emotional disturbance that could account for their reported experiences (Bloecher, Clamar, and Hopkins 1985; Parnell 1986; Parnell and Sprinkle 1990; Rodeghier, Goodpaster, and Blatterbauer 1991; Slater 1985; Spanos et al. 1993; Stone-Carmen, in press). My own sample demonstrates a broad range of mental health and emotional adaptation. Some experiencers are highly functioning individuals who seem mainly to need support in integrating their abduction experiences with the rest of their lives. Others verge on being overwhelmed by the traumatic impact and philosophical implications of their experiences and need a great deal of counseling and emotional support.

The administration of a full battery of psychometric tests is time-consuming and expensive. I have undertaken to have four of my cases tested by Ph.D. psychologists. One twenty-one-year-old man, who I knew was quite troubled—one of two of my seventy-six cases who had to be hospitalized for psychiatric reasons—revealed a complex picture of emotional disturbance and troubled thinking in which cause and effect in relation to the abduction experiences could not be sorted out. The other three tested in the normal range with no obvious psychopathology found.

The effort to discover a personality type associated with abductions has also not been successful (Basterfield and Bartholomew 1988; Basterfield, in press; Mack, in press; Rodeghier, Goodpaster, and Blatterbauer 1991). Psychologist Kenneth Ring has posited the notion of an encounter-prone personality (Ring 1992; Ring and Rosing 1990), a tendency of an individual who has been affected by unusual experiences to be more open to them in the future. But in this, as in any hypothesis concerning the personalities of abductees, it is important to keep in mind that the encounters may in many instances be found to have begun in early infancy, and children as young as two years old have talked of their abduction experiences. I have two boys under three in my own sample. Cause and effect in the relationship of abduction to building of personality are thus virtually impossible to sort out.

Similarly, there is no obvious pattern of family structure and interaction in the case of abductees. When I began this work I was struck by how many abductees came from broken homes or had one or more alcoholic parents. But some of my cases come from intact, well-functioning families. There also seems to be a "poor fit" between some individual experiencers and their parents, and a number of my cases complain about coldness and emotional deprivation within the family (for example, Joe, chapter 8). Some abductees experience having been told by an alien female that she was their true mother, and they even feel in some vague but deep way that this is actually true, i.e., that they are not "from here" and that their Earth mother and father are not their true parents. I have several cases in which the abduction-affected sibling seems to have fared better in life than the other siblings and attributes this to the warmth and love received during their life from the aliens themselves! It appears, as in the case of sexual abuse (see below), that the alien beings seem interested in human woundedness and may play some sort of healing or restorative role. Careful research to document this possibility is needed.

I have the impression that abductees as a group are unusually open and intuitive individuals, less tolerant than usual of societal authoritarianism, and more flexible in accepting diversity and the unusual experiences of other people. Some of my cases report a variety of psychic experiences, which has been noted by other researchers (Basterfield, in press). But here too biases related to the effects of their abduction experiences, the particular segment of the abductee population that came to me in the first place, and the results, in some instances, of our work together must be considered. Subtle measures, such as tests of openness, intuition, and psychic ability, that might distinguish abductees as a group from a matched sample of nonexperiencers, have yet to be developed or applied in the field of abduction research.

An association with sexual abuse has also been suggested in the abduction literature (Laibow 1989). But here too errors related to the misremembering of traumatic experiences, or the reverse—traumatic experiences of one kind (abduction) opening the psyche to the recollection of traumas of another kind (sexual abuse)—can lead to falsely overstressing the association. I have worked with one woman, for example, who went to a capable psychotherapist for presumed sexual abuse and incest-related problems. Several hypnosis sessions failed to reveal evidence of such events. But during one of her sessions she recalled a UFO that landed near her home when she was a six-year-old girl from which emerged typical alien beings who took her aboard the craft. For the first time, she experienced powerful emotions, especially fear, in the therapy hour. The therapist who referred the woman to me told me that he was "clean," i.e., was not directly familiar with the abduction phenomenon and did not suspect that she had such a history. There is not a single abduction case in my experience or that of other investigators (for example, Jacobs 1992, p. 285) that has turned out to have masked a history of sexual abuse or any other traumatic cause. But the reverse has frequently occurred—that an abduction history has been revealed in cases investigated for sexual or other traumatic abuse.

Sexual abuse appears to be one of the forms of human woundedness that, at least from the experiencer's standpoint, has led the aliens to intervene in a protective or healing manner. A thirty-five-year-old woman, for example, remembered consciously being sexually abused by her father at age four and weeping in the cellar afterwards. Several familiar alien beings—she recalled encounters from fourteen months old—"checked me to see if I was hurt, 'cause I did hurt," found underwear for her (not the "right ones") and "did up my sandals," she told me.

There have also been efforts to relate the abduction phenomenon to Satanic ritual abuse (Dean, in press; Wright 1993) and multiple personality disorders which, like sexual abuse, are related to psychological traumas in which the mechanism of dissociation is employed (Frankel 1993; Ganaway 1989; Spiegel and Cardena 1991). But it is important to realize that dissociation is a means whereby the personality copes with a traumatic experience by splitting off part of itself to keep disturbing emotions out of consciousness, thus allowing the rest of the psyche to function as well as it can. "Dissociation" per se tells us nothing about the source or content of the original disturbing experience. Abductees *will* use dissociation as a way of dealing with their threatening experiences, i.e., to keep them out of consciousness, and it may even be a prevalent coping device among abductees (Jacobson, in press). But the fact that they employ this defense mechanism does not tell us anything about the nature of the original traumatic experience. I feel sometimes that in the mental health profession we are like the generals who are accused of always fighting the last war, invoking the diagnoses and mental mechanisms with which we are familiar when confronted with a new and mysterious phenomenon, especially if it is one that challenges our way of thinking.

The first cases that were referred to me in the spring of 1990 confirmed what Hopkins, David Jacobs, Leo Sprinkle, John Carpenter, and other pioneers who were investigating the abduction phenomenon had already discovered. These individuals reported being taken against their wills by alien beings, sometimes through the walls of their houses, and subjected to elaborate intrusive procedures which appeared to have a reproductive purpose. In a few cases they were actually observed by independent witnesses to be physically absent during the time of the abduction. These people suffered from no obvious psychiatric disorder, except the effects of traumatic experience, and were reporting with powerful emotion what to them were utterly real experiences. Furthermore, these experiences were sometimes associated with UFO sightings by friends, family members, or others in the community, including media reporters and journalists, and frequently left physical traces on the individuals' bodies, such as cuts and small ulcers that would tend to heal rapidly and followed no apparent psychodynamically identifiable pattern as do, for example, religious stigmata.

In short, I was dealing with a phenomenon that I felt could not be explained psychiatrically, yet was simply *not possible* within the framework of the Western scientific worldview. My choices then were either to stretch and twist psychology beyond reasonable limits, overlooking aspects of the phenomenon that could not be explained psychologically, such as the physical findings, the occurrence in small children and even infants, and the association with UFOs—i.e., to keep insisting upon a psychosocial explanation consistent with the prevailing Western scientific ideology. Or, I might open to the possibility that our consensus framework of reality is too limited and that a phenomenon such as this cannot be explained within its ontological parameters. In other words, a new scientific paradigm might be necessary in order to understand what was going on.

WORKING WITH EXPERIENCERS

With this dilemma in mind I approached Thomas Kuhn, author of the 1962 classic *The Structure of Scientific Revolutions*, which analyzes how scientific paradigms change, to get his advice about my investigations. I knew Tom Kuhn since childhood, for his parents and mine were friends in New York and I had often attended eggnog parties at Christmastime in the Kuhns' home. I found the advice that he and his wife, Jehane, who is highly knowledgeable in the fields of mythology and folklore, gave me to be very useful. What I found most helpful was Kuhn's observation that the Western scientific paradigm had come to assume the rigidity of a theology, and that this belief system was held in place by the structures, categories, and polarities of language, such as real/unreal, exists/does not exist, objective/subjective, intrapsychic/external world, and happened/did not happen. He suggested that in pursuing my investigations I suspend to the degree that I was able all of these language forms and simply collect raw information, putting aside whether or not what I was learning fit any particular worldview. Later I would see what I had found and whether any coherent theoretical formulation would be possible. This, by and large, has been the approach that I have tried to follow.

When a possible abductee comes to see me, either referred through the UFO network, by another mental health professional, or self-referred upon learning of my work through the media, I explain that I regard him or her as a co-investigator. Although abductees understand that I am engaged in research about the phenomenon, I explain that my first responsibility is to their health and well-being. The overall investigative and therapeutic approach I use has evolved over the past three and a half years and is still changing (Mack 1992). I do an initial screening interview, which generally lasts about one and a half hours. During this session I obtain a history of possible abduction-related phenomena and learn as much about the person and his or her family as I can. Sometimes additional family members, who may or may not be experiencers themselves, will be interviewed.

Abductees may have a great deal of conscious recall of their experiences without hypnosis. One nineteen-year-old man remembered the details of an abduction at age four in our first interview. He told anxiously of being "picked up" from a clearing behind his home by gray aliens at midday, and taken into a spaceship. He was able to describe the saucerlike UFO and the beings themselves in great detail. On the ship he was unable to move and was forced to lie down in a cubicle where he was bathed in laserlike light and a skin sample was taken with a cylindrical instrument. After this he was returned and told to "run along now" down a path to the apartment complex where he was living.

But often abductees say that there are vast areas of their lives that they strongly feel are outside of conscious recall and yet powerfully affect them on a day-to-day basis. Although they generally know that these experiences may have been traumatic and that their recollecting them will be disturbing, the majority of abductees I have seen elect to investigate their experiences further. It is far more difficult, they have felt, to have major episodes of their mental lives and experiences unavailable to them than it is to confront what they sense has happened, however disturbing the events may prove to be.

The inducement of a nonordinary state, a modified form of hypnosis in my cases, seems to be highly effective in bringing abductees' walled off experiences into consciousness and in discharging their traumatic impact. I do not quite understand why this is so dramatically true. Abductees seem to move readily into trance, although I know of no study that has compared them for hypnotizability with other groups, especially other trauma survivors. Sometimes the simplest or most modest of relaxation techniques is all that is needed to bring back many memories. It is as if hypnosis undoes, in a kind of reverse mirror-imaging of the original altering of the psyche's consciousness, the forces of repression that were imposed at the time of the abduction.

These repressing forces are felt by the abductees to be much more than their own self-protective defenses. They may feel that as much as ninety percent of the energy that kept them from remembering was the result of an outside turning or switching off of memory by something the aliens themselves do. According to the abductees, the aliens will frequently communicate to them that they will not, or should not, remember what has occurred. Sometimes it is explained that this is for their own protection, and indeed, especially as in the case of small children, ongoing conscious recall of painful or traumatic experiences could interfere with daily life (for example, Jerry, chapter 6). The experiencers may feel that they are specifically disobeying the admonitions of the alien beings, with whom they often feel connected or allied on a very deep level, when they cooperate with me in recalling their abductions. This requires reassurance on my part that no harm, to my knowledge, has ever come from recalling these experiences when done in an appropriately supportive context.

It has been suggested that the experiencers' sense that they are "not supposed to" remember these events, and the alliance they often feel with the alien beings, are manifestations of the "Stockholm syndrome," in which a hostage or victim comes to sympathize with the perpetrator(s) as a means of retaining some agency in an intolerably coercive situation. This analogy is useful in facilitating the experiencers' initial expressions of outrage; however, it does not hold up as we move through deeper levels of uncovering. As I believe is clear in the case material, abductees come to feel a more authentic identification with the purposes of the whole phenomenon than occurs, for example, in hostage situations.

The economy and history of remembering in the abduction phenomenon is one of its most interesting aspects. Detailed recall of experiences that were never in conscious awareness may be triggered years, even several decades, after the event by something seen or heard which may bear only a minimal relationship to the actual abduction. What combination of abductee/alien factors determines the timing of recall, including when the experiencers elect to investigate their histories and who comes to tell their stories, remains to be understood further. The information presented in this book will necessarily be biased by these factors.

The type of hypnosis or nonordinary state I employ has been modified by my training and experience in the holotropic breathwork method developed by Stanislav and Christina Grof (Grof 1985, 1988, 1992). Grof breathwork utilizes deep, rapid breathing, evocative music, a form of bodywork, and mandala drawing, for the investigation of the unconscious and for therapeutic growth. In its emphasis on the breath, the Grof method has much in common with ancient meditative practices. I have found that maintaining a focus on the breath, as a tool for centering and integration in association with hypnosis, is invaluable in working with abductees. This is related to the extraordinary intensity of the energies involved—connected apparently with the power of the original experience—manifested in bodily sensations and movements and strong emotions, especially terror, rage, and sadness, that come up as recall of the abduction experience takes place.

After a simple induction that includes soothing imagery, a systematic relaxation of the parts of the body, and frequent return of attention to the breath, I encourage the experiencer to envision a comforting and relaxing place, to which the experiencer may automatically return at any time during the session. This enables the individual to mediate the pace of remembering, and reinforces the priority I place on his or her well-being. Because of the recurrent and unpredictable timing of these experiences, I have found that it is best not to use the word "safe" in describing this imagined refuge. For many experiencers, especially in the early stages of uncovering, there is no such thing as "safety," and to suggest it is to deny the full power of the experience.

As is often true of survivors of other traumatic events who seek to bring the events into full consciousness, abductees want to remember. Sometimes there is a danger that the unfolding of the narrative, the recall of the abduction events, will run ahead of the abductees' defenses once again, resulting in their becoming overwhelmed and traumatized. Through focusing on the breath during the induction process and the hypnosis session itself the experiencer is able to be grounded and to approach their experiences with greater strength. I explain to the abductee at the beginning of the session that I am more interested in their integration of their recalled experiences as we go along than in "getting the story." The story, I explain, will take care of itself in due time.

Having achieved together a relaxed (if often somewhat apprehensive) state of being and established skills for pacing and grounding the recall, we proceed to the process of memory retrieval. Numerous examples of this part of the session are detailed in the following chapters. It is useful in reading these accounts to note the way in which a return to focus on the breath in difficult moments often reduces fear by grounding the memory in raw perception and quieting the interpretive mind In addition, at moments of special distress during the session, I may place my hand gently on the abductee's shoulder to assure him or her of my presence. But in providing this reassurance one must be careful not to create a confusing replication of the original intrusion, which any physical contact with an experiencer who is in the depths of traumatic recall may create.

At the end of the session the experiencer may feel powerful tension or cramping in certain muscle groups, especially for some reason in the hands, and a tension-exaggerating approach, as developed by the Grofs, may be useful in discharging the remaining tightness or cramping. We also spend some time at this point discussing the material that has emerged. This conversation helps to bring the material more fully into normal consciousness and to further the process of integration. It is at this time that many experiencers begin to struggle deeply with questions of accuracy and meaning, and they often ask me how they should regard their hypnotically recovered memories.

This question has received a great deal of attention in both the UFO and therapeutic communities. Critics and skeptics cite work on the inaccuracy of recall with hypnosis and the possibility that the experiencer is developing recall of memory to please or comply with the hypnotist's expectations, to question the reality of the abduction phenomenon itself. I believe that these criticisms cannot be supported. Daniel Brown, a noted expert in the field of hypnosis research, determined after carefully reviewing the literature on recall among trauma sufferers under hypnosis that there are simply no studies of the accuracy of memory in this population, i.e., among individuals for whom the events in question are of core meaning or central importance. Rather, conclusions regarding the inaccuracy of recall under hypnosis have been based on studies in which an environmental context was created and memory was tested in relation to events that were of peripheral significance to the subject (personal communication, 18 October 1993). These studies, therefore, may not apply to abduction experiencers, who are highly motivated to remember accurately intense occurrences that are of the most vital importance to them.

If the abduction phenomenon, as I suspect, manifests itself in our physical space/time world but is not *of* it in a literal sense, our notions of accuracy of recall regarding what did or did not "happen" (Kuhn's advice about suspending categories seems relevant here) may not apply, at least not in the literal physical sense. Under these circumstances the reported experience of the witness, and our clinical assessment of the genuineness of that report, may be the only means by which we can judge the reality of the experience. John Carpenter's finding that experiencers who have been abducted together and hypnotized independently consistently provide accounts of what "happened" to them on the ships that correspond in minute detail thus becomes all the more remarkable (Carpenter 1993). Using the criteria of affective appropriateness and a narrative consistent with what I know about how abductions generally proceed, it is my impression that the reports provided under hypnosis are generally more accurate than those consciously recalled. We will see, for example, in the case of Ed (chapter 3) how his conscious memory of an abduction that occurred when he was a teenager contained bravado and pleasurable happenings consistent with his adolescent self-esteem. The same experience recalled in more detail with difficulty under hypnosis was humiliating and altogether uncongenial to his teenage self-regard.

The suggestion that the abductee is trying to please the hypnotist during the session and is making up the whole story—because, presumably, the hypnotist is there to discover an abduction—fails to take into account how disturbing abductions are to the experiencers and how intense the resistance is to bringing what they have gone through back into consciousness, or to accepting the reality of the phenomenon at all. As will be seen in the later chapters of this book, I sometimes need to invoke every morsel of alliance and cooperation to enable the abductee to go forward into the depths of the forgotten experience. Furthermore, abductees are peculiarly unsuggestible. To meet the above criticisms I and other investigators have tried repeatedly to trick abductees by suggesting specific elements—hair on the aliens, corners in the rooms on the ships, for example—only to be met with direct contradiction of these efforts. Proponents of the controversial "false memory syndrome" as an explanation for abduction memories need to account for this as well as the points outlined on page 43.

This discussion, like my conversations with the Kuhns, raises interesting epistemological questions that will be with us throughout this book, especially those concerning consciousness as an instrument of knowing. In this work, as in any *clinically* sound investigation, the psyche of the investigator, or, more accurately, the interaction of the psyches of the client and the clinician, is the means of gaining knowledge. But it must then be noted that although we analyze and formulate as objectively as possible afterwards, the original information was obtained nondualistically, i.e., through the intersubjective unfolding of the investigator-abductee interaction. Thus experience, the reporting of that experience, and the receiving of that experience through the psyche of the investigator are, in the absence of physical verification or "proof" (always quite subtle in the abduction phenomenon, as will be discussed later), the only ways that we can know about abductions.

When experiencers ask me about the status of their experience under hypnosis, I can only say that the elements of their story have appeared again and again in the stories of other individuals who are not crazy. I note that the feelings and emotions they have shown me seem quite real to me, and I ask them if they can find any explanation for feelings that intense. Finally, I tell them that I have no answers, and I ask them to rate the reality of their "memories."

At the end of the session I instruct the experiencers to call either me or my assistant, Pam Kasey, who is present during almost all the meetings, for a follow-up discussion. They usually do call, but if they do not we call them. We are interested in how the experiencer has dealt with the powerful feelings that came up during the session, additional memories that surfaced, and how they are managing what I call the "ontological shock" of the abduction events; for until the powerful reliving that has occurred during the hypnosis session, the abductees may have still clung to the possibility that these experiences are dreams or some sort of curable mental disorder. The denial never disappears altogether, and a shock may recur, even after several hypnosis sessions, especially if a second abductee reports independently witnessing or experiencing during a shared abduction precisely what the first one has reported.

Regular support group meetings, held in a friendly, private atmosphere where easy socializing is possible, are an important aspect of my work with abductees; for members of this population feel extremely isolated and unable to communicate, except with other experiencers, a central aspect of their lives without fear of rejection or outright ridicule. In the support group they find a community of individuals with similar experiences. In the group abductees can share what they have been through, or are still experiencing, can keep up with what is going on in the UFO/abduction field generally, and can explore the various possible meanings and implications of the experiences in their individual and collective lives.

Although one or more professional investigators are present during the support group, it is important that the abductees develop a self-help support network among themselves outside of the regular meeting times. Sometimes this involves small group meetings; at other times telephone contact is sufficient. As I have stressed, abductees are not, generally, mentally disturbed individuals. But they have undergone powerfully traumatic or confusing experiences, feel isolated from the mainstream belief structures of the society, and often need a great deal of support from people who know about or are familiar with the abduction phenomenon.

It is often useful for an abductee to have an ongoing relationship with a psychotherapist who is familiar with this phenomenon. When I began my work there were very few mental health professionals who involved themselves with this field, and some were doing considerable harm by trying to fit the experiencers into a familiar diagnostic category, most often of some other form of traumatic abuse. But that is changing, and in the Boston area and some other metropolitan centers there are increasing numbers of clinicians who are open to the reality of the abduction phenomenon and able to work with this population, although few are prepared to take on the powerful reliving of the abductees' experiences through hypnosis. Training programs, begun in 1992 with the leadership and support of Las Vegas businessman Robert Bigelow, and organized in various American cities by abduction investigators John Carpenter, Budd Hopkins, and David Jacobs, are familiarizing many mental health professionals with the abduction phenomenon.

In talking with people who work with experiencers, I have come to believe that what is most important during a regression and in all interactions with experiencers is the way of holding the energies of these experiences. This includes a degree of warmth and empathy, a belief in the ability of the individual to integrate these confusing experiences and make meaning of them for him- or herself, and a willingness to enter into the co-investigative process and risk being changed by the information. These are, of course, qualities that are important in any relationship, and they become critical in this work, where we are all pushed to our edge, experiencer, investigator, and therapist alike.

CHAPTER TWO

ALIEN ABDUCTIONS: AN OVERVIEW

INDICATORS OF ABDUCTIONS

Although some abductees may recall only a single dramatic experience, when a case is carefully investigated it generally turns out that encounters have been occurring from early childhood and even infancy. Indications of childhood abductions include the memory of a "presence," or "little men," or other small beings in the bedroom; recollections of unexplained intense light in the bedroom or other rooms; a humming or vibratory sensation at the onset of the experience; instances of being floated down the hall or out of the house; close-up sightings of UFOs; vivid dreams of being taken into a strange room or enclosure where intrusive procedures were done; and time lapses of an hour or more (Hopkins 1981) in which the parents may have been unable to find the child. Awakening paralyzed, with a sense of dread, and experiencing strange beings or a presence in the room, are common indicators in both children and adults.

Sometimes the alien beings are remembered as friendly playmates, or even healers (in the case of Carlos, for example, the abductee felt he was literally cured of life threatening bouts of pneumonia by the alien beings). Often, the aliens are protectors in early childhood, but the encounters become more serious and disturbing as the child approaches puberty. But even small children (as in the case of Colin, Jerry's son, whose history is in chapter 6) may be terrified by the experience of being taken from their family up into the sky against their will and subjected to painful procedures. Frequently the child will tell the parents of these encounters, which the child knows to be real, and are told by the parents that they were dreaming. They learn eventually to go "underground," and often resolve to tell no one until, as adults, they finally decide to investigate their experiences.

Abductions run in families, sometimes over three or more generations (Howe 1989). Here too the vagaries of memory—the peculiar mix of psychological defenses and an apparent control of recall by forces that the aliens command—make it difficult to develop meaningful statistics regarding the number or percentage of relatives involved. In the cases of Jerry and Arthur (chapters 6 and 15), for example, the experiencers contacted me after a conversation with an affected sibling triggered their memory. Parents who may eventually acknowledge close-up UFO sightings, or even actual abduction experiences, often initially deny their own experiences and even their children's, not wanting to be reminded of their own abduction traumas. Sometimes children will see a parent on the ship, but when the child confronts the parent with that experience the parent may not recall being abducted. Or the reverse may occur—a parent, as in Joe and Jerry's cases, or an older sibling, may recall being abducted with a child or a younger sibling, and feel deeply troubled at being unable to protect the child. Or, conversely, a child may resent an older sibling or parent, who may or may not recall the abduction, for not protecting him or her.

Although abductions or abduction-related experiences may recur throughout the experiencer's life, the pattern and timing of these encounters is not clear. Some abductees believe that they occur at times of stress or particular openness or vulnerability. But this is by no means certain. One of the most distressing aspects of the phenomenon to investigators and experiencers alike, although for different reasons, is the unpredictability of its recurrence.

There are other symptoms that are tied to unconscious association with particular elements of the abduction experiences. These may indicate a possible abduction history, but are not by themselves definitive. They include a general sense of vulnerability, especially at night; fear of hospitals (related to the intrusive procedures on the ships); fear of flying, elevators, animals, insects, and sexual contact. Particular sounds, smells, images, or activities that are disturbing for no apparent reason may later prove to be connected with the abduction experience. Insomnia, fear of the dark and of being alone at night, the covering of windows against intruders, sleeping with the light on (as an adult), and disturbing dreams and nightmares of being in strange flying craft or enclosures, are common among abductees.

Odd rashes, cuts, scoop marks, or other lesions may appear overnight, or unexplained bleeding may occur from the nose, ear, or rectum, which by itself might not draw attention, but attains significance in association with other abduction-related phenomena. Other symptoms, which later prove to be specifically related to aspects of the abduction experience, include sinus pain; urological-gynecological complaints, including unexplained difficulty during pregnancy; and persistent gastrointestinal symptoms.

For a clinician like myself, trained in the Western tradition, the investigation of abduction cases presents special challenges, since much of the information that is obtained does not fit within accepted notions of reality. The temptation is to accept some experiences, especially those that appear to make some sort of sense within our space/time paradigm, and reject others as too "far out," i.e., too far from what we know as possible from a physical standpoint. I suspect such discriminations are not wise or useful. For the whole phenomenon is so bizarre from a Western ontological standpoint that to credit some experiences because they appear, at least superficially, familiar to us and reject others on the grounds of their strangeness seems quite illogical. My criterion for including or crediting an observation by an abductee is simply whether what has been reported was felt to be real by the experiencer and was communicated sincerely and authentically to me.

THREE CLASSES OF INFORMATION

Applying the above framework I have found it useful to distinguish three classes or levels of information. First comes what might be called the nuts-and-bolts level. This concerns phenomena such as the visual sighting or radar spotting of UFOs, light and sound phenomena associated with them, the burned patches of earth that they sometimes leave, aborted pregnancies, and lesions on the surface of or implants left in abductees' bodies following their experiences. These are phenomena that appear to occur within the physical universe familiar to Western science and can be studied by its empirical methods. The field of ufology—the sighting of UFOs—was concerned primarily with directly observable phenomena until the discovery of the abduction syndrome.

Second are phenomena which *look* like they *could* be understood within our space/time universe if only we had the scientific and technological knowledge and ability to do so. These could be "extraterrestrial" phenomena which suggest technologies thousands of years ahead of us. These phenomena are, at least theoretically, not inconsistent with some sort of extension of the physical laws set forth by Western science. This category would include how the spacecraft get here (the "propulsion systems"); how they can accelerate at unbelievable speeds, virtually flicking across the sky or disappearing suddenly from a radar screen; the means by which the alien beings "float" people through doors, windows, and walls; the switching off of memory and consciousness of abductees and potential witnesses and other forms of mind control; the creation of alien/human hybrid fetuses, seen by or brought to the abductees on the ships; and the creation or staging of powerfully vivid images of landscapes, experienced by the abductee as real (for example, Catherine, chapter 7). Although we do not understand the mechanisms by which these effects are achieved, they do not, per se, require a fundamental change of paradigm. Spectacular advances in physics, biology, neuroscience, and psychology might, conceivably, shed light on them.

Finally, there are phenomena and experiences reported by abductees for which we can conceive of no explanation within a Newtonian/Cartesian or even Einsteinian space/time ontology. These include the apparent mastery of thought travel by the aliens and sometimes by the abductees themselves (as Paul, described in chapter 10); abductees' sense that their experiences are not occurring in our space/time universe, or that space and time have "collapsed"; a consciousness abductees experience of vast other realities beyond the screen of this one, beyond the "veil" (a word they frequently use); the deeply felt sense of opening up to or returning to the source of being and creation or cosmic consciousness, experienced by abductees as an inexpressibly divine light or "Home" (another word they commonly use); the experience by abductees of a dual human/alien identity, i.e., that they are themselves of alien origins (for example, Peter, Joe, and Paul, in chapters 13, 8, and 10); and the powerful reliving of past life experiences, including great cycles of birth and death. In addition, the aliens appear to be consummate shape-shifters, often appearing initially to the abductees as animals—owls, eagles, raccoons, and deer are among the creatures the abductees have seen initially—while the ships themselves may be disguised as helicopters or, as in the case of one of my clients, as a too-tall kangaroo that appeared in a park when the abductee was seven. The connection with animal spirits is very powerful for many abductees (for example, Carlos and Dave, chapters 14 and 12). This shamanic dimension needs further study. These phenomena cannot be understood within the framework of the laws of Western science, although as I have indicated, they are fully consistent with the beliefs developed thousands of years ago by other non-Western cultures.

PHENOMENOLOGY: WHAT DO EXPERIENCERS TELL US?

The summary of abduction phenomenology provided in what follows will be developed in more detail in the case examples.

How Do Abductions Begin?

Abduction encounters begin most commonly in homes or when abductees are driving automobiles. In some cases the experiencer may be walking in nature. One woman was taken from a snowmobile on a winter's day. Children have experienced being taken from school yards. The first indication that an abduction is about to occur might be an unexplained intense blue or white light that floods the bedroom, an odd buzzing or humming sound, unexplained apprehension, the sense of an unusual presence or even the direct sighting of one or more humanoid beings in the room, and, of course, the close-up sighting of a strange craft.

When an abduction begins during the night, or, as is common, during the early hours of the morning, the experiencer may at first call what is happening a dream. But careful questioning will reveal that the experiencer had not fallen asleep at all, or that the experience began in a conscious state after awakening. As the abduction begins the abductee may experience a subtle shift of consciousness, but this state of being is just as real, or even more so, than the "normal" one. Sometimes there is a moment of shock and sadness when the abductee discovers in the initial interview, or during a hypnosis session, that what they had more comfortably held to be a dream was actually some sort of bizarre, threatening, and vivid experience which they may then recall has occurred repeatedly and for which they have no explanation.

After the initial contact, the abductee is commonly "floated" (the word most commonly used) down the hall, through the wall or windows of the house, or through the roof of the car. They are usually astounded to discover that they are passed through solid objects, experiencing only a slight vibratory sensation. In most cases the beam of light seems to serve as an energy source or "ramp" for transporting the abductee from the place where the abduction starts to a waiting vehicle. Usually the experiencer is accompanied by one, two, or more humanoid beings who guide them to the ship. At some point early in this process the experiencer discovers that he or she has been numbed or totally paralyzed by a touch of the hand or an instrument held by one of the beings. Abductees may still be able to move their heads, and usually can see what is going on, although frequently they will close their eyes so they can deny or avoid experiencing the reality of what is occurring. The terror associated with this helplessness blends with the frightening nature of the whole strange experience.

When abductions begin in the bedroom, the experiencer may not initially see the spacecraft, which is the source of the light and is outside the house. The UFOs vary in size from a few feet across to several hundred yards wide. They are described as silvery or metallic and cigar-, saucer-, or dome-shaped. Strong white, blue, orange, or red light emanates from the bottom of the craft, which is apparently related to the propulsion energy, and also from porthole-like openings that ring its outer edge. After they are taken from the house, abductees commonly see a small spacecraft which may be standing on long legs. They are initially taken into this craft, which then rises to a second larger or "mother" ship. At other times they experience being taken up through the night sky directly to the large ship and will see the house or ground below receding dramatically. Often the abductee will struggle at this and later points to stop the experience, but this does little good except to give the individual a vital sense that he or she is not simply a passive victim. There is a debate in abduction work as to whether abductions can be stopped, or even whether to do so is a good idea (Druffel, in press). There are small variations in what is experienced during this phase of the abduction. Arthur (chapter 15), for example, described ascending to a UFO on a kind of threadlike arc that extended to the craft from the car his mother had been driving when the abduction began.

Independent Witnessing

Independent witnessing of an abduction does occur, but is, in my experience, relatively rare and limited in nature. As in so many aspects of the phenomenon, the evidence may be compelling, yet at the same time maddeningly subtle and difficult to corroborate with as much supporting data as firm proof would require. Husbands and wives, for example, are commonly "switched off" while the spouse is being abducted and have "slept" through the whole event. The abductee is sometimes highly frustrated when loud screaming fails to rouse the sleeping partner, who may seem to be in a state of unconsciousness deeper than sleep, appearing as if dead.

Hopkins has documented a case, now being widely discussed, where a woman made an unsolicited report to him that from the Brooklyn Bridge she saw his client, Linda Cortile, being taken by alien beings from her twelfth story East River apartment into a waiting spacecraft that then plunged into the river below (Hopkins 1992, in press). These observations corresponded precisely with what Mrs. Cortile had told Hopkins happened to her when he recovered her memories of an abduction that occurred in November 1989. This is, to my knowledge, the only documented case where an individual, who was not him- or herself abducted, reported witnessing an abduction as it was actually taking place. The witnesses of an abduction, it seems, are often themselves abductees, who may be involved in the same event, raising questions about the "objectivity" of the observer. Sometimes, according to reports, the abductee may be noted to be missing for a half hour or more or, in rare cases, for days, as in the famous Travis Walton case (Walton 1978; Tormé 1993), by family members or others. But in these instances no one has seen them being taken into a spacecraft, and there is no firm proof that abduction was the cause of their absence.

One of my first cases, a young woman of twenty-four, was abducted as a teenager with a friend after midnight from the basement den of her friend's home. The girls' fathers were frantic when they could not find their daughters during the night. According to both girls (I have spoken with the other girl, who confirms my client's account) the fathers checked the den during the early hours of the morning and found that the two teenagers were not there. By six o'clock they were both back in the den. In another case the eight-year-old daughter of one of my abductee-clients, who is herself probably also an abductee, observed her mother to be missing from her room when she looked for her during the night. The mother told me that she had had an abduction experience at the exact time the daughter told her she was missing. In the morning the girl said to her mother, "Daddy was there and the covers on your side were turned down but you were gone." Another client of mine was abducted with a college roommate from their dormitory. She actually saw her roommate being returned through the door through which she was taken. When the beings returned the roommate, my client observed, "her head was hanging and her hair was hanging down and I thought that she was dead." But she herself was then abducted, so that her credibility as an "independent" witness might therefore be questionable.

Independent observation of a UFO near where an experiencer reports that an abduction took place is another kind of corroborative evidence, especially if the abductee did not him- or herself actually see the craft. We will see in Catherine's case (chapter 7) that she was shocked to discover in the media the next morning that a UFO had been observed traveling the exact route north of Boston that she had felt compelled to drive during the night. Her drive culminated in an abduction, which took place in a wooded section of a suburb about fifteen miles northeast of Boston, but Catherine never saw the UFO itself, except in the air near her car as the abduction was actually taking place. Peter (chapter 13) reported being taken into a UFO from a Connecticut home while three of his friends who were walking outside witnessed a UFO close above the house. The case is weakened by the fact that the three witnesses failed to check inside the house to see if he was actually missing.

Inside the Ships: The Beings

Sometimes abductees will remember being taken into the ship through its underside or through oval portals along its edge, although often they cannot recall the moment when they entered the craft. Once inside they may at first find that they are in a small dark room, a sort of vestibule. But soon they are taken into one or more larger rooms where the various procedures will occur. These rooms are brightly lit, with a hazy luminosity from indirect light sources in the walls. The atmosphere may be dank, cool, and occasionally even foul-smelling. The walls and ceilings are curved and usually white, although the floor may appear dark or even black. Computer-like consoles and other equipment and instruments line the sides of the rooms, which may have balconies and various levels and alcoves. None of the equipment or instruments are quite like ones with which we are familiar (Miller, in press). Furniture is sparse, limited generally to body-conforming chairs and tables with a single support stand that can tilt one way or another during the procedures. The ambiance is generally sterile and cold, mechanistic and hospital-like, except when some sort of more complex staging occurs. Many more details of the inside of the ships and, of course, of the abduction processes themselves, will be provided in the case histories.

Inside the ships the abductees usually witness more alien beings, who are busy doing various tasks related to monitoring the equipment and handling the abduction procedures. The beings described by my cases are of several sorts. They appear as tall or short luminous entities that may be translucent, or at least not altogether solid. Reptilian creatures have been seen (Carlos, chapter 14) that seem to be carrying out mechanical functions. Human helpers are sometimes observed working alongside the humanoid alien beings. But by far the most common entity observed are the small "grays," humanoid beings three to four feet in height. The grays are mainly of two kinds—smaller drone or insectlike workers, who move or glide robotically outside and inside the ships and perform various specific tasks, and a slightly taller leader or "doctor," as the abductees most often call him. Female "nurses," or other beings with special functions, are observed. The leader is usually felt to be male, although female leaders are also seen. Gender difference is not determined so much anatomically as by an intuitive feeling that abductees find difficult to put into words.

The small grays have large, pear-shaped heads that protrude in the back, long arms with three or four long fingers, a thin torso, and spindly legs. Feet are not often seen directly, and are usually covered with single-piece boots. External genitalia, with rare exceptions (Joe, chapter 8), are not observed. The beings are hairless with no ears, have rudimentary nostril holes, and a thin slit for a mouth which rarely opens or is expressive of emotion. By far the most prominent features are huge, black eyes which curve upward and are more rounded toward the center of the head and pointed at the outer edge. They seem to have no whites or pupils, although occasionally the abductee may be able to see a kind of eye inside the eye, with the outer blackness appearing as a sort of goggle. The eyes, as we will see in the case examples, have a compelling power, and the abductees will often wish to avoid looking directly into them because of the overwhelming dread of their own sense of self, or loss of will, that occurs when they do so. In addition to boots, the aliens usually wear a form-fitting, single-piece, tuniclike garment, which is sparsely adorned. A kind of cowl or hood is frequently reported.

The leader or doctor is slightly taller, perhaps four and a half or five feet at most, and has features similar to the smaller grays, except that he may seem older or more wrinkled. He is clearly in charge of the procedures that occur on the ship. The attitude of the abductees toward the leader is generally ambivalent. They often discover that they have known one leader-being throughout their lives and have a strong bond with him, experiencing a powerful, and even reciprocal, love relationship. At the same time, they resent the control he has exercised in their lives. Communication between the aliens and humans is telepathic, mind to mind or thought to thought, with no specific common learned language being necessary.

Procedures

The procedures that occur on the ships have been described in great detail in the literature on abductions (Bullard 1987; Hopkins 1981, 1987; Jacobs 1992) and will be summarized only briefly here, although many of them will be described in detail in the case examples. They might be categorized as of two sorts, physical and informational.

The abductee is usually undressed and is forced naked, or wearing only a single garment such as a T-shirt, onto a body-fitting table where most of the procedures occur. The experiencer may be the only one undergoing the procedures during a particular abduction, or may see one, two, or many other human beings undergoing similar intrusions. The beings seem to study their captives endlessly, staring at them extensively, often with the large eyes close up to the humans' heads. The abductees may feel as if the contents of their minds have been totally known, even, in a sense, taken over. Skin and hair, and other samples from inside the body, are taken with the use of various instruments that the abductees can sometimes describe in great detail.

Instruments are used to penetrate virtually every part of the abductees' bodies, including the nose, sinuses, eyes, ears, and other parts of the head, arms, legs, feet, abdomen, genitalia, and, more rarely, the chest. Extensive surgical-like procedures done inside the head have been described, which abductees feel may alter their nervous systems. The most common, and evidently most important procedures, involve the reproductive system. Instruments that penetrate the abdomen or involve the genital organs themselves are used to take sperm samples from men and to remove or fertilize eggs of the female. Abductees experience being impregnated by the alien beings and later having an alien-human or human-human pregnancy removed. They see the little fetuses being put into containers on the ships, and during subsequent abductions may see incubators where the hybrid babies are being raised (as do Catherine, Jerry, and Peter, among my cases). Experiencers may also see older hybrid children, adolescents, and adults, which they are told by the aliens or know intuitively are their own. Sometimes the aliens will try to have the human mothers hold and nurture these creatures, who may appear quite listless, or will encourage human children to play with the hybrid ones as, for example, Catherine is made to do.

Needless to say all of this is deeply disturbing to the abductees, at least at first, or when they first remember their experiences. Their terror may be mitigated somewhat by reassurances the aliens give that no serious harm will befall them, and by various anxiety-reducing or anesthesia-like means they use. These involve instruments that affect the "energy" or "vibrations" (words that abductees often use) of the body. These processes may greatly reduce the abductees' fear or pain, and even bring about states of considerable relaxation. But in other cases they are incompletely successful and terror, pain, and rage break through the emotion-extinguishing devices used. As I will document in detail in several case examples, the traumatic, rapelike nature of the abduction memories, or even of the process itself, may become altered as the abductees reach new levels of understanding of what is occurring, and as their relationship to the beings themselves changes in the course of our work.

In sum, the purely physical or biological aspect of the abduction phenomenon seems to have to do with some sort of genetic or quasi-genetic engineering for the purpose of creating human/alien hybrid offspring. We have no evidence of alien-induced genetic alteration in the strictly biological sense, although it is possible that this has occurred.

Information and the Alteration of Consciousness

The other important, related aspect of the abduction phenomenon has to do with the provision of information and the alteration of consciousness of the abductees. This is not a purely cognitive process, but one that reaches deeply into the emotional and spiritual lives of the experiencers, profoundly changing their perceptions of themselves, the world, and their place in it. This information concerns the fate of the earth and human responsibility for the destructive activities that are taking place on it. It is conveyed by the direct mind-to-mind telepathic communication referred to above and through powerful images shown on television monitor-like screens on the ships themselves. The information may begin to be conveyed when the abductees are children or adolescents (see Arthur, chapter 15, and Ed, chapter 3), but its implications are not fully understood until much later. The investigator seems to play an important part in enabling the abductee to bring forth and realize the significance of the information they have been receiving during abductions that have been taking place over many years.

Scenes of the earth devastated by a nuclear holocaust, vast panoramas of lifeless polluted landscapes and waters, and apocalyptic images of giant earthquakes, firestorms, floods, and even fractures of the planet itself are shown by the aliens. These are powerfully disturbing to the abductees, who tend to experience them as literally predictive of the future of the planet. Some abductees are given assignments in this future holocaust as it is displayed, such as to feed the survivors, or are told, as in the prophetic books of the Bible, that some will perish while others will be taken to another place to participate in the evolution of life in the universe.

Some abduction researchers believe that these images are not shown for the purpose of altering the course of the planet's history in a positive way. Rather, they maintain, the beings are studying the experiencers' reactions and are deceiving them into believing that they are concerned with our fate while they proceed to take over our planet, their own having presumably been destroyed by an apocalypse of science and technology similar to the fate that might befall us (various personal communications 1990–93; also Scott, chapter 5). They argue further that if the aliens were truly concerned with our well-being they would manifest themselves more forthrightly and intervene directly in our affairs in order to make things better.

The aliens themselves, when confronted with this issue, say that we are not ready to acknowledge their existence, and would treat them aggressively as an enemy as we do anyone or anything different from ourselves that we do not understand. But most importantly, the aliens say, their methods are different. They do not wish to bring about change through coercion but rather through a change of consciousness that would lead to our choosing a different course. Some abductees receive information of battles for the fate of the earth and the control of human mind, between two or more groups of beings, some of which are more evolved or "good," while others are less evolved or "evil."

The abductees usually remember fewer details of their return to Earth than they do of their abduction. Usually they are returned to the bed or car from which they were taken, but sometimes "mistakes" are made. They may be returned quite a distance, or even miles, away from their home. This is rare, and I have seen no cases of this kind, although Budd Hopkins has told me of such instances. Smaller mistakes are more common, such as landing the experiencer facing in the wrong direction on the bed, with his or her pajamas on backwards or inside out, or with certain garments or jewelry missing. Sometimes the aliens seem to be making a point, or a certain humor is involved. One two-year-old among my cases was tucked into his bed tightly after an abduction, which the parents say that neither they nor his older sister had done; he, of course, was incapable of doing this. Hopkins tells of a case where two abductees were returned to the wrong cars. As they drove along the highway the drivers recognized each other's cars. They were "reabducted" and returned to the appropriate vehicles (personal communication, December 1992).

After the abduction the experiencer may have varying degrees of recollection of what occurred. Sometimes what happened will be remembered as a dream. The abductee may wake with unexplained cuts or other lesions (the mucus membrane taken from inside the nose and under the tongue in one of my cases), small lumps under the skin, a headache, or nosebleed. Generally experiencers are quite tired afterwards and feel as if they have been through some sort of stressful experience.

Physical Aspects

The physical phenomena that accompany abductions are important, but gain their significance primarily in that they corroborate the experiences themselves; for the effects tend to be subtle and would not by themselves convince a Western-trained clinician of their meaning. For example, even though the abductees are certain that the cuts, scars, scoop marks, and small fresh ulcers that appear on their bodies after their experiences are related to the physical procedures performed on the ships, these lesions are usually too trivial by themselves to be medically significant. Similarly, abductees will often experience that they have been pregnant and have had the pregnancy removed during an abduction, but there is not yet a case where a physician has documented that a fetus has disappeared in relation to an abduction (Druffel 1991; Miller and Neal, in press; Neal 1992). Many abductees have noted that electrical or electronic devices—television sets, radios, electric clocks, telephone answering machines, electric lights, and toasters—malfunction in relation to abductions, or simply when the experiencers are nearby. But it is almost impossible to prove that these disturbances are related to the abduction process, or even that they have occurred at all.

Abductees frequently experience that some sort of homing object has been inserted in their bodies, especially in the head but other parts as well, so that the aliens can track or monitor them, analogous, the abductees themselves will observe, to the way we track animals with various devices. These so-called implants may be felt as small nodules below the skin, and in several cases tiny objects have been recovered and analyzed biochemically and electromicroscopically. MIT physicist David Pritchard, who has also been analyzing an implant that came out of a man's penis, has written about the criteria for examining and determining the nature of such objects (Pritchard 1992). I have myself studied a ½- to ¾-inch thin, wiry object that was given to me by one of my clients, a twenty-four-year-old woman, after it came out of her nose following an abduction experience. Elemental analyses and electronic microscopic photography revealed an interestingly twisted fiber consisting of carbon, silicon, oxygen, no nitrogen, and traces of other elements. A carbon isotopic analysis was not remarkable. A nuclear biologist colleague said the "specimen" was not a naturally occurring biological subject but could be a manufactured fiber of some sort. It seemed difficult to know how to proceed further.

There is no evidence that any of the implants recovered are composed of rare elements, or of common ones in unusual combinations. In discussions with a chemical engineer and other experts in materials technology, I have been told that it would be extremely difficult to make a positive diagnosis of the nature of any unknown substance without having more information about its origins. Under the best of circumstances it would be difficult to prove, for example, that a substance was not of terrestrial or even human biological origins.

Assuming that, in fact, these objects were left in the human body by alien beings, which would be virtually impossible to prove, it would not be difficult for the aliens, in light of all the other seemingly miraculous things of which they appear capable, to adapt a small object to the human body by forming it along the lines of the body's own chemistry. If that were the case, the analysis would yield nothing unusual. This was actually my experience in the case of Jerry (chapter 6), who felt strongly that two small nodules that appeared on her wrist following an abduction experience had not been there before. She agreed to have these removed by a surgeon colleague of mine, but the pathology laboratory found nothing remarkable about the tissue.

There was considerable excitement among abduction researchers when the first implant was "discovered." Here, at last, would be the concrete physical proof of the reality of abductions, a real object recovered from the alien world, the smoking gun that would silence the critics. I am not now so sanguine that the phenomenon will reveal itself in this fashion. To hope so may even be a sort of "error of logical types." In other words it may be wrong to expect that a phenomenon whose very nature is subtle, and one of whose purposes may be to stretch and expand our ways of knowing beyond the purely materialist approaches of Western science, will yield its secrets to an epistemology or methodology that operates at a lower level of consciousness (the point is made in Eva's case, chapter 11).

A theory that would begin to explain the abduction phenomena would thus have to account for five basic dimensions. These are:

1. The high degree of consistency of detailed abduction accounts, reported with emotion appropriate to actual experiences told by apparently reliable observers.
2. The absence of psychiatric illness or other apparent psychological or emotional factors that could account for what is being reported.
3. The physical changes and lesions affecting the bodies of the experiencers, which follow no evident psychodynamic pattern.
4. The association with UFOs witnessed independently by others while abductions are taking place (which the abductee may not see).
5. The reports of abductions by children as young as two or three years of age (see Colin in chapter 6).

IMPACT AND SEQUELAE OF ABDUCTIONS

Needless to say, abductions profoundly affect the lives of those who experience them. These effects are traumatic and disturbing, but they can also be transforming, leading to significant personal change and spiritual growth. Whether this transformational element is intrinsic to the abduction phenomenon itself, dependent in part on integrative therapeutic work with the investigator, or is a by product of coming to terms with the traumatic nature of the experiences, is one of the questions that will be explored in this book.

Trauma

The traumatic aspect has four dimensions. First are the experiences themselves. To be paralyzed and taken against one's will by strange beings into a foreign enclosure and subjected to intrusive, rapelike procedures, some of which are especially humiliating to human dignity, is obviously highly disturbing. In this light, it is surprising that abductees as a group are not more emotionally troubled than they are.

Second, abductees experience a lifelong sense of isolation and estrangement from those around them. Whether or not they recall consciously many elements of their experiences, abductees feel that they are somehow different or "other," that they do not belong in this society even if, superficially at least, they seem to get along well. As children they have commonly been told that the abduction-related events they describe are dreams, or even that they are lying, so experiencers learn to keep these matters to themselves and feel very much alone with their experiences. One savvy eight-year-old abductee looked at me incredulously when I asked him if he told his friends about his "encounters," which he was able to distinguish sharply from dreams, even when they had to do with UFOs. "No, I don't tell anybody that I don't know that well," he said. "I just don't want them to know that I have encounters. I think that a lot of people I know get scared if they hear scary stories . . . I guess people are like, 'Hey! That's too weird!'" This boy is, in fact, popular with other children and his teachers and they find nothing unusual about him. As adults too abductees learn not to talk about their experiences, except under trusting circumstances, knowing that they are likely to be met with skepticism and false interpretation, if not outright derision.

Third, abductees experience what I have called "ontological shock" as the reality of their encounters sinks in. They, like all of us, have been raised in the belief that we on Earth are largely alone in the universe and that it would simply not be possible for intelligent beings to enter into our world without using a highly advanced form of our technology and obeying the laws of our physics. Abductees tend to persist in the hope that a psychological explanation for their experiences will be found, even when they tell me that what has happened to them is as real as the conversation we are having.

Finally, abduction-related traumas are unusual in that they can recur at any time. Most traumas, such as war-related experiences, rape or childhood abuse, are finite; they occur and then are over, even if they persist during a given period of time. But abductions are unpredictable and their recurrence in an individual's life follows no foreseeable pattern. Parent abductees will commonly first seek to investigate their own experiences when they discover that one or more of their children are having abduction encounters. The discovery that they cannot fulfill their protective responsibilities as a parent will breach their denial and motivate them to confront their own buried experiences so they can be more helpful to their children.

In addition to these specifically traumatic long-term effects, abductees may also suffer from a number of long-term symptoms that, though subtle, they relate to their abduction experiences. These include various fears, discussed earlier, such as of hospitals and needles, as well as symptoms such as headaches, nasal sinus pains, limb pains, gastrointestinal and urological-gynecological symptoms, and disturbances of sexual functioning (Jerry, chapter 6). It is somewhat ironic in view of these pathological sequelae that so many abductees have experienced or witnessed healing of conditions ranging from minor wounds to pneumonia, childhood leukemia, and even in one case reported to me first hand, the overcoming of muscular atrophy in a leg related to poliomyelitis.

It is interesting that not all abductees experience the intrusive, traumatic procedures that have come to be seen as characterizing the phenomenon (for example, Arthur, chapter 15). I do not think this is simply a matter of resistance or denial. Some individuals seem to be "selected" primarily to be instructed, even "enlightened," a kind of "reprogramming," as one woman puts it, by beings that are usually of the subtler or luminous sort. Perhaps these individuals, who seem to have spiritual leadership qualities, have a different consciousness, are more fearless—or willing to be out of control and move through their terror—than other abductees. It is a question that deserves further study.

As will be discussed in several cases in this book, an abduction history can place a great strain on a marital or other intimate relationship. This is especially the case when one member of the couple is an experiencer and the other not only evidently is not, but finds that he or she cannot accept the reality of the spouse's experiences. Relationships are also disrupted when one member of a couple undergoes significant personal development, directly or indirectly resulting from their experiences, leaving the spouse more or less behind (Eva, chapter 11).

TRANSFORMATIONAL OR CONSCIOUSNESS-ALTERING ASPECTS

I will devote more attention in this book to the transformational and spiritual growth aspects of the abduction phenomenon than has been the case in other literature on the subject. There are several reasons for this decision. First, I believe that this feature of the phenomenon has either been neglected or has been viewed as incompatible with the traumatic dimension of an abduction as it has most often been described. Second, it is my impression that this largely unresearched area is of considerable significance. Finally, and most interesting, I think, is my personal experience as a psychiatrist dealing with abductees: I seem to receive more information of this kind in my work with abductees than, apparently, do other investigators. It is not altogether clear why this is so. Perhaps my caseload is preselected, biased in the direction of individuals who, in seeking the help of a psychiatrist, deepen their understanding of their experience through exploration of their consciousness. Possibly, abductees sense that I am open to hearing about experiences or information that might be considered too "far out" for most investigators, and my own personal evolution may, in fact, have made me more open to the information they are seeking to convey. In any event, I try to be as scrupulous as I can not to lead clients in any particular direction, so that if information that is relevant to the spiritual or consciousness-expanding aspects of the abduction phenomenon emerges during our sessions, it will do so freely and spontaneously and not as a result of specific inquiries of mine.

Because so much of this book is concerned with the spiritual or transformational dimensions of the abduction phenomenon, here I will only briefly outline the types of experiences that could be placed in this category. Of overriding importance is the shift that needs to occur in the relationship between the experiencer and the alien beings before consciousness-altering information can be received. Although the relationship with the aliens may have been playful, even intimate, in early childhood, it tends to change to a more disturbing and traumatic one as puberty approaches and the reproductive hybrid "project" begins. As traumatic intrusions take place, the abductees tend to feel themselves to be victims of hostile beings who regard them coldly, or simply as specimens in a project that serves the needs of the aliens. They may feel betrayed by the alien beings as the nature of their interaction changes.

But as our work deepens, especially as the reality of the alien intelligence is acknowledged and the abductees come to accept their lack of control of the process, the frightening and adversarial quality of the relationship seems to give way to a more reciprocal one in which useful human-alien communication can take place and mutual benefit is derived. The abductees may even experience a profound love for the alien beings—in some ways more powerful than the love they experience in human relationships—and may feel that this love is returned. Connection through the eyes seems to play an important part in the evolution of this process. Whereas, for example, the abductees felt bitterly resentful about having their sperm and eggs used by the aliens in the hybridization project, they may come to feel that they are participating in a process that has value for the creation and evolution of life.

There are those who might argue that such a shift in stance by the abductees in the face of the ongoing helplessness of the abduction situation is in fact a defensive shift. It could be considered an attempt by the ego to retain a sense of mastery by giving away voluntarily what will be taken by force, or an attempt to reduce cognitive dissonance by believing that the emotional cost of such a traumatic experience can be balanced by providing something good and positive for the universe. On the other hand, it is possible that working through the shattering experience of the abduction may give abductees access to experiences of transpersonal meaning, universal love, and connectedness that make such compassion possible.

As is true of so many aspects of the abduction phenomenon, it is difficult in the area of transformational or spiritual growth-related experiences to separate cause and effect, or even to think in causal terms at all. Does an abductee, for example, receive (and communicate) information about a past life experience because his consciousness is open to the possibility of such matters? Or will the emergence into consciousness of the memory of a past life, itself facilitated by our work together, bring about an expanded personal horizon and a broadening of the sense of self in relation to the larger fabric of universal consciousness?

The fact that the relationship between the abductees and the aliens can evolve so dramatically over time makes me question categorizations of the beings into constructive, good, and loving ones and others that are deceptive and hostile, bent on taking over our planet—the idea that the light beings are good or caring, for example; the grays are businesslike and indifferent. This kind of taxonomy smacks suspiciously of the sorts of polarization that characterizes human group or ethnonational relationships and may have little to do with the way interspecies or interdimensional relationships work beyond the earth. Furthermore, it is common for abductees to experience, for example, both light beings *and* little grays (Arthur, chapter 15) or reptilian and other sorts of beings (Carlos, chapter 14), during the same abduction. It is possible that we are dealing with interconnecting or reciprocal relationship processes that are evolutionary in nature and not comprehensible in the linear terms of our familiar polarities.

Types of experience during abductions that appear to be related to personal growth and transformation are as follows:

1. "Pushing through" occurs, i.e., fully experiencing the terror and rage associated with the helplessness and intrusive instrumentation on the ships. When this takes place acknowledgment and acceptance of the reality of the beings becomes possible and a more reciprocal relationship follows in which personal growth and learning can take place. From the "ego death" follow other levels of transformation:
2. The aliens are recognized as intermediaries or intermediate entities between the fully embodied state of human beings and the primal source of creation or God (in the sense of a cosmic consciousness, rather than a personified being). In this regard abductees sometimes liken the alien beings to angels, or other "light beings" (including the "grays").
3. The abductees may actually experience themselves as returning to their cosmic source or "Home," an inexpressibly beautiful realm beyond, or not in, space/time as we know it. When this occurs during a hypnosis session, powerful, inexpressibly joyous, even orgiastic, feeling occurs. Conversely, abductees may weep with sadness when they experience having to leave their cosmic home, return to Earth and become embodied once again.
4. Past lives are experienced during the sessions with strong emotion appropriate to what is being remembered. This is most likely to occur when the investigator picks up on cues in the sessions during which encounters from infancy are being remembered. Complaints or simply observations of being here on Earth "again," of being "back" or having "returned," are voiced (about which I then inquire). The past lives that are recalled seem to have relevance to the personal development or evolution of the experiencer, as I have seen in the cases of Dave and Joe.
5. Past life experiences provide abductees (and the investigator) with a different perspective about time and the nature of human identity. Cycles of birth and death over long stretches of

time can thus be relived, providing a different, less ego derived sense of the continuity of life and the smallness of an individual lifetime from a cosmic perspective. Consciousness is experienced as not coterminous with the body; the notion of a soul with an existence separate from the body becomes relevant.

6. Once the separateness of consciousness from the body is grasped, other kinds of "transpersonal" experiences become possible; identification of consciousness with virtually endless kinds of beings and entities through space/time and beyond often occurs. Paul (chapter 10), among my cases for example, found himself during our sessions identified with dinosaurs or dinosaur-like reptiles from another era and experienced himself present at the site of a UFO crash several decades ago when alien beings were destroyed by human fear and aggression. Another abductee, a young Brazilian man, found that his alien encounters opened him to the identification with the myths and spirit entities of his culture's folklore, from which his Western scientific and intellectual training had cut him off.
7. A distinct but important aspect of this kind of transpersonal experience is an abductee's sense of possessing a double human/alien identity. In their alien selves they may discover themselves doing many of the things that the "other" aliens have done to them and to other human beings, such as studying their minds or even carrying out reproductive procedures. The alien identity seems to be connected in some way with the soul of the human self, and one of the tasks the abductee then confronts is the integration of their human and alien selves, which takes on the character of a reensoulment of their humanity.
8. The reliving of abduction experiences leads abductees to open to other realities beyond space/time, realms that are variously described as beyond a "veil" or some other barrier which has kept them in a "box" or in a consciousness limited to the physical world. When asked about these experiences abductees have trouble finding the words to describe what has occurred and speak of the "collapse" of space/time, of the nonrelevance of the notions of space and time, and of being in multiple times and places at the same moment.

The result of all these experiences for abductees is the discovery of a new and altered sense of their place in the cosmic design, one that is more modest, respectful, and harmonious in relation to the earth and its living systems. Emotions of awe, respect for the mystery of nature, and a heightened sense of the sacredness of the natural world are experienced along with deep sadness about the apparent hopelessness of Earth's environmental crisis. One of John Carpenter's cases described herself as having become a "child of the universe" after she had become conscious of her abduction experiences. The meaning and implications of these shifts of consciousness for possible human futures will be discussed more fully in the case examples and in the concluding chapter.

The thirteen cases presented in this book—eight men and five women—were selected from the seventy-six abductees I have interviewed on the basis of the following criteria:

1. Their stories, although in some instances complex, seemed to me sufficiently clear to permit a coherent narrative.
2. Each case appeared to illustrate in depth one or more central aspects of the abduction phenomenon.
3. Each person was willing to have his or her story told, with or without the use of his or her actual name.
4. I knew these individuals quite well. But there are abductees I have known longer or worked with in greater depth. If I have chosen not to tell their stories here it is because I could not do justice to the richness of their experiences in a sufficiently clear and concise manner.

The sequence of cases reflects generally a kind of progression from simpler stories to more complex multidimensional narratives. The last case suggests what the abduction phenomenon may hold for the transformation of our institutions and collective lives.

CHAPTER THREE

You Will Remember When You Need to Know

Ed is a technician in his mid-forties at a high tech firm in Massachusetts, married to Lynn, a writer with whom he shares an interest in science and technology. One day in the summer of 1989 Ed and Lynn were walking along the Marginal Way in Ogunquit, Maine, a cliffside path that weaves along the rocky shore for several miles. Suddenly Ed found himself becoming tense, moody, and withdrawn. He then became sweaty and worried and grabbed Lynn's hand tightly. He had no idea what his distress was about. Ed was practicing meditation and believes that this may have contributed to his eventual recovery of important memories. Ed has also had a number of frightening childhood experiences which are probably abduction-related. These will be discussed in the context of his hypnosis session. From early childhood Ed was unusually fearful of doctors' offices and operations—"anything to do with medicine"—even before his tonsillectomy at about age nine.

One day at the seashore, a week or so after the walk in Ogunquit, following a day of relaxation, he says, "it came to me." Ed began to recall an experience from the summer of 1961 when he was in high school. Over the next few months more details came to his mind through what he called "flashbacks." Ed had had some interest in what he calls "alien intelligence." As a result of his memories he became interested in the UFO phenomenon and attended a MUFON (the citizen-based Mutual UFO Network) conference in New Hampshire. Several people he met through this network suggested he contact me, and so he telephoned my office in July 1992. Since then I have interviewed Ed and Lynn for several hours and have hypnotized him with the aim of recovering more details of his high school experience. He and Lynn have also been attending my support group.

Ed's case is important for two principle reasons. First, the timing of his teenage experience and his recall of it indicate a process of information reception, storage, recovery, and integration of great purpose and potential power. Second, the narrative which Ed was able to recover in an altered state of consciousness appears, from what we know of the abduction phenomenon, to be much more plausible than the account he could provide from conscious memory. This supports the argument for the power of hypnosis to recover memories of abductions that are both meaningful and true to the actual experience (whatever the source of these experiences may ultimately prove to be), and suggests that, at least in the case of UFO abductions, hypnosis may be more of a clarifying than a distorting tool.

In what follows I will first tell the story of young Ed's abduction experience as he remembered it in our initial meeting on July 23, 1992. I will then bring in the rich details that he recovered in hypnosis on October 8 which give meaning and greater coherence to his experiences and subsequent life. Ed also has less clear recollections of earlier childhood visitations.

In July 1961, Ed, his friend Bob Baxter, and Bob's parents took a trip up the coast of Maine in the Baxter's Nash. One damp and foggy night they stopped at a place where the shore was rocky; Ed does not recall exactly where except that it was north of Portland. The Baxters stayed in a cabin while the boys slept nearby in the Nash, which had fold-down seats in the back. The car was parked perhaps a hundred yards from the sea. Ed and Bob had been talking about how "horny" they were and "speculated about the great encounters we were going to have at the beach." Ed believes he had been asleep when "the next thing I know I'm out on this precipice" in a "pod" that had "some sort of a glass bubble over it." He was naked in a small room with transparent curved walls. The room felt warm and safe, but Ed could see the pounding surf and "hear the wind buffeting around outside." He believes he could hear strains of melodious, light classical music coming perhaps from houses nearby. There is no question in Ed's mind that this really happened, although he calls the experience "beyond language."

With Ed in the pod was a small, slight female figure with long, straight, thin silvery-blond hair. Although Ed could recall no specific earlier abduction experiences, the figure "had familiar aspects to her," and he had vague, "very sinister," memories of "something out there" from his childhood. The female entity had a small mouth and nose, intense large dark eyes, and a "sort of triangular" shaped head with a "largish" forehead. "I had this uncomfortable feeling that every time she looked at me she could just see right into me." He found her "attractively unusual" and felt "a little self-conscious." The figure, perhaps sensing this, "gave me some sort of blanket or big towel or something." She seemed to sense his thoughts without his saying anything, reassuring Ed, for example, that they were safe and would not fall off the precipice onto the rocks below. Ed was sexually excited, and the female being "sensed my horniness." Although he was "hazy" as to how this came about, Ed said, "we had intercourse." According to Ed this act was "similar" to human sexual intercourse with "fondling of the breasts," insertion of the penis in the vagina, and active participation by both individuals. Interestingly, although Ed was a virgin at this time, he did not recall this experience and still felt himself to be a virgin when he had sexual intercourse some time later.

After the sex, which Ed said was "fulfilling" and "great," he sensed that the female wanted to get on with a "bigger agenda—You know, you take care of the immediate physical needs, and you get on to the lesson." Her attitude was "now we'll get down to business . . . like a teacher. You might settle your students by telling them a light story when they first come in the classroom, just get them settled, focused, and then kind of lead them into . . . Then she started explaining things to me." Ed wanted to write things down so he could remember later, but she would not let him and "just worked at my perception by awareness, sort of mind to mind." Sensing his frustration she assured him, "You will remember when you need to know."

In this interview Ed's recall of the content of the information this female entity gave to him was sketchy, but he remembers that he felt "dumbfounded" with "my mouth hanging open." He had had a traditional Roman Catholic upbringing, attending parochial school until the fourth grade, and nothing he had been taught prepared him for messages of such spiritual or cosmic import. Somehow the alien being opened Ed's consciousness. "She wired me into my emotions, and at some point early in the encounter, perhaps during the 'foreplay,' 'interplay,' orgasmic parts of it, like she sort of gains either a clinical judgment of my emotional/mental typography, or gained my agreement to go on to part two. She looked at me and said, 'Well, do you think you can handle the next part?'"

Some of the information concerned "the way humans are conducting themselves here in terms of international politics, our environment, our violence to each other, our food, and all that. And she kept explaining that the laws of the universe are this way, and, it's like, if you're driving on the wrong side of the road, what's going to happen inevitably, you know. It's like here are the laws, and here's the way you humans conduct your affairs, and slam, bang, you know, it's inevitable . . . " Ed's father was a machinist by trade working at a major firm in the New England region and Ed's idea, inculcated in his "flag-waving" family, was that he would be a "techie," go into electronics and help to "beat the damn communists . . . We gotta develop more, better technology to kill those damn Commies before they kill us."

Although the information was largely new to Ed, it somehow "made sense to me." He had "a scientific bent towards things," and "she explained things in "scientific, logical terms . . . elucidating a series of interrelated concepts, that these are the laws of the universe, specifying in detail these concepts. And you know, here, you geniuses on the planet doing this and that, and this is the way the thing should operate, and you're out of harmony, and at some point the sheets get balanced out. Sooner or later. And I'm standing there, oh my God. It's like she gave me this second, this safety mechanism of embedding it deep in my core . . . I just sat there, like, you know, like, oh my God, it's, it was like trauma time . . . deep anxious concern over the path I could see us humans taking . . . a world trauma."

Ed was told of the "heavily destructive" path we were taking, which was also destructive to the "humanoid's planet." He feels that his whole psyche was permanently changed by this encounter. "How do I put it? She said that with my acquiescence that I was, my emotions, my cognitions, my whole perceptions, were going to be modified, and that my style of functioning would be slightly different. The closest analogy I can give is like changing some of the software architecture and some of the hardware in the computer. At first when you sit down to work with it you may not see it, but then you say, wait a minute, now, the software works differently, faster. It's got more capability." She also assured Ed that he would no longer be traumatized by negative alien visitations. "She sensed in my mind an awareness about previous encounters with a negative type of alien and said, 'Yes, yes, yes, that's in the past now. They won't be around. They won't be able to mess with you anymore.'" She "made reference to . . . her own people, positive aliens, as a group that came from someplace else" and explained that the team that was with her served as her "supporting staff."

Subsequent to the encounter, Ed found himself making intuitive and somewhat impulsive statements on social, political, and scientific matters, and "other kids would look at me and say, 'Boy, is he weird.'" Although not especially accomplished in the conventional academic sense, Ed found that he had an instinctive appreciation of modern physics and modern chemistry, of such matters as Einsteinian relativity, micro- and macrorealities, the curvature of space, and the paradoxes in scientific laws. He also found he could talk of these things to other teenagers in his high school class so that they would say, "Yeah, that makes sense." These changes in Ed seemed to intrigue his teachers.

Ed's intention had been to be a flag-waving supertechie, but once in engineering school he found himself "frozen emotionally" and "angry, frustrated, and despondent." After less than a semester at an engineering college, he transferred to a small liberal arts college where he tried "to find out what makes civilization tick, trying to understand the nature and structure of human civilization." He became interested in "the pageant of history, Rome, Greece, and all" and what he calls "the bigger quest."

Although Ed is aware of no further encounters, he senses now that his teenage experience lingered within him, and he sometimes has had brief flashbacks or glimpses of the cove or other geography where the event took place. In his twenties and thirties he became somewhat of a loner. He would be stirred by pictures of blond women and sometimes when bicycling "whenever I'd see a woman of a slight build [with] long blond hair I'd try to pedal up to see her" and he'd think "Geez, is that her?" But he would be disappointed that "it's not her." Both Ed and his wife Lynn have Nordic roots. They met at a cultural organization where both were studying Northern European literature and history. Lynn "felt like I knew him already." In addition to their common interests in science and technology, nature, and the outdoors Lynn has blond hair. After five years of spending time with each other they were married in the late 1970s. The couple is childless, although they have been trying to have children. There have been a number of fertility problems, which may or may not be abduction-related, including three or four spontaneous terminations of Lynn's pregnancies. Lynn herself has had a missing-time episode and other experiences that make her suspect that she has also had encounters.

When I met Ed he was having trouble finding his proper "niche," was "lost in the desert," and "beating his head against the wall." Lynn believes that this may have contributed to their difficulty in having children, for "we've been in sort of suspended animation, waiting for the light to go on."

Ed has always felt especially close to nature, to woods, trees, and plants and feels that he can "talk to plants." He feels a desperate "race with the earth," a need to put together the pieces of what he knows "like an erector set," and that he has been challenged to do this. He has practiced meditation and studied Eastern philosophy in his struggle to find his authentic path. He and Lynn both feel that Ed's "time in the desert" may have been worthwhile and that out of all the things he has tried he may find a way to integrate his ecological and spiritual commitments with his technological and scientific abilities.

At the end of the first session we discussed the fact that Ed's teenage experience occurred two months before the Betty and Barney Hill case that began the history of modern abductions, and he expressed with some conflict a desire to gain access to further memories, especially of the information that the female alien provided him in the capsule, through hypnosis.

After our initial meeting, Ed found himself becoming increasingly troubled "in the name of the future" about "eco-instability" and "assaults upon the earth." He increasingly longed for guidance as to what he might do and felt pressed to discover more of "what was it she told me?" He felt awed by a sense of responsibility, as "I'm not like JFK's nephew or something . . . I have this intuitive feeling that what she told me is damn important, important for my unfoldment—I'm afraid to even say for the earth's unfoldment. I mean, I'm no messianic leader that thousands are going to follow." He was determined to go deeper into his experience from that night in Maine with the use of hypnosis, and a session was scheduled for October 8, eleven weeks after our first interview.

Before the regression Ed and I went over in more detail the misty quality of that July night, the location of the Nash in relation to the sea (about a hundred yards away on the other side of a road), the cabins where Bob's parents were staying, and the gently rocky nature of the coast where they were to spend the night (compared to the more rugged terrain and sweep of the coastal cove by the promontory where the pod was perched). I had him describe the car's fold-down seats and the two boys' bedtime preparations, which I went over in still more detail during the hypnosis session itself. As he talked of getting ready to go to sleep that night, he referred to nighttime fears from perhaps age four when he would wake up in terror, screaming for his parents, after having dreamed of going down a path "into a field on some railroad tracks, and then the railroad tracks would disappear into some sort of starlit, black night." During the period when he had this dream he had "deep anxiety about going to sleep at night" and was afraid to be in his room by himself in the dark.

After the induction of the hypnotic relaxation, I explored further details of the boys' bedtime preparations with Ed. He described Mrs.

Baxter's motherly fussing over the boys' warmth and comfort and remembered that he slept in a sleeping bag while Bob slept with blankets. He again recalled his and Bob's talk of girls ("we're still virgins"). He remembered the sound of cars going by on the road, the fog "shrouding in," and that he felt "a little uncomfortable deep inside of me about problems or something, I don't know." As Ed described "drifting off, going into my thoughts," he felt "some tingling" in the area of the base of his head or upper neck and then found himself "getting frightened" as the memories of that time began to come back to him.

The tingling sensations increased and Ed said, "I sense something around the car." He thinks he had been asleep but is not sure, dreaming perhaps of "getting it on" with a nice teenage girl from one of the area beaches. But then he saw one or two figures through the car windows, a "couple of human sort of things, but cripes, their eyes are big! You know, they're not spiders, and I don't think I'm dreaming." The figures "don't look like normal humans." They had "big black-gray, or black, eyes and intense, small mouths. Some sort of earlike thing . . . In this damn, this fog, I can't quite make out what you guys look like, you guys are clever. You know how to use camouflage." The beings were of a "slightly different type" than he had seen before. Ed's fear mounted in the session and he recalled feeling "just mortified, like somebody's ready to jump me, and I'm gonna fuckin' fight for my goddamn life."

As the tingling sensation "at the base of the skull" continued, Ed felt himself drifting out of the car. He made a growling sound as his anger came back, but then felt relaxed and even "happy," which puzzled him. Two or three of the beings were "looking at me" and Ed experienced "a sensation of floating, and my whole body is starting to float. I'm floating, I'm floating, I'm floating! Why am I floating?" At this point in the session Ed felt confused and tried to "keep control of what's going on." I encouraged him to stay with his actual experience. He then saw a gray fog around him, "the tingling sensation going further into all parts of my skull," and he was "vaguely aware of scenery changing . . . This is certainly many steps beyond the TV program *One Step Beyond.*" Ed noted. "I keep trying to control it, but I'm trying to not control it. I, my mind, doesn't want to go and get back there." Then he said, "I'm sort of like zapping through all this cellular, atomic-type structure, and I'm seeing, I'm just penetrating through it. I just penetrate. It just keeps coming and coming and coming."

Then Ed felt himself "going down this time tunnel" and literally "traveling" with "no reference point . . . I'm aware of some process of transporta—I'm aware of some movement of some sort." He continued fighting ("The rules of the street are you fight like hell"), but realized that the rules by which he operates did not apply and "they're overriding me somehow or other" so that "I have no choice [but] to let it happen" and "it did happen."

Next Ed felt himself to be "very stiff" and "out someplace along the shoreline, floating, someplace." He recalled hearing the "surf pounding" and became more perplexed about the fact that he was "moving above the water, somewhere along the coast, with no visible means of propulsion," although he did "have a funny feeling that they're around me someplace." He felt that the beings were "very gentle" without "mental intimidation" or "harshness to their physical treatment of me," but "definitely, we are in control, thank you." Ed had the sense that he was moving about as fast as a car, perhaps fifty or sixty miles per hour, and could make out "houses and streetlights, maybe a car here and there, a porch light, maybe a TV" below him. He was embarrassed at the thought that someone might see him. "Geez, if anybody looks up and sees some guy in his pajamas floating along, cripes . . . I might be embarrassed, geez, if my mother ever heard about it. She'd chew me out. Jesus Christ! Proprieties and everything."

Ed found these recollections so extraordinary that he wondered out loud, "Am I bullshitting you?"

"I don't know. Are you?" I asked.

"No, no," he said, "because this keeps coming back, and I sense, I, the sincerity of myself says it's not coming from just, it's not I'm just making this up." But, "It doesn't fit what's supposed to happen . . . This is what you submit to a script to *Twilight Zone* . . . It doesn't fit anything I would have seen on the tube or at the theater." Ed found that he was "approaching this outcrop of land, and the surf, and all that, and I'm seeing some sort of luminescent, domelike pod." Then "I'm being dragged in a sense that I have no control over my body . . . towards the bottom . . . somewhere on the bottom I feel like I'm coming through, and I don't know how we got through the bottom, but here I am." This was evidently physically possible because the craft seemed to be "just suspended there," protruding beyond the rocky outcropping. "Parts of it are near the rock, and other parts of it are just, they're kind of just over the rock."

Inside the vehicle Ed noted a bluish-silver glowing light. He felt his vision was being limited and he was angry. "I hate being in this position of no control. I hate this. I hate this, damn it! I hate this! So stupid, so goddamn stupid, for Crissakes! How will I explain this to anybody?" He had a pain in the front of his head and was confused, "disoriented mentally and physically . . . My wallet's in the house. I have no ID with me." He also had an erection which embarrassed him—"this is not part of my upbringing."

There were "at least half a dozen" beings in the room which he called "an amphitheater" and "sort of like a surgery theater" or "an O.R." and "there's these white lights around." One of the beings seemed to be "the head doctor, in charge." This one "gives off all the vibrations of being a female." The others he called "drones" or "staff people just milling around doing whatever." The female being had long, silvery hair with large, black eyes without pupils or irises. She was looking at Ed "with those loving, big, sensuous eyes and they exude a loving sexuality, like she's a very wise, mature woman conveying telepathically to me—'I have control of the situation.'" She was wearing a "dresslike, long smock," open at the neck that covered her arms and shoulders. He noted that "She has breasts . . . Maybe there's some sort of a pendant or medallion thing around, dangling from her neck on some sort of a chain, or pin, can't quite make out."

The female being "thinks" his name to him and Ed asked, "How do you know me," observing, "You look very sexy." Although she was staring at Ed ("she looks right into me"), he felt she was not letting him look at her. He seemed to be trying to bring forth a sound from his throat but "it won't come," and he re-experienced the sense of being "totally shut down" that he had felt originally. He found himself "sinking into a gray fog" as if "enshrouded." The being told Ed, "You're okay. Don't fight it. Don't fight it." She knew his fear and was "reading my mind like an open book." She said, "I know what happened to you when you were younger" and assured Ed that "these things will not happen again." His fear diminished and he said, "Somehow or other you're doing a great job of convincing me that you're here for my well-being. I'm not just your laboratory rat, guinea pig" (as he had felt in the past).

Ed recalled that he experienced some sort of forced arousal. "She's filling my mind with all sorts of erotic escapades . . . She's just raising, somehow or other, she's just giving me, giving me a hard-on." The female being knew that Ed wanted to have sex with her—"She reads my mind like a book. She can just thumb through the pages at her will." But with a "mirthful expression" she told him something like, "Yeah, you'd like that wouldn't you, but that's not the way it's going to be." Instead, she explained that they needed his sperm for "their needs . . . to create special babies" and "for work we're doing to help the people on your planet." Ed continued to struggle with his helplessness and lack of control but was somewhat mollified and persuaded that he was being used for something worthwhile.

Some sort of "tube" or "container" was placed over Ed's penis and he felt "very relaxed now." He experienced a rubbing sensation or friction, which built in intensity. "It's a very smooth, handlike thing. I want to believe it's her hand." After he ejaculated the female being "thinks it to me, 'That was good. That was great,'" in appreciation that they "got a good sample." He then felt hot and sweaty. This completed the first part of the experience. Although Ed felt "at first they manipulated me to put me in the position where they could tell" him what they wanted, he soon felt, "Okay, you convinced me that what you did was for some greater good that I don't fully understand." Of the female alien he said, "There's something very trusting about you, loving, caring, wanting to help." Then he added in some awe, "I've never been through anything like this in my life."

At this point the scene changed, and Ed felt as if "I'm in a different space . . . Before was more like an O.R.," but "now, all of a sudden" he was in "the pod with translucent walls" that he had remembered consciously before the hypnosis. He did not recall "how I got from there to here." The female entity "wants to talk to me" and Ed felt scared about what was coming next. "A seriousness comes over me. It's like when the doctor or the teacher comes in and says 'now the heavies . . . now the serious part . . . now we think we can tell you things.'" He noticed that the female being, "the spokesperson," was wearing a "silvery, somewhat metallic tunic. It's sort of a weave, woven/knit sort of thing with some sort of shimmery stuff on it, and I can see her exquisite breasts jutting from beneath it." She had a "gentle grin" and was "looking at me very lovingly." Ed felt that the inside of his head and his eyes were "burning and spinning" as she began to talk to him and he was shown different things.

The remaining forty to forty-five minutes of our hypnosis regression was taken up with Ed's recollection of the information he received during the abduction. The sequence of our dialogue may not reproduce altogether accurately the order in which these thoughts and images came to him at that time.

The narrative was filled with apocalyptic images. The being communicated to him telepathically in what Ed calls "allegorical terms" a message of "instability on your planet, eco-spiritual, emotional instability . . . Volcanic eruptions are a sign . . . It's allegorical towering plumes of eruptive rage. Not ejaculations of ecstasy, but eruptions of anguish. Be careful. Pounding waves of eruption, watering, rushing, and engulfing about and around you." Ed protested, "Why do you talk to me in allegory? I'm no poet."

But the relentless communication continues. "Towering, pounding surf, shifting plates, instability, Earth shuddering in anguish, crying, weeping at the stupidity of humans losing contact with the inner soul of their being." She said to him, "You have a chance, Ed. You have an inner sensitivity." Again he protested, "But the teachers always pound me for my grammar and spelling." But she insisted, "You have a sensitivity, Ed. You pick up on things. You can talk to the earth. The earth talks to you." He affirmed to me that, indeed, as a small boy he would struggle with his mother because he loved to go into the woods (for which she would sometimes punish him by making him go to bed without supper or spanking him). "Things talked to me," he said, "The animals, the spirits . . . I can sense the earth. I can sense the interplay of nature." The battle with his mother continued ("Only bad boys go in the woods," she would say), but Ed was undeterred.

The female being, whose name he now recalled as something like "Ohgeeka" or "Ageeka," picked up on these qualities and underscored the responsibility Ed has for his gifts and powers. "Listen to the earth. Listen to the earth, Ed. You can hear the earth. You can hear the anguish of the spirits. You can hear the wailing cries of the imbalances. It will save you. It will save you . . . Things are going to happen," she said, but he must "listen to the spirits," even if he is taunted, and not feel overwhelmed. "She gave me a flash . . . she opened up that channel and turned up the volume. Some of [the spirits] are crying; some of them are mirthful. She just ran me through the whole thing in a couple of seconds. 'All this you can see, hear, and feel. Other people may think you are crazy.'" The earth itself, the being told him, is enraged at our stupidity, and "the earth's skin is going to swat some bugs off" that do not know how "to work in symbiotic harmony" with it.

I asked Ed how this swatting off was going to happen. "Convulsions of the earth," he said, "almost like puking us off, or shunning us off . . . They will get rid of parts of us." Meanwhile, the being kept telling Ed that he has a "greater agenda," and must "Listen to the music of nature, the exquisite sounds of nature. 'The music will make sense to you, and it will leave a harmony with your emotions, your intuitions, your hunches. You have a gift with hunches, intuitions, and emotions.'" Ed, imbued with the doctrine of his hierarchical Roman Catholic upbringing in which "God talks with the Pope, and the Pope talks to the priests, who then talk to my mother, my father," reacted cynically. "Yeah, I'm gonna open up a tea leaf parlor. Put a turban on my head. Ten bucks a pop." But the being cut through his sarcasm and insisted that he own up to the gifts he was born with.

At this point he recalled that he actually saw the spirits in the form of "mirthful little playful creatures, just kind of, just bounding around." I asked him to describe them. "They're like energy forms . . . of many different kinds. There are many different shapes, colors" (he giggled at this point). He found "hilarious" the "contortions" and "mirthful" things that they can do. I ask for an example. They can "float around," he said and "change the laws of nature," by which he seemed to mean they can change shape and form. One of the spirits "is talking to me." He had long, silvery hair and an oversized head and is only a foot or two tall, like a "micromidget." The spirit said, "Well, see, I put myself like this so you could sort of look at me and relate to me. But I don't have to be like this if I don't want to be like this, and I can change myself into a multitude of forms . . . I put myself into this semihumorous shape so that you would feel comfortable and mirthful with me, because I know certain other types of creatures you're afraid of, like spiders, in particular, and snakes. So I'll make myself like this!"

Then Ed realized that in some sense he and other human beings also "have the power to change the laws of nature if I access certain parts of my being," which frightened him. I asked what scared him. "Well, Jesus, if I blow it and use this power at the wrong point . . . "

"You are right, Ed. You are right, Ed," the female being told him. He continued with her words, "'You've got to learn how to use this, when to use this, where to use this, or else they will take you away! And they will label you, and they will put you away, away, away. And they will tell your parents, "poor Ed, he had such promise but something happened, and Mrs. M. [his mother], we're terribly sorry."' She's running this in my mind. 'The authority figures, "we'll have to put him away . . . "'" The being spoke to him further about the task of cultivating his mind, warning, for example, about pursuing science in an orthodox academic setting and fashion at the expense of his intuitive skills. "The process of cultivating your mind in the traditional, rational way will obliterate these other things, and you have seen all the infinite possibilities," she said.

I asked Ed to say more about what he was told or shown about these possibilities. He spoke of "seeing how the laws of the universe are made" and "something about the point where the universe comes into birth." Again she warned him about misusing his understanding.

"'You see the ambush of the planet now,' she said." At this moment in the session Ed felt "blocked from seeing" further. I asked him what he saw of the universe being born.

Ed: An incredibly blinding, searing white light.
JM: She showed you that?
Ed: Yes.
JM: What was that like for you?
Ed: Almost too much. But it was like, holy shit. There's a particular chord or passage in Mahler's tenth symphony. Goes like, it just opens up and there it is. It's like a galaxy being born. It really is, really is, really is this. "But [she says] I don't want you to see too much of it. You have to know. You must be wise in how you talk about this, and when and where you talk about this. There are those that will use your mind for asinine purposes."

I took him back to the glimpses he had of the anguish of the spirits. In addition to the "innocent play with nature, the way we're supposed to be," Ed remembered seeing "distorted forms of entities, of spirits out here now because man has done so much damage, hurt and damage both to himself and to each other and to Mother Nature." He was shown "grotesque forms . . . Horrific. There's dark, gray, malignant forms that they were trying to heal, or rebalance. It took a lot of effort [of the healthy spirits] to keep these malignant forms from growing in magnitude and malignancy." The female entity continued, "You are distorting, growing greater malignancy, trying to get yourself . . . We're trying to keep them under control and bring back into play. You see how malignant and deformed they are, Ed. They're gray, they're horrific, distorted masses. You see how these energy forms, Ed, are so nice and mirthful, and they look so healthy, and then there are these others over here."

Ed spoke further of the communication he received about the consequences of "our pillaging on the planet." The malignant, destructive forms, were created by the imbalance of "the collective human psyche . . . dark gray things that just come along and suck up and destroy anything. They just go berserk, and they just zap anything in their path, in their range." I asked what he was told the ultimate result would be. "They run their course," he said "until they work out all this negative energy and return to normal, happy state. They've got to work it out of their system, like getting the pus out of your wound. It won't heal until you get the pus out of it."

I asked him what is called for, what he has been told is to be done. He answered personally, in terms of what he has been told about his own survival in the face of the "cataclysmic earth changes" that are to come. "She's telling me, showing me that I have the tools within me to survive. I have this extra dimension. I have the choice of listening to it or not . . . I am to listen to my inner, something deep inside myself, and listen to the earth." Ed's partner, Lynn, he realized, has known intuitively what the being has told him and "it doesn't faze her in the least." I asked if the female in the pod knew that Lynn would someday show up for him. "She knew," he said. "She gave me a sense that a certain unspoken thing would resonate," and that his and Lynn's task would be to "to teach those human beings who will listen . . . There are those who will listen before it happens, and prepare themselves."

I asked if he was given any information about whether it was still possible to prevent the cataclysm. "No, no, no," he said. "There are not enough. Too few will listen, but those who will listen and can work with the laws of nature, will survive to teach others on the other side who then will listen and say, 'Oh, geez, we were screwed up back then.'" I expressed my lack of clarity as to whether he meant "cataclysm" in a literal, physical, or metaphoric sense. He said there would be "a series of geological and meteorological convulsions." Finding this all rather depressing I asked how this information about spirits would help him or anyone else survive. Undeterred he said that "the spirits of the earth" will "make safe havens" for those who will survive. I wondered what was the use if everything was going to be destroyed. He said that it was more a reconstitution—not just destruction—a rebalancing, and repeated that "humans have to learn how to work on this planet with the laws of nature and not plunder the earth," to use the "raw materials" in "the way they should be used." Then "the earth will rebalance itself."

I remained confused about the literalness of all this and of the various distinctions at work between spiritual and physical catastrophe. Ed had some difficulty understanding my confusion and said "A person has to spiritually rebalance themselves. I have to balance myself out spiritually to hear these messages. When I hear these messages I will know on a physical level where to go to work, those spots on the earth which will still be sacred and accessible." At this juncture we were both becoming somewhat tired and agreed to end the session.

After I brought him out of the regression, Ed observed that this information has always been "before my eyes" as if "on a page," but that he had "looked right by it . . . She had it there for me all the time," but "I was afraid to look at it." Ed felt awed by what he had told me and spoke of the responsibility of translating his "access into their dimension," his knowledge and concern for the imminent breakdown of planetary systems, into usable terms for the larger society. He experiences himself as a "Joe average" person with extraordinary information. "It's like trying to be Superman and Clark Kent. You can't go walking around in your cape and tights all the time. You have to be a kind of Clark Kent."

"Love is the key," Ed said, "love and compassion for the earth or the beings on the earth, be they corporeal or incorporeal—not love in the mush and gush sense, but there is a deeper sense of love." Both Ed and Lynn understand now why the female being "put the dampers" on his remembering and talking about his experience. "If I had come back and . . . started talking" about his knowledge of the laws of nature he might have been "whisked off to make the bomb of the millennium." He still faces the dilemma of how to be effective. "Who's going to listen to a mild-mannered . . . technician?"

Ed did not experience his encounter and the recovery of memories related to it as markedly disturbing. "I don't find what happened to me traumatic, like in the case of the other people in the encounter group [he had attended my support group for other abductees a few weeks before] who were used like laboratory rats." Instead he felt as if "a great cloud, a shroud has been pulled away from my awareness that has been sort of just there." We talked further together about how, as Lynn put it, "to responsibly take action." One step that we could agree on was the value of his talking with other experiencers in order to share information and to build the affinities of their growing community.

Discussion

Although it is not uncommon for an experiencer to recall a single, major encounter, it is interesting that Ed's occurred when he was a teenager and was not recalled for nearly thirty years. The forces involved in the implanting, storage, and recovery of information remain among the central mysteries of the whole abduction phenomenon. As we have seen, Ed's teenage abduction seemed to have worked subtly in his psyche throughout his life, making him somehow different, perhaps more intuitive or aligned with nature, than his contemporaries. Yet we do not know what is cause and effect, whether his openness itself may have predisposed him to being chosen for abduction, whatever that may mean. We also do not understand completely why it should suddenly happen that many of the memories of the earlier experience returned. Yet it does appear that as a mature man he is in a better position to apply his knowledge in some sort of earthly calling, integrating the abduction-related information with his psychological gifts and professional expertise. It is not yet clear how he, with Lynn, will do this.

The forced taking of sperm for some sort of poorly understood interspecies breeding program is characteristic of male abductions. Information about ecological disaster with powerful apocalyptic imagery is also commonly transmitted by the aliens to human subjects. What is somewhat unusual in Ed's case is the extent of the detail he was given in a single abduction, and the fact that he originally received this information in the summer of 1961 two months before the Barney and Betty Hill abduction in September of that year. Much of what Ed was told or shown about the breakdown of planetary life systems and human violation of nature's laws is well known, at least within the environmental movement. A well-researched book like Donella Meadow's *Beyond the Limits of Growth* (Meadows 1992), makes clear the consequences, if not the causes, of our continuing destruction of the earth's environment. What is of particular interest is the power of the information for someone like Ed. His learning was not simply intellectual. The realities transmitted to him by the female alien, of impending ecological and spiritual disaster as the result of human disharmony with nature, were felt deeply in every part of his being. As he said himself, his whole self, his "entire perception," was changed by the encounter.

It is not clear how literally to regard the apocalyptic visions Ed has received. They have the quality of prophetic tradition, warning of disaster and of the need for fundamental change. Whether they are concretely predictive, or calls for change and action, cannot, of course, be answered. That Ed has taken the information seriously enough to change his life and commit himself to communicating what he has learned to others who will listen is certain.

Ed's direct encounters during the abduction with shape-shifting spirits that have actual form is interesting, as is the actualization of negative emotional forces in the shape of malignant or demonic spirits. The Western worldview has no place for such entities, tending to regard them as the products of fantasy or the projections of the psyche, although they are believed in or their existence acknowledged by other cultures throughout the world. For example, in a meeting in India in April 1992, in which I participated with a small group of professionals who were invited to discuss the abduction phenomenon with Tibetan leaders, the lamas saw alien beings as spirits that have become upset by the invasion and destruction of the earth's environment, which they inhabit as well. A Tibetan physician explained that as a result of our ignorance, material attachment, and aggression, manifested by the desecration of the planet, these spirits have become annoyed and irritated and are causing "negative disturbances." A leading spiritual figure among the Tibetans also saw the aliens as spirits that have become so troubled by human destruction of the realms they inhabit that they have been forced to come among us, seeking our compassion and transformation.

A final word needs to be said about the use of hypnosis in Ed's case. Before my first meeting with him, Ed had recalled a great deal about his teenage abduction. But his conscious memory before the regression tended to simplify the experience and, more significantly, to gloss the narrative in ways that were more syntonic with the self-image and desires of a young adolescent than what he recalled painfully during the hypnosis session would be. Many embarrassing details relating to powerlessness and loss of control were not available to him except under hypnosis. In particular, the happy outcome of pleasurable sexual intercourse with the cooperative, sexually active, female alien gave way to the forced, quite humiliating, taking of a sperm sample as the being watched approvingly. This second scenario, which is obviously more disturbing, is far more typical of male abduction experiences and, therefore, more believable.

All this suggests that, at least in Ed's case, the information recalled painstakingly under hypnosis is more reliable than the consciously recalled story, which seems to have been unconsciously adjusted to be compatible with Ed's wishes and self-esteem. There are other details obtained during the hypnosis session relating to the transport to the craft, the numbers of beings (the female leader and her "staff" rather than the single alien "woman"), the two chambers (an O.R.-like room and the podlike briefing room) rather than the single pod, and the great amount of information transmitted by the alien female, which make the story obtained during the regression more believable, or at least more consistent, with other abduction accounts.

CHAPTER FOUR

"Personally, I Don't Believe in UFOs"

Sheila N. was a forty-four-year-old social worker when she was encouraged to contact me in the summer of 1992 by a psychiatrist at the hospital where she had done her internship not long before. She was seeking understanding and relief from the stress of what she called "electrical dreams" that had begun more than eight years earlier, following her mother's death. The psychiatrist Sheila had been seeing for seven years encouraged her to see me, but it was the hospital psychiatrist's knowledge of and interest in my work that brought us together. Sheila's case illustrates some of the issues that psychiatrists and other mental health professionals confront when working with abductees.

Sheila had been very close to her mother, and so her death and the events surrounding the seven-day hospitalization in January 1984 that led up to it were deeply disturbing to Sheila. Sheila's mother had had a heart attack five years earlier and in 1984 was hospitalized for an endarterectomy, a surgical procedure aimed at clearing coronary artery blood flow. She did well at first, but then suffered a cerebral hemorrhage that led to her death several days later. Sheila was unable to find out what the relationship was between the surgery and its fatal complications, and felt that the doctors were abrupt and uncaring toward her mother. She also felt that her mother was unnecessarily maintained on life support after there was no hope and that thus she was robbed of her dignity. This insensitive treatment was especially troubling to Sheila in view of the history of incest between her mother and her mother's father. In Sheila's words, her mother had been "robbed of dignity in her childhood because of her father's sexual demands." Sheila was also angry and sad that her mother's grave was still open with the cover to the vault left exposed and only covered with earth three days after she was buried. Following her mother's death a division developed between Sheila and her husband, who she felt was unable to support her adequately in her grief. "Jim can't deal with illness. He has to be happy," she said.

In the days after her mother's funeral Sheila was in great pain. She would walk the streets at night, feeling very irritable, as if "nothing could hurt you any worse," but unable to cry. Sheila wrote in her journal on February 9, four weeks after her mother's surgery, of activity in the night sky—"more planes than cars." She also began to have recurring dreams in which she would experience terror, be unable to move, and her body would feel as if it were vibrating or "full of electricity." At first she called these "spiritual dreams," and they made her feel like someone or something else were controlling her body, as if she were "possessed" by demons. Later she thought of the dreams as seizures. "At this time [just before we met] I describe them as if electricity is traveling throughout my whole body. Regardless of the name, the experience of these dreams has not changed." Ordinary dreams, according to Sheila, are "more fragmented," whereas in a dream in which she saw alien beings, there seemed to be "a natural progression in a certain direction."

One of the dreams, which Sheila believes occurred in March 1984, about ten weeks after her mother's death, was different from the others with respect to certain remembered details. In anticipation of our first meeting she wrote me about it.

> I woke up to a very loud noise with flashing lights. The noise was a high pitch sound and remained at that tone for the duration. I was struck by the precision of the red flashing lights. Other bedroom doors and the bathroom door were open. I could see down the length of the hall and it appeared that the lights were coming through the windows on all sides of the house at the same time.
>
> At this point in time I was lying on my back. I was very frightened. I eventually raised myself up on my elbows. I saw *several*, small, peoplelike things walking down the right side of the hall, one behind the other. They looked like their whole bodies were silver in color. I noticed some blue on the first one in line below his right shoulder. It appeared to be a reflection of something, although there is nothing blue in the hall. They were short with thin arms and legs. As they approached the bedroom, the third or fourth one in the line raised his right hand up. I knew they were coming to me, but I had never seen them before. They appeared to be waddling [later this proved to be an artifact of the flashing light pattern]. They were very awkward on their feet [also related to the flashing lights].

By October 1984, nine months after her mother's death, Sheila's estrangement from her husband had reached the point where she felt she had to move to a separate room. "I attempted on several occasions to discuss the basis of my sadness with my husband. He would not listen." She also did not feel she could tell him of her strange dreams and moved out of the bedroom in part to "protect" him and "allow him to sleep." He did not protest when Sheila moved out, and they have not slept in the same room since.

Shortly after this Sheila asked the pastor of her Methodist church for a referral for psychotherapy, but difficulties arose when the therapist insisted, against her wishes, on talking to the pastor about her. Frustrated with the lack of progress in her weekly appointments and feeling unable to trust her therapist, Sheila's despair deepened.

In July 1985 she learned that the pastor, whom she felt close to, was told by his doctors that his life expectancy was three to five years due to cancer diagnosed four months previously. In addition to the loss of her mother and the pastor's fatal illness, since November 1983 Sheila had also experienced the deaths of other close friends and family members. Refusing to honor Sheila's requests, her therapist still insisted upon talking with the pastor and her husband of her work with him. Feeling desperately bereft and alone, on July 17, 1985, Sheila bought a bottle of aspirin and ingested twenty tablets with "every intention of taking them all." Except for general physical discomfort and buzzing in her ears she suffered no untoward effects. Shortly before I first saw her, Sheila wrote to me that "Suicide is not my typical coping style" and "I want to reassure you it is not my intent to give up under any circumstance."

Soon after this episode Sheila began to see a psychiatrist, Dr. William Waterman. Although she seemed to have resolved her grief over her mother's death, Sheila still felt no closer to understanding or obtaining relief from the electrical dreams, which continued to plague her. A particularly disturbing one occurred on New Year's Eve of 1989. She had been sleeping downstairs, while her daughter, Beverly, and Jim were in their rooms upstairs. As in 1984 she heard a loud noise and sat up with her body feeling "full of electricity . . . Something forced me down," but she did not recall seeing "the small peoplelike things" again. Six months after the New Year's Eve episode Sheila wrote to Dr. Waterman, "Prior to January 1, 1990, I thought all of those 'whatevers' entering my bedroom were only symbols in a dream. Since that time, I have come to recognize the hostility and aggression I experienced with it as a connection to the recurring 'spiritual dreams.' In my dream, those 'whatevers' meant business, and so do I."

An article that Sheila saw in her local newspaper in 1985, which she later retrieved, described UFO sightings in the town where her mother was buried. As she wrote to Dr. R., one of the doctors she saw later for hypnotherapy, "This article made me question any involvement my mother had to do with all of this." Later she would piece together the sightings with what she had heard, read, or seen about abductions and began to question if, indeed, her electrical dreams were dreams at all and whether it was really true that her continuing fear of the night was related to delayed grief over her mother's death. On July 14, 1990, Sheila wrote to Dr. Waterman, "I told you long ago that it is my firm belief that these dreams go far beyond loss, but I simply had nothing more to add to that."

Determined "to conquer my fear of the night whatever it takes" and wanting desperately to "put an end to these dreams," Sheila and Dr. Waterman explored other approaches, especially ways of uncovering buried memories. They considered and rejected the possibility of an Amytal interview. Finally, and with reluctance on Sheila's part, in the summer of 1990 at Dr. Waterman's recommendation she contacted a psychiatrist, Dr. G., at a Boston teaching hospital, who specialized in the therapeutic use of hypnosis, in order to explore the source of her symptoms and to obtain relief from her anxiety. He asked another psychiatrist, Dr. R., who had an interest in learning more about hypnosis, to call Sheila. A careful workup of Sheila's case included, in addition to the history, a neuropsychological evaluation of possible temporal lobe epilepsy and an all-night sleep study. In his note referring Sheila to behavioral neurology, Dr. R. wrote that in one of Sheila's dreams, "She is sitting on her bed and sees several little figures: the first one stood at the foot of her bed and the second one by the door. They walked in an awkward way and had the shape of a small humanoid body." In another dream, he continued, "She is lying down with a blanket over her and saw two of these small beings standing over her."

The psychologist who did the neuropsychological evaluation described Sheila's sleep disturbance but noted that she "denies other current symptoms of depression." The psychologist put together a high school incident in which Sheila sustained a minor injury on the right side of her head that led to nausea and light sensitivity for several days with some difficulty concentrating during the testing, "variability of attentional functioning" and "motoric restlessness" and suggested "a possible diagnosis of Attention Deficit Hyperactive Disorder." Sheila wrote in the margin of a copy of the report she sent me beside these words, "I will *never* believe this is true of me." Noting Sheila's tension during the examination, in addition to her history, the psychologists also considered the possibility of an "anxiety disorder" and a post–traumatic stress disorder "given the trauma of her mother's death, her experience of nightmares, and her exaggerated startle response." They concluded, however, "Intelligence testing found Ms. N. to be a very bright woman functioning in the above average and superior range." An all-night sleep study found "nothing remarkable except anxiety and poor sleep."

According to her records, between August 1990 and July 1992, Sheila had twenty-four appointments with Dr. R. and/or Dr. G., which included at least seven hypnosis sessions. The appointments were generally one hour and the hypnosis ranged from fifteen to twenty-two minutes.

In a letter to me, Dr. G. wrote of Sheila's concern when he first saw her about the vault being left open overnight and said she expressed her sadness over her mother's death and other aspects of the history recorded here. In her journal Sheila wrote that Dr. G. cautioned her in 1991 that although hypnosis could be of benefit in "producing new material," it does not "guarantee accurate recall" and "can be an extension of a personal fantasy or experience." The focus of the treatment was on the impact of her mother's surgery, death, and funeral, and on cognitive/behavioral approaches aimed at reducing the distress of her "dreams," especially by asserting to herself they were "not reality." "I never felt safe," Sheila said of this treatment process. In May 1992, Sheila accepted a prescription for Klonopin, .5 mg, to combat her anxiety, and in June for an antidepressant, Wellbutin, 100 mg, to be taken at bedtime. She continued both medications until early August. In Sheila's view Dr. G. continued to connect her dreams with depression. In his letter to me, Dr. G. stressed that hypnosis may not lead to accurate memories and also wrote that "whether or not she is subject to delusion or deeply religious beliefs is similarly difficult to answer." He said that my willingness to accept what people like Sheila say was "much more than neutrality," and he argued that although my kindness, sympathy, and support toward Sheila may have led to therapeutic improvement, that "does not necessarily confirm the theories that surround UFOs and abductions." I discussed all of these issues in a meeting with Dr. G., who generously agreed to schedule a Grand Rounds at his hospital to discuss them further in a medical academic setting. During the well-attended rounds, the issues raised by Sheila's case were discussed. Dr. G., on the basis of his research in other types of cases, again doubted that hypnosis could lend credibility to Sheila's abduction reports. The complex questions surrounding memory and hypnosis in abduction cases are discussed on pages 24–25.

Sheila told Julia, another abductee with whom she talked at length before I was able to see her, that under hypnosis she saw "a skeleton with no nose or mouth," a "curling iron" with a handle and a rotating pencil tip "like a drill," a twelve-by-eighteen-inch-long rectangle the color of piecrust with slits cut in rows that created a great amount of fear in her, and she also remembered her arms being stretched out and tied down with rubber tubing. In notes to me in early January 1993, Sheila wrote, "I did not discuss these with Dr. G. as they occurred because I was afraid he would reject these and I would be left feeling alone with them. I did tell him about these collectively well after the fact. He asked me how I responded to them at the time. I told him that I immediately replaced it with a vision of a beautiful flower garden full of hummingbirds. He commented favorably on this and I perceived his comment as encouragement to turn away from fear."

At some point during the course of the evaluation and treatment, in response to showing him the article about the UFO sighting in 1985, Sheila told me that Dr. G. said, "Personally, I don't believe in UFOs." On two occasions when they were alone according to Sheila, Dr. R. said, "You don't really believe in Martians, do you?" She says she never used the word "Martians" and found these comments "condescending" and they undermined her trust in him. (Dr. G. and Dr. R. say that they did not make these statements.)

Sheila wrote to me of how difficult it was for her to challenge "a professional" with what she felt were errors. "I wanted help so much, and I was always fearful of losing it," she wrote, "even when I recognized I wasn't really being helped at all." On July 31, 1992, she wrote to both psychiatrists terminating her work with them. She says she would have preferred to speak with each in person but "we were having difficulty scheduling time when I assumed full employment." Both letters, which she has given me copies of, are appreciative, courteous, and frank. To Dr. G. she wrote of "moments where I have felt a lack of understanding and acceptance . . . Over time," she stated, "I have come to think that there is a relationship between the single dream and those that reoccur. I desperately wanted to believe otherwise. It simply does not make sense in the world as I know it to be. It is my feeling that a clearer understanding of this relationship is the only way I can experience freedom from it all." She also noted a paradox in relation to the dream of March 1984 and asked, "Why did I fall asleep when every other terrifying dream wakes me up? Why was the setting of it in my bedroom when that is where I was when I had this dream?" And, finally, the ultimate question, "Was it a real experience and, if so, what happened that is so painful for me to remember?"

In her letter to Dr. R. she thanked him for the feeling of security that his presence during the hypnosis sessions had given her, and then explored further than in the letter to Dr. G. the sense of actuality she had of the beings in her room in the 1984 episode and the seemingly purposeful pattern of their deployment about the room. She also echoed the experience of so many abductees when she wrote, "What a horrible and terrifying thought that you can't protect your own child [in reference to her suspicion that Beverly has had abduction experiences as well] in the privacy of your own home." In saying "Good-by" Sheila concluded, "I do think that our work really ended the day Dr. G. said, 'Personally, I don't believe . . . ' I know what I saw, and I can only say that this experience changed my life . . . One of the first questions Dr. G. asked me," she continued, "was if I thought I have a creative imagination. It took me eight years to get this far, so I guess my answer is 'NO' . . . I will figure this out," she ended, "or die trying." In replying thoughtfully to her letter Dr. G. wished Sheila well and urged her to forget the most disturbing dreams as best she could and to take a mild tranquilizer so that she could gain "some respite" from her distress.

Shortly after she wrote these letters Sheila spoke with Dr. T., the psychiatrist and friend who encouraged her to contact me. In May she had taped the CBS miniseries on abductions, *Intruders*. Although she had been unable to watch it herself, she brought the tape with her to the meeting with Dr. T., which was supposed to be about a project which was an extension of her master's thesis. Putting aside the discussion they had planned, Sheila declared that she had another agenda, namely to connect, possibly by writing a book, her abduction-related experiences with health care professionals to the question of patient satisfaction. She told Dr. T. about her experiences and her struggle to get help in dealing with them and gave him the tape of the miniseries which he much appreciated. Of the March 1984 experience she told him, there was a "possibility that it wasn't a dream anymore." On August 12, Dr. T. and Sheila met again, by which time he had compared Sheila's story and the *Intruders* cases. He told Sheila, "Writing the book is one way, but this is another way." And he gave her my name and phone number. The next day she called my office.

I was unable to see Sheila for several weeks and asked Julia to speak with her. They met on August 27, and Sheila found their conversation greatly comforting; it helped her confirm the possibility of a relationship between the electrical dreams and the March 1984 experience. Sheila said that the night before their meeting, Beverly reported she had awoken about five o'clock in the morning "incredibly scared," with an unexplained green light shining into her bedroom through a large window fan, creating a flashing effect and illuminating the wall opposite her bed. The meeting with Julia was the first time Sheila felt that someone had heard and understood her.

On September 1, Julia told me that Sheila had called her to tell her that the night before she had had another electrical dream in which she had been paralyzed with great pain in her hip where she felt a needle had been inserted into the bone. She felt intense fear and desperation and I agreed to meet with her as soon as my schedule permitted. We talked at length on the phone a few days later, and our first meeting—a four-hour session which consisted of a review of Sheila's history and a long hypnosis session—was scheduled for September 21.

Sheila grew up in a small city west of Boston. She is the middle of five children. "My family of origin is close," she wrote in one of her communications to me. Her father was a commercial airline pilot who was away a great deal when Sheila was young. Although retired, he still travels a lot, mainly now between Maine and Florida. He does not admit to seeing a UFO, Sheila said, "but he wouldn't believe them either." "My mother never missed an opportunity to instruct us on proper etiquette in our youth," Sheila wrote me after reviewing the first draft of my write-up of her case. "She was a true lady and always strived to set a good example for us." In Sheila's view, her mother "protected us a lot" when the children were growing up, which may be related to the history of incest.

As a small child Sheila attended the Presbyterian church, then went to the Congregational church, and in high school began going with friends to the Methodist church, where she became close to the pastor who first referred her for counseling after her mother's death. She attended Sunday school as a child, and the Bible became important to her. Beverly now attends a Christian school. Sheila always viewed God as a "source of energy," but her abduction experiences, or, more accurately, the way these experiences complicated the grieving process after her mother's death, challenged her faith. According to the Bible, she told me shortly before her first regression, "you should love God more than your parents. I was angry with myself because that wasn't how I was feeling."

As a child and adolescent Sheila was musically and athletically inclined and socially active. "I like to be with people," she said. She met her husband, Jim, who is now a school teacher, when she was a sophomore in high school and he was in college. After college Jim enlisted in the army for three years and encouraged Sheila to date other men, which she did, "but my heart wasn't in it." She actively pursued Jim, writing to him every day while he was in the army. "He tried to pawn me off on his brother," but "I just hung in there." Rather self-disparagingly she says that after Jim came out of the army "all his friends from high school and college were scattered so I was the only one left." They were married in 1970 and Beverly was born in 1975. Sheila says that the marriage was good until after her mother's death and the electrical dreams began. Sheila attended college and social work school at Massachusetts institutions, initially obtaining her license in 1980. With characteristic persistence, when circumstances permitted, Sheila returned to graduate school and finished in 1991. At the time of our first meeting she was working in the adult psychiatry unit of a general hospital west of Boston.

The first recollection Sheila has of experiences that might be related to the abduction phenomenon is an incident that occurred before she was six. She and her brother saw something during the night they called the "Gaw." Sheila wrote me later that "The 'Gaw' without a doubt had the same appearance as the beings that entered my bedroom" [in 1984]. Another incident occurred between ages six and eight. "I saw someone walk out of the closet" in her bedroom, she said. She screamed and her parents came in, checked the room, and reassured her that she had had a bad dream, but "I guess I had my reservations . . . I felt it was real." The being was quite tall and shadowy in appearance. It walked toward her window and disappeared. To this day Sheila keeps her closet door closed. Even before our first regression Sheila associated this being with the entities she saw in her room in 1984. "It was the same as coming into the bedroom eight years ago."

When she was in her early teens, Sheila was driving in a car with her mother and one or two of her siblings while the other children and her father were in a second car (with five children they had to take two cars) on a visit to her paternal grandparents. Each person in the car that Sheila was in saw a bright light which looked like a bolt of lightning but was straight, just above and parallel to the horizon. "It was just one of those things you couldn't explain." Before she began therapy in November 1984, Sheila shared her March 1984 experience with her sister Melissa. Melissa took it seriously, gave Sheila a hug, asked for details, and told Sheila of her own related experiences. But until 1989, Sheila still dismissed the possibility that it could be real. As she pointed out, "It's no wonder that those who have not experienced have so much trouble accepting."

What Melissa told Sheila was of an incident when she was seven or eight and Sheila was thirteen or fourteen, and she saw "something big and silver in the sky" that so disturbed her she was shaking and crying. When she told her father, he dismissed her with, "It's just a blimp. Don't worry about it." Melissa continued as an adult to be troubled by this memory and sought help from a hypnotherapist, but whenever she would get close to bringing up the experience, she would start to shake and cry again. Melissa, to whom Sheila feels very close ("she's wonderful"), has told Sheila that under hypnosis she has seen different colored reflections in association with the object, blue on one occasion and "orangy-pink" on another, but because of her terror the hypnosis has not been productive.

When Melissa was in her early twenties she saw a ball of light come through the sliding doors of her apartment, "bounce around the room," go down the hall and into another room and "through a wall." She was with a friend, and they both ignored the phenomenon until Melissa said, "Wait a minute, the curtains are closed. It's not a light from a car going down the street." As Sheila related, "They followed the light around the room before it went back outside through the same sliding doors."

When Sheila told a cousin of her father's about coming to see me, the cousin told her that she had seen a UFO in her neighbor's backyard and that Sheila's older sister, Laura, had had an experience "with a loud noise and red lights" resembling Sheila's. But Sheila and Laura are quite different and Laura never told her of this incident.

Sheila's daughter Beverly's apparent involvement in the abduction phenomenon is an important element in Sheila's determination to explore her experiences. She, like many abductees who are parents, is deeply troubled that she cannot protect her child. When Beverly was fourteen or fifteen months old and still sleeping in her crib, Sheila remembers waking in the middle of the night and going downstairs in the dark, which was very unusual for her. She saw "something white across the stair" and wondered what she might have left there. When she turned on the light she saw it was Beverly, sound asleep in her pajamas without a blanket. When Beverly was about eight Sheila took her to the pediatrician for a possible ear infection. The doctor removed an object about the size of a pencil eraser with "junk all over it" and discarded it. Beverly insisted, weeping, that she had not put the object in her ear, yet she told Sheila that as far back as she can remember she covers this ear with her sheet and blanket to keep it from being exposed. As is characteristic of child abductees, Beverly had frequent unexplained nosebleeds when she was little.

Sheila and I met on September 21, October 12, and November 23 for hypnotic regressions. Dr. Waterman was present during the second and third sessions. As I had not seen her before, half of our first meeting was given over to filling in details of Sheila's family background, personal history, and the experiences she and her family had undergone that might be abduction related. Initially she appeared to be a rather timid, anxious, and soft-spoken woman who was clearly determined to forge through considerable distress to meet with me. Her anxiety, I later discovered, was increased by the troubling elements of her previous experiences with doctors. Her concern with control and the fear of losing it were evident from the outset.

Before beginning the first regression, whose objective was the recovery of memories related to the March 1984 experience, Sheila and I reviewed her conscious memories of the episode and I asked her to draw the layout of the rooms and sleeping arrangements in her house. Just before the start of the regression Sheila was talking with me about the changes in religious faith and feelings of isolation she had been experiencing. In the hypnosis sessions Sheila spoke very softly, while her body movements reflected intense energies that correspond to the "tremendous" pressures and other forces she was experiencing. Before the second session I asked Pam Kasey to take notes describing these movements and will report them in that respect.

Under hypnosis Sheila immediately spoke in a very soft tone of feeling scared of "the noise and the lights." I took her back to an earlier point in the night and asked her to describe the process of going to bed. She said that she had gone to bed about eleven o'clock, after her husband, and tried unsuccessfully to get him to give her a hug (they were still sleeping in the same bed at that time). She believes that she went to sleep lying on her left side, but the next thing she actually recalls was being awake on her back, hearing a very loud, high-pitched noise—"I can't scream louder than that noise"—and seeing an evenly blinking red light "coming in all the windows all over the place . . . Jim looked like he was dead," she said, "His mouth was open and the lights made him look like a funny color." She became so frightened at this point that I had to reassure Sheila that the beings would not come into the room we were working in. Later she wrote me, "The greatest benefit in this was that I was certain that I was not alone. I knew you were there."

Feeling confused, Sheila pushed herself up on her elbows and saw several of "those things" coming down the hall. One of them raised his hand as if to signal to others. With her breath coming in gasps, and with much encouragement on my part to breathe deeply and to recenter herself, Sheila described three figures with "skinny arms and legs" that came in a straight line into her room. Two of them came to the end of her bed and the blinking lights and noise ceased as the beings at the bottom of her bed stared at her, "just looking." Her fear mounting, Sheila told me that she wanted to lie down again so the beings would not see her, but realized this was to no avail. "I know they're going to come to me. I feel that," she said. Two of the beings came beside her and one stood over Jim. She looked at their eyes and saw "power" but could not speak of this further.

Part of her terror, Sheila then realized, was that she had seen the beings before ("I know them"), and "I don't want to see them." Her fear seemed to reach a crescendo as her body writhed in awful contortions and her limbs twisted about, tensing and relaxing. Their eyes are "so ugly," she said, and reported that she feared the beings' touch. Thrashing and turning, Sheila now saw a white light above her and felt her arms being held to her sides.

"This is not my bedroom," she announced, and moaned, "I want to go home. I don't know how I got here." There were now a lot of beings "coming and going. It's hard to count them because they're all over." Sheila felt she had been forced to "lie down" and said that the beings "took my energy . . . There is power in those eyes," she declared. Something touched her on the abdomen and "they won't let my arms go. They always do that." Then she felt the "horrible pressure" and pain of "something square" pressing into her body through the lower abdominal wall. "The scary part is you're not in control," she says. Sheila did her best with my help to overcome her generally ladylike ways and to express the rage and humiliation she was obviously feeling. But all she could do was say she would like to kick the beings, tie them up, and "send them home." She described their "funny-shaped" and hairless "fat heads," and pronounced, "They don't look like us."

The scariest thing about the beings is the eyes, Sheila said. "They're so big . . . They're different." Tiring now, giggling and punning weakly, she said, "Maybe they're from God, with big temples!" Toward the end of the regression I asked Sheila to describe where she was. "It doesn't look like my bedroom," she said. "I feel like I'm on a table," which "feels hard." Her body "feels hot" from the "fighting" and "trying to get away." She returned to "the power of the eyes," especially of the one she called "'the leader' . . . He controls with his eyes. Everyone respects him." The experience of "just their eyes and my eyes" was like "neurolinguistic programming," Sheila added. They "take control and then you don't have the energy to fight." Before stopping I asked her to give her perspective on the reality of what she had gone through. "I know it happened," she said somewhat sadly.

I spoke with Sheila the next day as part of my follow-up routine. She felt "confused" and "vulnerable now," for the experience had seemed "so real." She was determined to prevent these "dreams." I said that I did not know how to help her do that, but might be able to assist her in coming to terms with her experiences. She agreed to "join in the mystery" with me.

Dr. Waterman had heard me lecture on the subject of abductions and was open to the possibility that Sheila's case might be an example of this phenomenon. He cleared his schedule to come to our next regression. Before inducing the hypnosis, in Dr. Waterman's presence, I explored with Sheila how she thought he regarded her abduction accounts. "I guess he thinks it could be true," she said. Then he spoke of his own fascination and limited familiarity with the phenomenon, and related his curiosity to his own Methodist upbringing. He resonated with something that I had said to the effect of how "society has tried to snuff out [his words]" the "spiritual side" of our lives, and said that he felt "endorsed by this. It has an important personal meaning."

Sheila seemed more self-confident when she arrived for the second regression. Despite her fear, her recollection of the disturbing emotions of the first session, and a feeling of "tremendous personal violation" that "someone could enter your home and invade your space," she was determined, almost eager, to continue. "I've lived with it long enough," she declared. Later Sheila wrote me of how our first session broke her isolation, validated her experience, and bolstered her strength and determination. In the course of reviewing what had happened during the first session she said, "I don't care what anyone thinks; I know what I saw." She spoke disappointedly of a friend whom she generally trusted and had told of her abduction experiences. But the friend "doesn't believe me," Sheila said, and lamented that "people think that we have a discrete universe that we know about, and they don't want to work with anything beyond those boundaries."

We reviewed the first session, which she recalled quite accurately, except that the square instrument that was pressed through her abdominal wall she now remembered as "rectangular," approximately one by two and a half inches. Typically conscientious, Sheila assured me she had not meant to "lie," but she was so disconcerted by her pain that she could not speak clearly. She asked for "some direction, some goal," and I suggested we explore further the March 1984 incident, picking up where we had left off.

Once again Sheila described the bright lights, the noise, her fear and confusion, and how frustrated she felt when she could not wake her husband. She spoke of beings coming into the room and to her bed and again focused on the leader's eyes. "He scares me," she said, but "I've got to see him better." Now the eyes appeared black, not brown, and staring into them gave her the feeling that "there's this black all around me." She felt she "couldn't breathe," "squished [unable to move]," and "covered with black stuff . . . I felt like I was in a black box." She had the sense that she fell asleep for "two seconds" and "then I saw a light, just white, on top . . . it's too bright." Then she was on the table, "feels great" for a moment before becoming intensely afraid as she was shown "needles . . . by my eyes."

Sheila's body movements and other behavior as described in Pam Kasey's notes tell the story of her relived experience more powerfully than her words. She clenched and unclenched her hands and stretched her arms. Her legs twitched and her eyebrows became furrowed. Her shoulders tensed and she shuddered almost convulsively. Her breathing became labored, then quiet. Sometimes there were long silences and small restless movements.

The needles were stuck "right in my forehead." At first this was painful, but "then I just relaxed." The needles in her head seemed to cause her right hand and arm to become "numb." Then the leader walked up to her with something that looked like a fan with "a needle in it . . . I tried to stop him because I'm afraid of it," she said. Then her legs tensed up and twitched as she described his approaching them with the fanlike instrument. Her legs were "a little bit afar [*sic*] apart," the left straight and the right bent, as one of the figures stuck the fan's needle into the side of her left leg. Sheila wanted to scream as she said, "Take it out of me. Take it out of me," but could only moan softly, even with my encouragement. Even after the needle was removed from her leg, it still felt "stiff and sore."

After this she saw "a lot" of the beings "standing over me." She was embarrassed to have no clothes on. Then there was "something comin' over me" that looked rather like an electric shaver with "somethin' black underneath it . . . I can tell they're holding it," she said, as one of the beings "keeps putting it on me" and dragged it across her abdomen from the left to right side, causing her to feel cold. This seemed to be the black rectangle she has spoken of before. Her body now became more tense and agitated and Sheila groaned as she told of the shaverlike instrument being pressed "on my right side, very low, where I imagine my right ovary or appendix is" while "they're trying to hold my [right] leg down." She felt intense pain as the instrument was held here. Sheila said, "I just have to balance myself . . . It feels like they're going to go through the other side! . . . Can you imagine an elephant balancing on one leg?"—i.e., standing on one leg on her abdomen. Then she had the feeling that an instrument outside her body was "suckin' stuff out from the inside of my body."

After a few minutes Sheila's body became more still and she pulled restlessly at her shirt as she declared, "I'm hot." She did not "take the time to pay attention" to what was removed through the wall of her abdomen. She was clear, however, that she "saw the black rectangle." She spoke again of her fear, anger, and pain and then said, "I see their eyes. I don't want to see them anymore."

At this point the scene changed and Sheila was shown something that looked like a huge, red stained-glass window with brown and gold lattice work separating the panes that curved vastly toward a dome; such windows lined the entire length of a long wall. The feeling she had was, "I really am there." The cone-shaped dome was so awesome in its towering depth above her that "it scares me to look at it." At the same time, the display seemed very beautiful, like "the northern lights . . . a fountain of pure gold." The sense of depth seemed to be created by recordlike grooves that spiraled upwards. When I asked Sheila what made this so frightening she could only say, "It's like power. I can't tell you any more." She said, "I want to go home," and recalled nothing further. "They took my memory away," she concluded.

After coming out of the regression Sheila said she felt sad, which she attributed at first to feeling "like I wasn't dressed" and being so out of control. Trying to return to her ladylike manner Sheila said, "My deodorant failed me under these circumstances. I'm soaked." But more significantly she observed how troubling it was to find out that "my body is not my own."

Dr. Waterman was impressed with the power and apparent authenticity of what Sheila had gone through in the session, and over the next few weeks struggled with his shifting views of her case so that he could be more helpful to her, "joining in the mystery" with us. In a follow-up telephone call, Sheila expressed deep gratitude for our work, spoke of the isolation she had felt as a result of not being believed, and expressed her determination "to force a sense of accountability" in the psychiatric profession in regard to abduction experiences. She spoke of her desire to do research that would integrate her experience with caregivers in relation to abduction, with patient satisfaction.

Julia and Sheila had a long telephone conversation several days after the regression. Sheila wanted to know that I believed her and had many questions about hypnosis and the processes of remembering and forgetting. She wondered if there is distortion in hypnosis, and what prevented her memories from surfacing on their own without it. She speculated that it could be fear. Julia concluded, "I was very impressed with the strength Sheila is exhibiting. This Sheila seems like a different person to me. She can hold her own in a conversation where before she was too timid and insecure, always holding back tears. To me this is the most fascinating aspect of the whole abduction phenomenon, how coming to terms with what is buried in memory can affect your whole life."

Before beginning the third hypnosis session, which was also attended by Dr. Waterman, we reviewed what had occurred during the six weeks since the last meeting. Just three nights before at about three o'clock in the morning, Sheila had had another dream "like the electricity" in what she called a "semiconscious" state. "I woke up real quick, and I could hear someone breathing funny and I felt a hand on my side." She jumped up, screamed in terror, turned on the light and saw no one there. Yet the pressure of what felt like a hand on her hip and a "short click" sound that she took for breathing seemed altogether real. Although sometimes now she feels strong, even at night, on some nights, like the one above, she is frightened by the sense that "there's someone in the room with you and you can't see anybody." She is especially troubled by the lack of control and knowing she cannot protect Beverly. She captured the feeling of most abductees when she said, "You're always living with a certain amount of fear that it's going to happen again."

After this discussion we talked about Sheila's first experience in my support group ("people are at different levels," she accurately observed). The experiencers are "all in the same boat," she noted, "a boat at this point in time I can't deny." She spoke then of the "overwhelming" impact of acknowledging her experiences, and the difficulty of finding anything positive in them. "I don't know if I'd want to live forever under someone else's control," she said. She was having difficulty living with the implications of what she was discovering in the hypnosis sessions, "just that there are greater powers . . . how insignificant we are as human beings," and "that you have maybe transformed [*sic*] some barrier." Of the beings themselves she wondered, "Where do they come from?" Most disturbing to Sheila, "worse than the fan and the needle and the tube and electric razor and the needles in my forehead, was the blackness . . . It's a terrifying experience of the black," she said, "looking in their eyes to figure out how they got there, and then it gets covered with black."

Sheila had resolved to explore the experience of New Year's Eve 1989, in which she had experienced "the electricity," and felt that "something forced me down," but did not see the beings. Before we began the hypnosis, she recalled waking at about midnight and again at a quarter of one, hearing a noise, and feeling "like someone had their hands on my arms and my legs. They stretched me face down on the couch, they stretched me out. I sat up and didn't see anything. I was scared out of my mind. I was terrified to be alone."

In the beginning of the regression we went over the events of that New Year's Eve. In a journal entry of January 12, 1990, Sheila had written, "I have been truly frightened at night since. It was the worst one [dream] I have had in a long time." This evening Sheila's father, her sister Melissa, and Melissa's, daughter, Kimberly, had come over to join Sheila, Jim, and Beverly. They left about eleven and Sheila went to bed at eleven-thirty, "because I was tired." Beverly had just persuaded Sheila to switch bedrooms, as Sheila's was larger and had a telephone. Explaining that she did not have the time to prepare Beverly's former room for her own sleeping, Sheila decided to sleep downstairs on a long sofa in the sunroom. She took a pillow from upstairs and covered herself with an afghan, feeling sad as she thought about her mother's not being there.

Sheila wrote in her journal that as she prepared to go to sleep, "I was scared out of my mind. I was terrified to be alone." The sunroom, which is off to the side of the house, was "so far away from Jim and Beverly." She recalled the humming sound of the humidifier and hearing the swinging of the pendulum of a downstairs clock. She believes that despite her fear she was able to fall asleep. She was awakened about midnight—the clock on the VCR said 12:02, according to her notes of January 12, 1990—and frightened by "fireworks" that seemed to be coming from across the street where the people "have company. They're entertaining tonight." Then she was frightened by a light that came into the room, and said, "It's too bright." Then the light is "gone," and Sheila said, "I got to watch for that light." She was still awake when a clock struck twelve-thirty.

I took her back to when she looked at the VCR clock and asked her what she did next. She said that she stood by the window and looked out at a light on a pole in her neighbors' driveway and attributed the light in her room to this source. Then, despite fear of such intensity it caused her to shake, she recalled lying down again on the couch to escape her fear. She remembered looking at the large organ in the living room nearby and seeing plants in the room. Her fear was mounting at this point, both in the experience that she was recalling and in the session itself. Lying on her left side Sheila tried to close her eyes. Despite her terror, Sheila said she was able to fall asleep. "It took a little while. I was very tired."

Stressing again how alone she felt so far away from Jim and Beverly, Sheila said, "I saw that light again. I tried to find it." It seemed to come from the side of the house. "It's real bright. Then I looked at it; then it was real black." She was afraid of the light and asked how to make it go away. Then, in the light, she saw something orange and pink with a "dark spot on it." Now on her back instead of her side she experienced a light so bright that she asked *me*, "Did you just flash a light?" Next she was in the middle of "gray and stuff all around me that is like mist." In her journal entry of January 12, 1990, Sheila had written "gray and V-shaped . . . I couldn't see them," the entry continues, "But there were two of them at each point—my neck, each upper arm, and about 6″ above my ankles—five sets of two in all. The only way I can describe them is that they hurt me and these 2 sets were perfectly symmetric—any other way I could feel them even after I woke up. They seemed so real." Her breathing was labored and gasping now and she felt as if she were standing but would like to lie down. She spoke of feeling "not cold," as if she were in "some kind of gray bubble that was room temperature" with "no defined walls or a flat ceiling."

Then Sheila said bluntly, "I just saw their eyes. I want to get away from them. They're right in front of me."

"Where?" I asked.

"By the gray stuff." A great struggle now ensued in which Sheila was drawn to look at the eyes, but also avoided them and wanted to make them "go away." She noted how "big" and "intense" the eyes were, and "I never see them blink." Compelled to look into them she acknowledged, "I see the eyes." Although this admission, or the act of looking into the eyes, or both, made her feel more relaxed, Sheila felt at the same time that this made her "feel like I could be crazy, like you don't know what you're doing, like you could be like psychotic or something, like you're not in touch with reality."

Virtually whimpering now, Sheila described the terror of giving up control. "They're in control," she said. "I have to surrender it." She felt "exploited" by them, but at the same time "that we depend on each other." Her thoughts returned to the "too bright" flashing light which terrified her. Then Sheila saw something "orange" outside the window, "very low to the ground." With enormous difficulty, despite much encouragement on my part, she was able to say only "I saw a big, orange, oval mass."

Sheila asked "to go back to the eyes and talk about depending on them . . . They just told me that," she said after she surrendered control. Acknowledging that neither she nor I really understood what it meant, Sheila said, "we're both depending on each other. I have to accept his [the being's] presence in my life," or at least that he "comes to me in the nighttime." She does not believe he would "come to me in the daytime." When I asked her how this information was communicated to her she said, "I just know. I know what he's thinking. He communicates, but he can't tell you how." She was not happy with all this—"I don't like him there"—but accepts the truth of what she has acknowledged.

Sheila had considerable difficulty steadying herself to sit up after the hypnosis was terminated. She connected the eyes clearly with the blackness she had experienced in the previous regressions. "The eyes are scary. Well, I was looking at their eyes and then I was surrounded with black. All I could see was black." This time she did not feel surrounded with blackness when she looked into the eyes; she even found that she relaxed, but found the idea that "we depend on each other" frightening because "they are not friendly. You wouldn't invite them over for the holidays." She wondered if they were "deceptive . . . You can't depend on someone who's deceptive. You can't rely on them." But she was not sure about this.

We talked together about how interdependence occurs when you surrender control. "When I looked in their eyes—we talked about depending on each other—and I started thinking about systems theory, you know, ecology. But now that I'm awake I think that, well it's very difficult to think they're around . . . I think they're around in the daytime even though I don't see them." Sheila struggled further with the problem of control. "I don't see the balance, like when you look in their eyes, and then they're in control and you surrender the control then that's when you depend on each other. But I don't see the balance in logical thinking." We talked further about the human desire "to be in control" and the destructive consequences of this. "You have to surrender to achieve balance," she said. Human beings have been "socialized to be in control" in the daytime. "That's the daytime control," she added. "Nighttime you surrender for that perfect balance." We ended on the question of whether the nighttime surrender of control could have any payoff for the day. "I think it could," Sheila said.

"How?" I asked.

"Well, that's what I don't know the answer to," she said.

Dr. Waterman was impressed with what happened in this session and the days that followed. Sheila continued to integrate what had come up in this session, especially in their meetings. Troubling details returned that she had actually set down in her journal on January 12, 1990, but which had not come up in our session, especially concerning the V-shaped points. Still, Sheila seemed to Dr. Waterman to be a "different person," smiling, worried for him (his father had recently died), and much more self-assured. Later she wrote me that she was impressed with major changes in his outlook, "shifting views—Something was different." I saw her at the support group on December 14, three weeks after the last session. She seemed more energetic, with a bright, direct look in her eyes, and said she felt more hopeful. We talked about her efforts to help another abductee who was struggling with feelings of hopelessness in the dawning realization of her experience, just as Sheila had done.

Discussion

It is difficult for us to admit when we do not know something. In psychiatry there is a tendency, natural enough perhaps, to try to fit psychological data or emotional phenomena into familiar categories. Total uncertainty is very uncomfortable. In Sheila's case the emergence of her "electrical dreams" and other features of a traumatic condition following her mother's death created a certain logic that argued for an explanation of her case on the basis of unresolved grief, depression, or a post–traumatic stress disorder related to the death of her mother, to whom she had in fact been close. Yet various therapeutic efforts that followed this direction failed to relieve Sheila's distress, and by the end of the summer of 1992 she had become increasingly desperate.

In retrospect there were several features of Sheila's case that do not fit the diagnosis of delayed grief reaction or depression alone. Although anxious about the intrusive and disturbing electrical dreams, her principle symptom, there was nothing in them that pointed to a preoccupation with loss, separation, or other characteristics of grief, nor was there the deep loss of self-esteem nor the self-reproach that is likely to accompany clinical depression. Even the impulsive suicide gesture in July 1985 was in response to a genuine problem of confidence in a therapist at a time when she was feeling particularly desperate and alone.

Sheila did in fact show features of a post–traumatic stress condition with general anxiety, troubling dreams, and difficulty sleeping. But the question to be answered concerns its source. The death of her mother was troubling to Sheila, as was the estrangement from her husband. There was little, however, in her reaction to these events—to which she seemed to have adapted reasonably well—or the content of her dreams to suggest that they were the principle source of her ongoing traumatized state. A neuropsychological evaluation early in 1991 documented Sheila's anxiety but did not show evidence of depression and described her as "functioning in the above average to superior range." No other cause of her trauma, aside from the abduction experiences, has been uncovered.

Sheila's case demonstrates typical features of the abduction phenomenon. These include frightening dreams that seem more real than ordinary nightmares, memories—some available consciously with others emerging under hypnosis—of intrusion into her bedroom by humanoid beings, and being taken into a strange enclosure and subjected to intrusive surgical-like procedures. In three hypnosis sessions we were only able to scratch the surface of what Sheila seems to have undergone. Yet joining her in exploring the mystery of these experiences, and giving her the opportunity to express the powerful repressed affects that are associated with them, has been therapeutically effective.

One could argue that Sheila's clinical improvement was the result of confirming a set of false beliefs or delusions. But there is nothing in Sheila's tough-mindedness to indicate a proneness to delusional thinking, or even suggestibility. Furthermore, psychotic individuals with delusions do not generally improve when their delusions are reinforced, as too much psychological energy must be invested in the belief system at the cost of other functioning. We might also consider the benefit that accrues when one becomes part of a community of belief, as in certain religious groups, but the abduction phenomenon runs counter to contemporary social belief, and Sheila, like almost all abductees, finds the idea that these intrusions, whatever their source, exist in reality, to be altogether unwelcome. If anything, she is additionally traumatized by her acknowledgment of the actuality of the abduction experiences. Finally, there is the witnessing of Dr. Waterman, who was initially a skeptic about abductions but open to working with me. Having known Sheila for more than seven years, he saw her responses under hypnosis as authentic, reflecting powerful traumatic experiences with no apparent source other than what Sheila reported during the sessions.

The abduction phenomenon runs counter to the notions of reality of the Western scientific worldview. We believe it is simply not possible for these events to be taking place. Yet we have, so far, no conventional explanation for what individuals like Sheila are experiencing. Sheila herself wrote of her case to Dr. G., "It simply does not make sense in the world as I know it to be." But as Freud once said, theory does not prevent facts from showing up. All that those of us in the mental health professions can ask of ourselves at this time is that we keep our minds open when dealing with phenomena like the alien abduction syndrome that we do not understand, and resist providing explanations prematurely. We would do well to follow Sheila's lead, as she wrote to Dr. Waterman in 1990, "I have left my DSM III-R behind." Listening without knowing, but with a willingness to explore, can in itself be helpful.

Although Sheila has had more difficulty than many abductees in recovering the memories of her experiences and moving through their traumatic content, she shows the beginnings of a transformational process that has become familiar to me. In association with her own surrender of control, she is beginning to recognize the negative consequences for herself as an individual, and for the ecological balance of the planet, that our struggle for dominance and control have brought about. We do not know whether this shift of consciousness is simply a by-product of her working through the traumatic experiences or is intrinsic to the abduction phenomenon itself. It is interesting in this regard that Sheila's experience of acknowledging her interdependence with the alien beings, followed by her concern for the earth's ecology, occurred when she felt she had to look into the alien leader's eyes and surrender control.

The alien abduction phenomenon is a potentially rich source of information for our understanding of ourselves and the surrounding universe in which we participate. But to make such knowledge available we need first to admit our great ignorance of nature and nature's secrets. As Sheila wrote to Dr. R., "Some day, you may hear someone else telling you about a similar experience. I do not have a 'scientific' explanation for this either, but that does not call for an ignorant stance. We can admit that psychiatry does not have all the answers to understanding mental disorders, so why should we believe that science is prepared to explain everything that happens in this world?"

CHAPTER FIVE

SUMMER OF '92

Scott was twenty-four when we first met in November 1991 after he expressed interest in joining my monthly abductee support group. He was seeing a psychotherapist at the time because of anxieties related to his abduction experiences, and she thought the group could be helpful in giving him the opportunity to meet other abduction experiencers and in allowing him to share the conflicts that had grown out of his encounters. My policy then and now has been to meet personally with abductees before including them in the group. Scott's case demonstrates the dramatic personal transformations that are possible when an abductee directly confronts the reality of his abduction experiences and the powerful emotions associated with them. Scott also is one of an enlarging group of abductees who discover a dual human/alien identity in the course of their exploratory work.

Scott is a tall, husky, forthright young man whose slightly breezy manner belies his underlying thoughtfulness and sensitivity, qualities which have expanded in the time that I have known him. Although Scott has resisted formal education, he reflects a strong, untutored intelligence. Scott works as an actor and filmmaker and with his father in his auto mechanic business and is a talented builder, capable of repairing pianos as well as cars. He has played the piano since childhood and is an aspiring songwriter. He also wanted to be a pilot, but "all the medical stuff" he was put through as a result of his abduction experiences made this difficult. "I've always kept busy," Scott says, "to keep my mind off what's been 'happening to me.'"

In the summer of 1992 Scott went through a period in which his customary vigilant, animalistic, and fear-laden defensiveness (calling himself a "security freak," and fearing each night that he would be abducted, Scott hardwired the house where he lives alone with a radio alarm he activated at night, mounted surveillance cameras in several locations as a "deterrent," and a microphone by the front door with a speaker next to his bed for night monitoring) gave way for a time to more intense feelings of vulnerability, helplessness, and separateness from his family. "I felt completely open to anybody to strike me down, or do whatever they wanted to." Instead of the controlled person he had been, Scott discovered that "the uncontrolled person was the 'real' thing . . . I was scared," he said. "I mean I felt I could be destroyed. I did not feel safe, at all . . . It was completely unpaved ground for me." It was this opening, the surrender of control, that paved the way to the transformation of Scott's relationship to his abduction experiences and to profound changes in his experience of his own consciousness and identity.

Scott's sister, Lee, nineteen months younger than he, is also an abductee, although she has been slower to recover the memories of her experiences. For many years she clung to the possibility that her fears of sexual intimacy were related to abuse by her father or someone else. A careful history failed to substantiate a story of abuse that could account for her fears, while a powerful hypnosis session that we did in November 1992 revealed a disturbing, invasive early teenage experience in which she was taken aboard a UFO by alien beings, a probing instrument was inserted in her vagina, and some sort of tissue, perhaps an egg, was removed. Ten days later Lee embarked on a previously planned trip to India for several months to pursue her spiritual development, especially to study Tibetan Buddhism.

After reading my account of her brother's case, which their mother had sent to her in India, Lee was concerned that my brief summary of her experiences would make her appear as too much of a victim. "I do wish to help by having my story known, to inform people." She would like to see "a more well rounded" account "to portray a series of encounters which not only produced physical and sexual trauma, but provided a priceless opportunity for spiritual growth and sensitivity to all sentient beings, ranging from insects to those of other dimensions and planetary systems . . . This adjustment," she continued, "would make me feel less an intergalactic rape victim and more like what I view it as (as of yet)—an experience of something which has nearly blown my head off with expansion of consciousness. I am strangely grateful." Earlier on in the letter Lee had written, "Tibetan Buddhism as a philosophy recognizes much of the spiritual encounters and 'awarenesses' abductees have had."

Scott and Lee's mother, Emily, age forty-eight, works in real estate and supports her husband Henry's business. She may also be an abductee, but what is most remarkable about Emily, and is an important aspect of this case, is the extraordinary steadfastness and support that she has given to her children. She is the only parent who comes regularly to my support group meetings, and though she has suffered deeply over her children's abduction-related distresses, Emily has fully accepted the reality of what her children report they have been through. In addition, she feels a deep, intuitive sense that the process that they are undergoing is one of personal growth and ultimate enlightenment. This attitude, whatever its ultimate truth may prove to be, is unique in my experience among the parents of abductees.

Scott's father, Henry, has been a mechanic for twenty years and has recently started another business. Henry is cautious about talking of his feelings and views, but is also supportive of his children. He believes what they have reported, but has more of a "show me" attitude toward UFOs and aliens. Scott has a brother, Robert, who has reported no involvement with abductions. Emily describes Robert as listening in a "detached" but supportive manner when the subject comes up at home. Robert is married and has three children, twin girls age three, and a baby boy of one and a half (as of January 1993), none of whom appears to be involved in the abduction phenomenon. Scott feels grateful for his generally positive family life, and cannot connect his abduction experiences to any hidden traumas or other aspects of it. "I look at my family, and I look at the way I've grown up and it doesn't coincide at all," he says.

When I first met Scott he had been coping for several months with trauma related to an abduction experience that occurred in April 1990 in which he consciously saw small beings ("the short guys") in his room. He connected the experience with a memory of seeing the same beings in his room and a "flying saucer" outside when he was ten. Through UFO organizations and a long chain of referrals, Scott finally was referred to a therapist. She was helpful to him, and in their work, which included several hypnosis sessions, he recovered memories of abduction experiences going back to age three. Scott has regularly attended the support group meetings since November 1991, and he and I have kept in contact outside of the meetings. We did two hypnosis sessions in March and December 1992, which Scott sought in order to discover and express his buried emotions more intensely and to explore a more co-investigative, less therapeutic, healing model.

Details of Scott's early history were obtained from medical records at Children's Hospital Medical Center in Boston from when he was fourteen and evaluated for "confusional episodes previously labeled as seizures." At six months his mother reported that he had a seizure in association with a fever, and she stated that on his fifth birthday he had "a generalized seizure" in the absence of a fever but accompanied by ear pain. He was not evaluated by a physician at that time, but his doctor was called and attributed the seizure to the "excitement" of the day. Scott now sees that time as a "post-abduction panic attack."

The first abduction experience that Scott has recollected occurred when he was three. In the summer of 1991, with the help of hypnosis, he and his therapist were exploring events related to the period when he was about nine when "I jumped back to when I was three years old playing in the dirt outside . . . and all of a sudden, boom. I turned around, I was playing with my trucks, and they were there." Out of the corner of his eye he saw two beings appearing from nowhere and then some sort of rod "put me under." He remembers running to his mother. After being returned he was frustrated because he was unable to tell what had happened. "I saw big ants out there," he said. Remembering this experience so alarmed Scott ("I jumped clear off the couch") that he discontinued the hypnosis sessions until his first regression with me.

Beginning when he was eight, Scott was taken repeatedly to physicians, especially neurologists, for the evaluation and treatment of frequent throbbing headaches that had begun when he was six, and some sort of "spells" or "seizures" that were poorly described as attacks of "strange feelings," "spacing out," or "confusional episodes." He was initially described as "a restless eight-year-old boy." The headaches were diagnosed as "atypical migraine" and treated with mild analgesics (painkillers). An initial electroencephalogram (EEG) during this period was read as mildly abnormal, followed by others that were normal. But over the next several years Scott was treated with substantial doses of several anti-convulsant medicines that had little effect. An outpatient note from when he was fifteen records "visual hallucinations" from age twelve or thirteen in which Scott reported seeing a spinning, colored triangle and "images such as a woman ['feminine figure,' Scott says] leaning over his bed, cars, and outdoor scenes etc."

By the time Scott was sixteen or seventeen the seizure diagnoses in his record had given way to "psychoemotional components," the headaches had become "'tension' in origin," the hallucinations were described as "paroxysmal feelings," and the EEGs were normal. At eighteen he was described as somewhat depressed and "listless." By nineteen the anticonvulsant medications were discontinued and the medical visits ended. Scott resented what he later came to feel were uninformed and unnecessary medical procedures. "It's just incredible the amount of medical bullshit," he said when I first met with him, and in the support group nearly a year later he objected to what he called "hit-or-miss drugging."

Except for night fears, moodiness, difficulty concentrating, and the other symptoms that led his parents to bring him to so many doctors to try to understand what was going on, Scott felt that his childhood was happy and full of friends and activities.

Scott's layer symptoms bear a complex but not altogether clear relationship to his childhood abduction experiences. Scott thinks they were "flashbacks," re-evoked memories of his earlier abductions. Emily has asked repeatedly "Where was I?" when Scott and Lee were undergoing the abductions; yet compared to most parents of child abductees both she and Henry have been particularly supportive.

"It's bewildering," Emily wrote me in February 1993, a week after her own first hypnosis session with me in which the depth of her commitment to her children was affirmed, "that this was all happening right under our eyes so to speak and we apparently were not aware of it—consciously at any rate—and to recall Scott's remarks referring to fear of seeing them in his room—flying saucer outside—dog put to sleep—running up to our room—Henry going outside to see what was out there with his gun. We do recall this, but [it] was at the very back of our memory until this all came up a couple of years ago when Scott said, 'Remember when I was a kid?' and we said, 'Oh yeah!'" Later Emily wrote that she and Henry were fearful of a robber or intruder and also thought that Scott had had a bad dream.

Scott recalls that his childhood encounters tended to occur when he was outside with Lee, while Lee remembers a "little gully" near the house where she and Scott played a lot and which she has come to believe was one of the sites of their abductions. Lee says that "we used to love it," but when she was a teenager she stopped playing there. "I used to think about the place as a special place." When Pam Kasey visited the family at their Massachusetts home in March 1992, Scott and his parents talked of various UFO sightings that the extended family had experienced over the years. Scott recalls "seeing a ship" at age eight or nine while riding his bike and reporting it to his uncle. But an abduction experience he describes as "a biggie," which had remained "buried" in his mind until it surfaced in a hypnosis session with his therapist, began in his room when he was ten.

Scott saw "a flying saucer outside," and then he saw several beings come into the room. They put the dog that was in the hall to sleep, "somehow with the rod . . . After they were done with me" Scott became afraid, as "I knew they were going upstairs to my parents' room." Scott recalls, "I ran upstairs—this was after [the] event—and I told them what had happened and I said there was a flying saucer outside, and my father got his gun. He was scared shitless—everybody was—got his gun, and went outside and there was nothing but nature." Scott recalls, "When I was a kid, I was scared to death they were going to kill my parents." The beings seemed to him "like a greater power than your parents." Despite his fears Scott also felt that the beings were somehow "more wisdomful than my parents," although he is not sure whether this is "them themselves" or "the wisdom created by the whole experience." Scott describes the "telepathy" he experiences during the encounters as "a two-way channel. They know your thoughts and you can see theirs. It's quite traumatizing because of its unfamiliarity."

The next abduction experience that Scott has recalled relates to a feminine figure leaning over his bed when he was twelve or thirteen, mentioned above in an outpatient note. At about this time Scott was referred to a psychologist to see if there was an emotional cause of his distress. But even with lengthy psychotherapy, little progress was made toward uncovering its origin. The encounter with the woman figure, which was part of an abduction experience, will be reported in detail in connection with his second hypnosis session with me.

Scott does not recall any further discrete abduction experiences until April 1990 when he consciously saw several entities in his room after first sensing their presence in his mind. "Whoever these people were they were not from around here," he recalled in his first meeting with me. "It was the same people. I knew it," he said, that had been present in his room when he was ten. Alarmed by the experience, he sought help as described above. In several hypnosis sessions with his therapist, Scott recalled that during this abduction he was terrified as a faucetlike device was placed on his penis, "wires" or "leads" applied to his testicles, and a sperm sample taken as he lay terrified and paralyzed on a table in a UFO.

After our initial visit and his attendance at several support group meetings, Scott's curiosity about his abduction experiences deepened and he wished to explore them further, because "they had affected my life so much." Meanwhile his personal life was becoming more complicated. When I first met him, Scott told me of strain in his relationship with his girlfriend, a tendency of his to "grab on" and not be able to "let go." In the January 1992 support group he said that although she had been initially, "when it really came down to it" she "wasn't very supportive" in relation to his abduction experiences. At about this time Scott was given the opportunity to share his firsthand knowledge of abductions on the set at CBS in Los Angeles, where the two-part docudrama dealing with the phenomenon, *Intruders*, was being filmed for airing in May. For two weeks in February, Scott was on the set each day, which he found highly stimulating. He made a valuable contribution to the cast and crew's understanding of abductions and became close to an actress whose daughter might have had encounters.

At our support group meeting on February 24, Scott, just back from Los Angeles, spoke of the feeling that we are being "prepared for" something, that there was perhaps a "plan" of some sort, that we are not in control and "somebody else" is "running the show . . . Getting through the trauma part," he said, "has opened up the real stuff, the spiritual behind it," and he spoke further of a "bigger power" at work in his experiences. Scott recalled that even "when it happened to me as a kid" he felt he had to work on being "able to stay in the same room without panicking, without fear." The April 1990 episode, he said, "was a step up in intensity, majorly, and almost like testing, but there's definitely that anger, God, anger just to lash out from being touched, being under somebody else's control."

Scott spoke further in this meeting about breaking through "the trauma stuff" and of personal growth. His abduction experiences, he said, had made clear to him that "there's a massive amount of information in my head that I can't even understand." The aliens, he suggested, are "helping us grow so we can comprehend them . . . They're getting us trained to get us to a point where we can deal with them." Scott elected after this meeting to undergo his first hypnosis session with me in order to move further through and, hopefully, beyond the traumatic dimensions of his abductions and to discover their deeper meaning for him and for others as well.

Scott arrived at my house on March 16 with Ann, the actress he had gotten to know in Los Angeles. Before Scott came with Pam and me to the upstairs room where I was then doing the regressions, we chatted a bit in the living room about what the experience of playing such a role had meant for Ann, her objections to the inaccurate and sensationalized parts of the script, and her sensitive efforts to maintain the integrity of the role she was playing.

Before beginning the regression we talked about Scott's apprehensions and his possibly abduction related experiences since the April 1990 event, which we agreed would be our focus. He had no recollections of discrete abductions, but spoke of vaguer "cloudy kind of stuff," a blue light coming into his room one night, unexplained needlelike marks that had appeared on his arms several times, and how on some mornings his left sock would be mysteriously missing from his foot. Scott spoke of fear of death and aloneness and of his feeling like "something in a cage, an animal, being a specimen." We reviewed the details of the April 1990 experience and Scott went back briefly over a few of the more frightening details of his previous hypnosis sessions with his therapist. I reassured him that I would not leave him feeling caged and alone during our session.

After the hypnotic induction Scott spoke almost immediately of feeling "mad." We then reviewed the events of that April evening before the abduction began. He had drunk a couple of screwdrivers, played the piano, and talked generally about his life in the living room with his mother and father (he was still living with his parents at the time) who were watching television. He went up to bed a little earlier than usual—at ten o'clock—feeling tenuous and vulnerable about the course of his existence. As he prepared for sleep and "jumped in bed," Scott felt some anxiety about a new film shoot that was planned for the next day.

Scott recalls reading a magazine, and before he could fall asleep he felt that the beings were "there, in my mind." As his fear mounted in our session, Scott spoke of the loss of mental privacy and of the familiarity of these feelings. His room had no door, and unexplained light was coming in from the direction of the adjacent clothes washer/dryer room. Scott's breath was now coming in loud, short gasps as he spoke of "six" of "them" with "boxy" and "angular" heads that were "after me." Then he saw a "round-tipped rod" pushing toward him, which Scott related to how he was anesthetized. "They know I'm aware," Scott said, and "they put me under" so that "I couldn't move" by touching him with a rod behind his ear. At this point a "buzzing" in his right ear changed to a ringing sound and "I lost control of my body." Then all Scott saw was a screen like a TV monitor that was "fritzing." Memories of his life flashed before him, as he felt had happened "so many times" during abductions, and he felt himself struggling to protect his mind "so they couldn't touch it." After this he quite literally lost consciousness, although he had been saying "as fast as I could 'I've got to remember, I've got to remember.'"

Next Scott recalled he was on a table in the presence of two doctor-like figures with odd, tan- and white-tinted skin, wearing "glasses" and white coats, and several shorter beings in "army suits." The beings had deep, black, slightly slanted eyes with gray borders around them. "I hate them" for "taking me from my mom when I was young," Scott said, and "for not telling me who they are." "They're curious about me," and "I'm curious, but I hate what they've done."

"What did they do?" I asked.

"They've used me."

The beings then placed a "faucet thing, like a suction" over Scott's penis. This device was connected by a tube to a box at the side of the table. At this point Scott had a kind of out-of-body experience from fear, as he looked down on himself and saw his head on a blocklike pillow and four prongs being pressed into his neck, high up just below the scalp, which he also felt pushing against him. Scott believed these were like "electrodes" that were used to manipulate and control his movements and feelings. At this point in our session, and also at the time, he felt calm, although he is angry when he thinks back about what was done to him.

I encouraged Scott to center himself through his breathing and to express any feelings that were near the surface. He gave forth a loud growling sound as he spoke of his naked terror, his sense of violation, and his fear of bodily injury. He noticed how quickly he tends to "build the walls" to protect himself. He saw more light in the room now, and for the first time in this session spoke of the "wires" that were applied to his testicles. It was these wires, Scott observed, in combination with the suction device over his penis that stimulated his erection and were "making it happen" and "taking things out"—i.e., his "sperm." The whole experience, Scott said, "just seems unbelievable."

The beings communicated telepathically to Scott that they were "making [really taking] more white stuff" for a purpose. They were using him "as a father . . . taking my whatever, my babies." "All the stuff" they took from him was being used, Scott knew, to "make babies." Strong feelings of shame came up for Scott at this point, and I explained that he had no reason to be ashamed as he had been confronted by powers or energy forms against which he was altogether powerless. "I'm mad," he said, growling again, but "I can't fight . . . They know exactly what they're doing," Scott remarked, "that's why they cover it up. They don't want us to remember."

I took Scott back one more time to the traumatic, shame-filled aspect of his experience. Once again he balked at the full reliving of his humiliation. "I don't remember. Too painful," he said, "too emotional . . . I had no choice," he allowed. "It's not my fault." But he quickly added, "I should have been able . . . "

"Nonsense," I said and reiterated what I had told him about powers in the universe beyond our control. Again Scott expressed his anger, and I assured him that he "couldn't do a damn thing."

After this Scott remembered being "dropped in bed" in his room feeling very frightened and also angry, but had no recollection of how he was returned. He had the sense the beings had been "messing around with my head," leaving information of some sort that he could not access. After coming out of the regression Scott was struck by the power of the emotions he had experienced. "I've never had those emotions before, never, never." It "felt good," he said, to give expression through his voice and body to such strong, bottled up affects. The intensity of his anger bothered Scott some. "I'm scared to death of what damage I can do," he said. The "whole experience," he said, "when it is brought back into the body it releases these things. We are the emotional patterns that structure things and our reaction." He also was awed by the intensity and brilliance of the light that he had seen while on the table. As a result of the regression he felt he had more access in this ("normal") reality to the experiences undergone during the abduction. Scott was also left with the feeling, as is common with abductees, that his mind had been "electrically" manipulated or tampered with. He was aware that there were still "walls up everywhere," and that there was much more inside him that he wanted to remember.

The nine-month period between our two regressions was a time of rapid change for Scott. He brought Ann, who was still in Boston a week after the first regression, to the March 23 support group meeting. They updated the group on the progress of the miniseries. During the meeting Scott reflected on an increased preoccupation with philosophical and religious concerns, such as "who's in control" and the possible views of God. Around that time Scott also made several television appearances, including an awful show on a Boston channel in which he was humiliatingly, but not untypically, introduced as a young man who had had sex with aliens. As the spring continued he had increasing difficulty integrating the stimulation and stress related to his high public exposure, and had more frequent meetings with his therapist related to this discomfort. His therapist and I discussed his case, and I referred Scott to a psychiatrist at my hospital for prescription of a mild tranquilizer, which helped to reduced his tension. She described Scott as initially depressed, anxious, "very vulnerable," and confused about what had happened to him. He seemed to her to be a traumatized person who had experienced "a different kind of trauma," manifesting the hypervigilance and difficulty relaxing "you see with other trauma survivors." As far as the abduction story was concerned, "I don't know what to make of it," she said. "Something bad clearly happened to him."

One of the effects of Scott's crisis of helplessness and vulnerability in the summer of 1992 was to rally the support of his family, especially his mother and sister, who began coming to the support group meetings. By

September he was clearly feeling better, spoke in the support group of the need for a sense of humor, and continued to complain about the constant intrusion of the alien presence into his mind, a kind of loss of privacy. Emily told the group movingly of how "two of my children have been affected," and spoke of how little understanding she had had of the "extreme terror" to which Scott had been exposed in his abductions. By October he was speaking more boldly of pushing through his fear and of his struggles "to integrate" his experiences.

Scott spoke increasingly through the fall of 1992 of the spiritual dimensions of his abduction experiences. In the November 9 support group meeting he told of how "exposure to them" had "opened up something in me . . . It's almost like you're given an intense jump into a spiritual realm you're not even ready for—like Yogis go through tons of work to do to get to a certain point." Lee, who was about to leave for India, talked of the "hardships that people often suffer at the hands of their spiritual teachers." The body's instinctual, fearful reaction to the alien encounters Scott felt was "a natural reaction" on "a species level" when confronted with something so deep and unfamiliar. He could not "imagine anyone reacting kindly or feeling safe," at least initially. But toward the end of the meeting Scott asked, "What are my choices?" and told the group, "Even though the way I think a lot of times is how mad I am, and how upset I am, and how bad my ego's been damaged or wiped out, there's only one way to think about it if I want to live, and that is to look or find whatever there is positive in it which, God, is very difficult for me at this time . . . But that seems like the only thread that will keep me alive."

On December 16, 1992, I met with Scott at his request to review his course and, as it turned out, to plan another regression. In that session he told me that one night about ten days before, as part of his increasing openness to the alien presence, he asked the beings to "show me a sign" of their actual existence. At about two or three in the morning he experienced in a partially awake state the feeling of "somebody touching me from behind." He became extremely frightened, but the touching continued—"it was almost like teasing me." The concreteness of the response to his request alarmed Scott. "I asked for them to show me something and they did . . . in a way," he said. We talked—conversations with abductees often move in this direction—about whether human beings generally were ready to perceive the alien presence. Scott felt, as many abductees do, that our destructive attitude toward anything unknown or foreign would make it dangerous for the aliens to manifest more obviously before us.

Many abductees begin to pursue a more explicit spiritual path as they open themselves to the depth and meaning of their experiences. Scott himself, in addition to his increasing curiosity about the spiritual dimensions of the phenomenon, had begun to meet with an acupuncturist, and, more recently, with a shamanic healer. He was also increasingly challenging the traditional treatment model. Scott said of some of the therapists he had seen, "I feel I could heal, I could help them more than they could help me, and that sounds completely arrogant, I know." His request to undertake another regression was part of Scott's desire to move beyond the traumata of his abductions to a more reciprocal, mutually communicative, relationship with the alien beings. We scheduled the session for five days later.

At the beginning of the session we reviewed how frightened, needy, and vulnerable, yet also more alive, Scott had felt during the summer. Although he had had a recent abduction experience, we decided to do an "open ended" regression. In recent months I had found that the psyche's own wisdom would take the experiencer where he or she needs to go in the trance and that the healing, integrating, and information-gathering process is better served by not "targeting" a specific abduction event. Before the regression Scott spoke of his "fear of letting go" and his determination not to "hold back" in this session.

At the start of the regression, after several thirty- to sixty-second pauses, Scott spoke of feeling the presence of "one of them" standing by a table on which he was lying on his back. He was thirteen and said that he had never faced or even recalled what happened to him at that age. He perceived a cylindrical tube he estimated to be four inches in diameter that was part of a machine near a wall. The image of the tubing, which seemed to be pointing at his chest, was disturbing and faded in and out of his consciousness. He also had images of other "tools," like a curved banana-shaped instrument, on another nearby table.

Soon he recalled seeing a nonhuman female figure carrying a tray with several cylinders, each containing a little baby "in glasses. . .I'm really mad," Scott said, but "I don't know what they're doing." The "woman," who had come quite near him (recall the "hallucination" he had at twelve or thirteen years old of a female figure leaning over his bed), left the room and Scott realized that the aliens—probably this figure herself—had been taking his "seeds" for the purpose of making the babies he had just been shown.

Scott realizes now that his fear prevented him from looking directly at the beings, though he had attributed this to their elusiveness. He also speculated that if he had remembered seeing the beings during this (or these) experience(s) he might have told his parents, which the aliens told him not to do. For he was "part of their family," one of the beings explained.

"If I'm part of their family, why am I here?" Scott asked. I encouraged him to explore that question. He kept getting images of an empty cylinder, about six inches in diameter and about a foot long, with clear fluid inside. "I want to be one of them," and "I want to be one of me," Scott said, "but I can't be both."

"Why not?" I asked.

"Then I'm never home either way."

Next Scott recalled being taken down into a huge underground, rock-walled place by one of "a bunch of" fast-moving elevators. It was hot there, but "better than family here," for "They know everything about me. There are no secrets." Nevertheless, "it's scary" and "just feels weird." At this point I felt that Scott was judging the truth of what was happening with his analytical mind and I encouraged him just to report his raw experience, saving the judgments for later. "I just can't believe they're here," Scott said. "When they come for me they know everything I know." He said it made him feel bad that they would not let him talk about these experiences. He wondered, "Why don't they stay?" He received no answer to this question, except that they and we "aren't ready." He said the beings are in the process of changing themselves physically "so they can breathe here. They don't breathe the same as we do."

Scott revealed other problems for both of our species should the aliens' presence be manifest on a large scale too soon. "We're not up to their speed," he said. "They think much faster than we do," and "They're going to make it so they don't hurt us."

"How would their thinking faster hurt us?" I asked.

It is "confusing when they talk to us with their minds," he replied. "Too much information. Our minds are not used to such contact—it's a sensory overload."

At this point the session took an interesting turn. Scott acknowledged that he himself had persisted in denying the existence of the aliens, and I asked him to explore what it was exactly that he was denying. To my surprise he replied, "denying that I am one of them." To acknowledge the beings' existence has meant that he would have to experience a kind of "empty" feeling, a nostalgia for another domain. "I've always known," he said, "that I was different, that I wasn't from around here." When he was a child, Scott recalled, "I always wanted to run away. I couldn't figure it out. I could run anywhere, but I couldn't get there." He knew that the beings did not reside in our solar system.

Scott grasped then why he had never wanted to look directly at the beings. With some struggle he said, "My humanness doesn't want to see this."

"What is this?" I asked.

"Them . . . The human side," he continued, "cannot handle the other side." The human being in him reacts with fear, "like an animal . . . They appear to be animals, and you act like a scared animal. It's instinct." Nevertheless, he emphasized, humans must "stop" and realize that the aliens, whom as a child he called the "inkies" because of their large, black eyes, like ourselves, "are alive." We need to learn that even though "we look different" and "we think different . . . we're all life."

Scott's memories moved then into the apocalyptic vision I have heard increasingly from abductees. Major changes in the world are coming. The aliens will only come "when it's safer." But that will not occur until there are "less and less" of us as we die off from disease, especially more communicable forms of AIDS that will reach plague proportions. This material was frightening and very sad for Scott, and he also felt that he was not "allowed" to speak of it. Although he was positive in his conviction about this, he said, "I just hope I was wrong."

At this point in the session Scott shifted to perceiving from the alien perspective, and he saw the earth as a blue body below him. He had chosen to come here from another planet because it was "closest to where we're from." He did not know the name of that planet, but it was yellow, mostly desert, and lacking water. Once there had been trees and water, but something having do to with "science"—he does not know just what—"went wrong" and his people "went underground." Scott felt "sick" inside and sobbed as he told of how science "destroyed our planet." Naturally I was curious to know if Scott had any further information about how this had come about. But he did not, except to observe that somehow the alien species "knew before" the destruction occurred but seems to have been powerless to prevent it. After the regression he recalled that the destruction had occurred because of "something they made they couldn't stop," and that on their planet the aliens live in an "artificial environment."

With considerable resistance Scott admitted that the intention of the aliens was to "live here" (on Earth) but without us, unless "humans change," in which case "we might be able to live together." Then he contrasted the ways of humans with the aliens. Human beings "are alone" and "they don't share." In the alien realm "nobody's in their own world" and "everybody knows everything. There are no secrets." I asked him about himself. "I'm one of them," he said, but in his human identity he imposes limits on his ability to love and share because of "my own ignorance."

I asked, "What else?"

"Tradition," the "whole focus of my life, my independence," he said. Because of "fear of being hurt, of not getting what you want, fear of not receiving" human beings have trouble "opening up and trusting that it's okay" to give and feel love.

Change has "got to start somewhere," Scott said, and I asked him about his leadership role as a kind of intermediary between the two species. "There's gonna be so much work," he said, and it is "gonna take a long time." I asked if he thought there was time. "Yes. I think so," he replied. He was getting tired, so I asked if there was anything else he wanted to say before sitting up.

He said, "It's got to be done one way or the other."

"What's got to be done?" I asked.

"If we don't change it's going to change for us." Then he added rather sadly, "I don't think we can live with them."

After the regression Scott felt awkward about what he had revealed. He had difficulty trusting the information he had received because "there's nothing that reinforces it when you're growing up." One realm "has nothing to do with the other," he said, and we are rarely, if ever, "exposed" to the existence of the "aliens' side." Fear simply does not exist in the "consciousness" of that "side," and so there is greater freedom there. Yet it is difficult for Scott and makes him sad and afraid to "acknowledge anything about" the alien world, especially that he is part of it. For that means "that I'm not one of us [humans]." I spoke then with Scott about the possibility of integrating his alien and human identities, and he recalled how that "just didn't work when I was a kid . . . That's just not the way people live," he said. "People're just different." I told him of four or five other "double agents" I was working with and of the possibility of their getting together as a group, which he thought would be a good idea.

After this session Scott felt great relief, as if a huge "weight" had been removed. He recalled that since early childhood he experienced himself as having "two personalities," and spoke of how "crazy" that had always made him feel. He now believes that the doubting and denying of his alien experience has been a destructive process in his life, and wonders what part telepathy plays in the existence of the dual identity.

Before concluding Scott, Pam, and I talked further of what the alien/human project could be about. "I don't think they're disposing of us. I think they're taking part of us." Then "they'll have everything we'll have, and they'll have everything they have." But there are problems with the integrating of our species, for "you and I as we are maybe won't mix."

We speculated then about the relationship of the aliens' active presence on our planet and the accelerating, catastrophic destruction of the earth's living environment. "It's not just coincidence," Scott said. From the information he has received Scott doubts that we would survive "our catastrophe" as well as the aliens did theirs. "For them it was not the beginning of science. I mean, they were well into science before this happened, whatever happened. It was a lot further ahead than we were . . . They had the resources" to survive. I pressed Scott to say more of what he knew of the relationship between our two species. "It isn't just black and white," he said, "the two sides. There's a correspondence between the two."

My last question had to do with his reluctance to look into the eyes of the aliens. He replied that when he was experiencing the alien perspective he felt he was viewing reality through their eyes. But as a human "I was scared" to "because I'd be looking at myself."

"Yourself as what?" I asked.

"As one of them," he replied. I pressed him to say what was so frightening about that, but he did not know.

He simply added, "My whole life has been useless. I mean, everything I've done has been insignificant."

"Compared to what?" I asked.

"If I had realized that [his complex double identity] a long time ago," he replied.

The day after the regression Scott told me that he felt "at peace" and that "all my questions just disappeared very quickly. That's amazing." He told the support group on February 8 that he felt quite "self-sufficient now." On December 23 he wrote me a letter which accompanied a Christmas card. After writing empathically about the "immense" weight he suspected went along with "what you [referring to me] know" he shared further information that had come to him since the regression.

"Success on earth would take an incredible shift," he wrote, "a shift from ego gratification to aspiring to achieve, but aspiring to rid ourselves of the human flaw." The difficulty, he continued, is to "eradicate the human flaws without destroying the machine itself. They are glued together very tight. The growing pains are extreme but necessary." Communicating with his alien voice he then wrote, "Our intellectual abilities and the scope of our view is too much for humans to understand. The translators, as I am, are necessary in order to make contact . . . I've always known. I've always denied [his alien identity]. I always wanted to forget, but that's not who I am. The reality comes through the thick screen of human defenses. The study continues of the human-alien consciousness struggle. They are integrating, each learning from the other . . . I am at peace now. I understand I realize the conflict will continue inside me, but I have reached the turning point where my power of uncontrol has overcome that of my human side.

"I fear humans more than anything else," the letter went on. "We have tried to change you many times. Many members of our species have been destroyed in the process . . . I must say the human being has very heightened emotions, too much for me to process at times. We are very sensitive, but our emotions are not as primitive as your own. Your emotions are recreation in a sense. We are happy to be able to feel more than we normally feel. Our fascination [with humans] revolves on this. Our evolutionary process has deemed emotions less important than understanding, but it's like candy to a child your emotions to us. It is like a drug that we enjoy very much.

"It is interesting," the letter concluded, "that this is the very thing which also makes you so dangerous to us. I do not feel it is safe for me to come out yet. It will be a few years. I feel there is much I wish to convey, and I feel at a time very soon there should be a meeting of the high powers of your world with us." Although he had a few anxious nights following the regression, in the next few months Scott made rapid strides toward achieving greater peace of mind, a heightened sense of energy and purpose, integration of his human/alien identity, and deepening understanding of the meaning for him of his abduction experiences. He was confident about the information he had received and conveyed in our regressions, and felt that for the first time he could face its implications honestly and realistically.

Discussion

Scott's case illustrates the multiple levels on which we can think about the abduction phenomenon. At one level he is, or has been, a typically traumatized abductee. He has been through the terror, helplessness, paralysis, and instrumentation—especially the humiliating forced extraction of sperm for making babies (which he later saw during a regression)—that have followed upon several recalled childhood abductions in association, on at least one occasion, with a close-up UFO sighting. But in addition to this nuts-and-bolts, or clearly physical, dimension, Scott has also undergone an important personal transformation that has been the result of a shift in his attitude toward his experiences. Of inestimable importance in this process has been the support of Scott's parents, especially his mother, Emily (herself a possible experiencer), who has attended conferences on the subject of abductions, come regularly to my monthly support group, and volunteered to undergo hypnosis with me in order to understand more deeply her own experiences and the ways that she can more fully support Scott and his sister, Lee, who is also an abductee.

Through his constant attitude of inquiry, his search for spiritual meaning, and, above all, his willingness to confront and move through his terror repeatedly, Scott has been able to achieve considerable peace of mind and a deeper sense of understanding of the abduction process. By overcoming his denial and accepting the instinctual, natural basis of his bodily terror and resentment, Scott has been able to open his psyche to important information concerning a widening sense of his own identity and to take responsibility for his role as a "translator" between our two worlds. A crucial period for him in this process was the summer months of 1992 when he was twenty-four. He was then able to acknowledge deep within himself his vulnerability, helplessness in the face of the power of the alien energies, and the stark fact of his lack of control. He discovered then, as he later wrote in his Christmas letter to me, his "power of uncontrol." Scott feels that his psychic powers have increased as a result of his experiences.

As has occurred in the case of several abductees with whom I have been working recently, Scott's full acknowledgment of the reality of the alien presence has led him to the realization that he has always had a kind of dual identity, and is capable of experiencing himself as both human and alien. The alien perspective, which apparently has always been imbedded in his consciousness, was not available to him until he surrendered the illusion of control. From this point of view Scott, like many abductees, has been able to grasp fully what a dangerous species we are, not only to the aliens themselves but to the living forms of the earth, especially as we apply destructive technologies so mindlessly. In his alien identity he comprehends how fear and anger, which are not part of the alien experience, constrict our capacity for love and connection. Knowing himself to be "one of them" has allowed Scott to experience the ways in which our two species are in some way linked (there is a "correspondence" between us, he says), the basis for which we are only beginning to understand.

It is hard to know what to make of some of the information Scott reported in his second regression with me. Like other abductees he speaks of another planet from which the aliens have come, one that has been made arid and lifeless by "science," and he warns of the depopulation of the earth through natural catastrophe, especially a more communicable form of AIDS. This kind of apocalyptic vision is common among abductees, but we have no way of knowing whether it is authentically predictive in the physical world—it certainly is not inconsistent with what we know to be occurring on the planet—or represents some sort of metaphoric prophecy or wake-up call. The question is made easier (or more difficult, depending on one's point of view) by the fact that in the realms of consciousness and of existence to which abductees travel during their experiences the distinction between the literal and the metaphoric, or the objective and the subjective, seems to lose its power.

Finally, there is a poignancy for Scott and his family in the vain and intrusive search that was made during his childhood and adolescence for a conventional medical explanation for his abduction experiences. Countless hours of medical examinations, tests, and procedures resulted in wrong diagnoses and inappropriate treatment. I suspect that even as these words are being written, a child abductee somewhere is being taken by anxious parents to a physician who is steadfastly ignorant of the abduction phenomenon, as Scott's parents were when he was a child. Hopefully through the "translation" of experiencers like Scott, and parents like Emily and Henry ("and physicians," Scott added) who are willing to consider the possibility of realities of which in Scott's words there is "little knowledge," other children may eventually be spared the compounding of the trauma that ignorance and denial bring.

CHAPTER SIX

AN ALIENATION OF AFFECTIONS

The intrusive sexual and reproductive procedures that are a central aspect of the abduction phenomenon can profoundly affect the intimate life and general well-being of abductees. If the source of this "alienation" is unrecognized, and conventional psychosexual explanations are actively pursued, the problems may deepen and the stresses that abductees and their loved ones experience are likely to increase. On the other hand, important therapeutic gains become possible when the source of the experiencer's dysfunction is discovered. This problem is well illustrated by the case of Jerry.

Jerry, who describes herself as "an ordinary housewife," had just turned thirty when she called my office in early June 1992. When I first met with her she consciously recalled a struggle with many UFO dreams, abduction encounters, and related experiences dating back to age seven. At her mother's insistence Jerry had reluctantly dismissed these as "nightmares" until she saw my name and "Harvard University" on the credits for the CBS miniseries on abductions, *Intruders*, and "figured, well, that person might be a little more trustworthy and I jotted your name down." Also, at a friend's recommendation, her mother had read one of Budd Hopkins's books and said to Jerry that the accounts of abductions there sounded like her experiences.

Our meetings have included four hypnosis regressions. In addition, Jerry has shared hundreds of pages of journal entries with me, which she began writing several months before she contacted me. These include details of her abduction experiences, poems, and the discussion of extensive philosophical ideas related to the profound transformational process she has been undergoing.

Jerry is the second of four children and as a child lived in a rural area near Kansas City, Missouri, where her father worked at a dairy processing plant. Her older brother, Ken, also had peculiar childhood experiences, including seeing unusual white and blue lights outside his window and terrible "nightmares" of "someone" entering his room while he was awake. Shortly before she met with me, Jerry and Ken talked about their experiences, and she discovered that "he's been plagued with them his whole life." In addition, in her first regression Jerry saw her younger brother, Mark, being abducted with her when he was an infant and she was seven, but she has not discussed her experiences with him.

Jerry's parents were divorced when she was eight. Her father remained in Missouri after the separation, and for many years Jerry had little contact with him. Recently she has had long conversations with him and feels that they are now becoming closer. After the divorce Jerry's mother, who has worked consistently as a social worker, moved with the four children to Macon, Georgia. Jerry remained close to her mother and over the years has consistently confided important experiences to her. During Jerry's later childhood and adolescence the family moved about a good deal in Georgia. "Perhaps we were gypsies at heart," she suggests. She joined the Brownies and then the Girl Scouts and went to summer camp where she took horseback-riding lessons and enjoyed riding and being with horses. Later, we discovered, Jerry identified the colts and their "dark, almond-shaped eyes" with hybrid alien beings. Although her teachers told her that she was a student capable of college work, Jerry left high school at the beginning of the tenth grade when an English teacher imposed college-level assignments she could not handle and the school refused to transfer her to another class. After this she held various cashiering and clerical jobs.

Given the fact that she has only a ninth grade education, both Jerry and her friends were surprised at the "flood" of poems and complex information that she first began to write down five or six years ago. Her writing intensified greatly in November 1991 following a powerful abduction experience. "I don't know where it's coming from," she said. Indeed, the sophistication and articulateness of her writings do seem beyond her educational level. Words whose meaning she does not know will come to her as she writes, but she discovers when she looks them up that they contribute to a consistent set of ideas. Jerry felt that many of her ideas did not come from within herself but from some other source. She was so shocked by communications she received from the beings themselves just after the November 1991 abduction that she burned her first notebooks.

Jerry's first marriage was to Brad when she was nineteen and pregnant with her daughter, Sally. She never loved him, and they were divorced in 1986. Jerry says that her ex-husband played sexual "games" with the children involving oral sex but not penetration. Jerry once thought that this was the result of her own aversion to sexuality. "I don't really think that is true any longer," she wrote in a journal entry in January 1993. "He could have had an affair instead. It was something about him that made him choose to do what he did, and I probably chose him because subconsciously I may have picked up on something in him that worked for me. Maybe he was not a threat to me sexually. I could easily get away with not having to deal with my fear of sex with him. He actually accepted the idea of having a sexless marriage."

In 1989 Jerry married her second husband, whose name is Bob. He works as a carpenter. She loves Bob, and longs to have a normal affectional and sexual relationship with him. In her journal she wrote, "Now I find myself in a much better marriage with a man who has a normal sexual orientation who desires to have a normal sexual relationship with me." But her abduction memories have made this impossible. "I keep telling myself that my husband is innocent and is not going to hurt me like the beings do," Jerry wrote in January 1993. "I keep telling myself that it is different, that he loves me and is not going to hurt me. I try to keep positive thoughts, but when it comes time to have sex, forget it. All of that goes out the door, and I am back being afraid. My feelings during sex are like the feelings I have when I am abducted. I feel frightened, used, and feeling that I have to endure this [at other times she has said that having sex is like "going to the gynecologist or being raped"]. Also, I think that I will be hurt at anytime. A feeling of powerlessness, and the inability to have any say in the matter. I think that I get a safe feeling when I say no to my husband and he respects that. I am desperate to resolve this problem. I just don't know how to do it."

Jerry's fears of intimacy have extended to being touched at all, and she often would drown her sorrow and frustration in alcohol. "I only drank when I thought I might have sex," she wrote in September 1992. Assuming that her sexual problem was rooted in early incest or sexual abuse, Jerry and her ex-husband went to three different marriage counselors. On one occasion her "nightmares" were interpreted as "something trying to work its way up to the surface," but nothing useful emerged and Jerry broke off the counseling.

Whereas her first husband was frightened by anything out of the ordinary and would not have listened to her abduction experiences, Jerry felt that her current husband and his family were supportive and understanding, at least at first. Bob was present during our first regression and was powerfully affected by the obvious authenticity of his wife's experience. But disbelief on the part of Bob's family seemed to close in around her so that Jerry has felt increasingly isolated and alone with her experiences, relying almost exclusively on other abductees, Pam, and myself for support. The pulling away of her in-laws has been particularly painful. "His family doesn't socialize with me like they used to," Jerry told me in March 1993. "So, you know, it's painful, because they're the only family I have and I don't like them to think of me, you know, as the eccentric one, the crazy one." But "there's no going back. I have to learn to live with it," she says.

All three of Jerry's children appear to be involved in the abduction phenomenon. From the time Sally, who was born in 1981, was six she has had severe nightmares and will scream out "Don't touch me. Leave me alone." When she was nine or ten she had frequent unexplained and severe nosebleeds. She also has seen UFOs filling the sky in her dreams or imagination and has remarked to Jerry that maybe the aliens choose specific families. Sally has "dreams" about the family going onto a deck and seeing a spaceship coming and of "a lot of little creatures surrounding her." In another dream an alien girl with no hair and a red bow "stuck to her head" came to her window and asked her to come out and play. She said she went to play with the girl and was shown around a spaceship. Following one of Sally's more recent nightmares, Jerry found her on top of her blankets with her nightgown twisted up and her underwear missing. Sally was groggy and Jerry could not wake her. In June 1993, Sally became frightened when she had an unexplained time lapse of nearly an hour while she was timing herself reading a book for school. She looked at the clock which said 6:02, read for what seemed like a couple of minutes, looked again and saw it was 6:58. "How could that be?" she asked her mother in alarm, and Jerry groped for some explanation such as she had fallen asleep. But Sally insisted this was not the case.

Matthew was born in 1983. He was frightened of the puppets he called "wo-wo's" from *Sesame Street* that came through a window. When the alien puppets were being shown, Matthew would cry and scream and tell his mother to turn off the TV. Bert, one of the puppets, had "scary big eyes," Matthew said. He was also frightened by a TCBY yogurt commercial in which a UFO flew down and landed. When this was shown, Matthew ran out of the room and again screamed for his mother to turn off the TV. He spoke of a dream of a pyramid-shaped flying saucer that talked to him and had eyes. Both children reacted strongly to the picture of an alien when Jerry showed them the Hopkins Image Recognition Test cards (HIRT) she obtained from a friend who was also an abductee. Sally "lost her breath" and put her fingers in her mouth. Frightened, Matthew asked, "Has Sally seen that?" and "What did she do?"

Colin was three in February 1993. His involvement has been intense and is well documented in Jerry's notes, her conversations with me, and a careful evaluation by another child psychiatrist. Jerry has also witnessed his presence during her own abductions. In a journal entry dated August 14, 1992, when Colin was two and a half years old, Jerry wrote of hearing him crying and talking to himself in the night. She went into his room and found him sitting up in bed. "He seemed very awake." He asked for juice, which she brought him, and then "started to ramble on about lights outside and owls with eyes." He pointed out the window and said, "See the eyes." Jerry felt "so weird because earlier that night I had the strong feeling that they were around." She took Colin upstairs to tell Bob what he had been saying, but "he got angry and said he must have had a nightmare." Colin is generally a sound sleeper, Jerry noted, and "never asks to sleep with us or even has a habit of waking up at night." But that night, for the first time, he would not sleep in his own bed and insisted upon sleeping with his parents.

This behavior continued for several nights, and Jerry wrote in her journal on October 29 that Colin talked often and consistently about "these things." When Jerry and Colin were outside together he would look up into the sky and ask about the stars and the moon and then talk about the "scary owls with the big eyes" that "fall down out of the sky" or "floated" down. A few times he demonstrated what the eyes looked like by circling his eyes with his hands curved in the shape of a C. One time he went "into a lot of action like running and screaming and saying that they make me eat some food and they attack [a word he actually uses] me," especially hurting his toe.

Colin also talked about spaceships, planets, and stars. One night he climbed in bed with his mother and noticed a small picture of the earth on the binding of a book. "That's the planet Earth," he said, and "it go away" and "the house go away." Pointing at the ceiling he said, "They say bye-bye see ya." Then he jumped out of bed and enacted a scene, talking anxiously. "The owls with big eyes fall down and jump and I jump," and "there's a spaceship and I come out of the spaceship . . . My toe hurt," he said, and "the big eyes are scary, Mommy." After this Jerry actually found blood at the end of one of Colin's toes and a torn toenail.

On November 8, I met with Colin, now two years and nine months old, and his parents in my home while his brother and sister played in the backyard. He impressed me as a sweet, lively boy, but he revealed few of his fears. He called the alligator puppet a "tiger" that liked to bite. (Later he would ask his mother, "Why do the tigers go and get you?" and seemed to have replaced the owl with the big eyes with the tiger.) He was particularly interested in the large globe in my office and wanted to locate himself on it. I went through the HIRT cards with him, and he reacted strongly only to the alien card, which he called a "scary man," and became more anxious after this. In her journal on November 15 Jerry wrote that Colin had cried out "Ouchy! Ouchy!" several times during the night. When she and Bob went in to see him he was sound asleep, but the next morning he said that the "monster owls" hurt his leg. Climbing into his parents' bed and pointing to the ceiling Colin said, "What's that big boat, that big boat in the sky."

In her journal entry of January 28, 1993, Jerry wrote that Colin's distressing experiences seemed to be occurring every week or two. On January 25, when she and Colin were in the bathroom as she was preparing to meet Bob for lunch, he said several times in an angry, frightened voice, "I don't want to go back to the spaceship!" Then, standing on the toilet with his teeth and fist clenched and obviously distressed, he said several times, "I get lost. I don't like it." Calming down, he said, "I was born there and fell from the stars." When Jerry asked him to repeat what he had said he added, "I born on the spaceship and it was dark." Then he became tense again and wiggled around. She asked him how he got onto the spaceship and he circled his eyes with his hands and said, "The eyes." When Jerry asked him if there was anyone on the spaceship with him he answered, "Yeah, I see the King. I see the King and He is God." Jerry wondered where he obtained the verbal skills, seemingly beyond his years, to say these things.

On the night of January 27, Colin came into his parents' room and climbed into bed with them. Jerry found this unusual as the gate on his bedroom door had been locked and no one remembered having opened it. Meanwhile a monitor that picked up sounds from the toddler's room started making strange, loud noises, clicking on and off so loudly that Colin asked Bob to turn it off.

In light of his continuing distress, and because I wanted to see if a conventional psychopathological explanation for Colin's symptoms might be found in an independent evaluation, I asked a capable and—I thought—open-minded child psychiatrist colleague who was not especially familiar with the abduction phenomenon to evaluate Colin. This doctor met with Colin and his family in February and sent me his report in March. Dr. C. did not find much that was remarkable in Colin's history other than the story of his encounters, saw him as a "very cute and engaging boy," and discovered little marital tension between Jerry and Bob, except in relation to Jerry's abduction experiences. Colin played with puppets and became interested in a rubber snake that ate fingers and toes. He referred to his own toe being hurt, but evinced little distress about this.

Although Dr. C. found no explanation for Colin's problems, he wondered if they might be tied to an as yet undiscovered incident in the family, perhaps related to interactions with his brother who had a history of sexual abuse and with whom Colin had shared a room for a time. He speculated also whether Colin's symptoms might be related to TV images of spaceships and the planet Earth, though his TV watching was restricted. Colin's distress seemed to abate some after this evaluation, and Dr. C. recommended no further intervention at this time, although he offered to see Colin further if his fears persisted. One effect of this evaluation seemed to be to divide Jerry and Bob a little further in their perception of the source of Colin's problems. Dr. C.'s failure to discover a more conventional explanation of Colin's symptoms affirmed Jerry's view that they were related to UFO abductions. But Bob found Dr. C.'s search for a more conventional traumatic source within the family reassuring, as he has resisted the idea that abductions have a reality, at least where his little son is concerned. In June, upon seeing a book with an alien on the cover Colin remarked, "He's a Rocketeer. He goes up and comes down."

Jerry has the sense that abductions and related phenomena have been happening throughout her life. She has always known that the experiences that were so readily labeled nightmares by her mother and others were powerfully real for her. Thus she has always lived with a strong feeling of isolation and the sense that she had no choice but to deny a major "part of my life." The appearance of scoop marks, scars, bruises, and other small lesions following abduction experiences helped Jerry affirm the actuality of what she has undergone throughout the years before she found a community of experiencers and investigators who were familiar with the phenomenon.

Jerry's first consciously recalled abduction experience occurred when she was seven and still living in Kansas City. This episode will be discussed in detail in relation to her first hypnosis session. Before this she remembered seeing some sort of unusual light, a spaceship, and small, thin, gray beings outside her window. When she told her mother of seeing these things she was told it was a nightmare, but Jerry "told her adamantly that I was not imagining or dreaming it, that it was real."

"I saw lights, I saw the ship, I saw them," Jerry said in our first meeting, and "never once did I say it was a dream or a nightmare." The perennial insistence by her mother that these experiences that she felt were real were just dreams made Jerry doubt her own sense of reality. During the first regression, in the course of talking about her abduction at seven years old, Jerry indicated that she had had earlier encounters. She could not recall how old she was, but remembers being small and that "I wasn't afraid before when I saw them the first time. I thought they were cute . . . Lots of them" were outside the window, "just real happy," and encouraged her to "come and play." When she was nine and staying in a motel just before moving to Georgia, Jerry remembers feeling a presence in the room and the frightening sense that "someone had just sat on my bed." At age eight she had an important traumatic and intrusive experience which we explored in our fourth hypnosis session.

A still more disturbing episode, which we explored in detail in the second hypnosis session, occurred in Georgia when Jerry was thirteen. She woke up terrified and remembered pressure in the abdomen and genital area and that she could not move. "In my head I was screaming," Jerry remembers, but does not know if any sound came out. "Somebody was doing something," she recalled, but it was "something alien." Although she recalls wondering to herself, "Is that how sex is done?" she knew with great certainty that "it wasn't a person."

Jerry wrote in her journal about two weeks before we explored this episode in the regression how her difficulties with intimacy and sexuality had begun soon after it. She was dating her first "real boyfriend," who was about two years older than she. Jerry found that she was "terrified by the idea of doing anything more than just kissing," whereas she had dated and "experimented with petting and none of it bothered me in the least" before this. Her parents were asleep, and Jerry and her boyfriend were in her bedroom. He suggested that they "do more than just kiss and hug." Wanting to "be rid of" her "fear of being touched anywhere on my private areas" she allowed him to "touch me in between my legs, to put it nicely." But then "I freaked out. I was completely tensed up. My whole body was stiff as a board. I went into a kind of panic attack. I was sweating and shaking and my heart was racing. I looked at my hand and all of a sudden it started to shrink and shrivel up. It started to turn grayish. I was petrified. I don't know what I did after that, but whatever it was it scared my boyfriend enough to go and wake up my mother. She came in and calmed me down." Jerry did not talk of this incident with anyone, but wrote, "Ever since then I have had an aversion to sex."

In the years that followed, Jerry had a number of "nightmares" in which she would awake paralyzed, hear "buzzing and ringing and whirring" noises in her head, and see humanoid beings in her room. "They were really causing me to lose a lot of sleep," she wrote. In an episode in 1987 she saw beautiful "glitter and sparkles" that seemed to have been thrown into the room, but she screamed in terror as two small beings wearing some sort of shiny "outfit or uniform" floated above her bed. She believes that she screamed and tried to wake Bob, who was then her fiancé, "but he did not budge." As the beings came closer to her, Jerry became still more frightened and "then a blank" and "I don't remember anything after that." In a later regression with another therapist, in which she investigated this episode, Jerry was deeply moved when twin girls were shown to her which she feels were her own hybrid offspring.

In 1990, Jerry experienced the most traumatic of her abduction episodes. We have not yet investigated this one under hypnosis because of the intensity of the terror and pain associated with it, but Jerry consciously recalls many details of it. She and Bob had just bought a duplex apartment in Plymouth, Massachusetts. She does not remember how the episode began, except that she felt a presence and a tapping on her shoulder. She was taken into a circular room which was shiny and metallic-looking and contained what looked like equipment. As she was suspended in a standing position and tests were being performed, Jerry recalls that her necklace flew off and fell to the floor. She communicated telepathically with a tall being with "blondish" hair who appeared to be the leader. When she told him that her necklace had come off, he said he saw it and motioned to a smaller being to pick it up. Jerry was told she could not have it just then because it was "contaminated" and the beings put it in a "plastic-looking pouch." The leader promised that it would be returned to her some other time. Months later her mother found (Jerry had told her about the necklace in the episode) what Jerry believes was the same necklace in a box in Georgia.

At first Jerry was not frightened during this episode, and was pleased that she was able to converse with the beings. The leader asked "how the medication has been so far," and she made the mistake of saying "fine." For after this a procedure was done to the back of her head above the neck that caused the most excruciating pain she had ever experienced, "even worse than childbirth . . . I thought they were killing me," she said, and remembers screaming, "How could you? You asked me how my medication is." In addition to the raw pain, Jerry felt muscle spasms that were out of her control and extended in rapid succession from her legs to her facial muscles. She screamed for them to stop and was filled with hate and rage. "Here I thought they were somehow perfect and loving beings. How could they have done that to me? I was so terrified. I blanked out after that. The next thing I was back in bed, waking up." Although she usually goes to sleep curled up in a self-protecting position on her side, Jerry awoke on her back. Her body was very rigid and straight, her hands were folded on her chest and her feet were pointing straight up and close together. Still panicked, Jerry tried to wake her husband but could not do so. She then telephoned her mother in Georgia because "I needed to tell someone what had just happened."

In one of the three episodes in 1991, Jerry recalls being taken by taller, more human-looking, fair-skinned, blond beings to what seemed like the top of a very large building with illuminated equipment in it. She had the sense that she was at a beach or a seashore, as she heard the wind and the water breaking, felt a breeze, and smelled the sea. High up in this building, Jerry was shown scenes of missiles and other weapons. She felt this was very important. They also showed her some sort of triangular machine that became circular when it spun and "had to do with flight maybe." Jerry was assured that she would not ever forget what she was shown on this occasion. The next day she found herself making triangles out of paper or with pencils and toothpicks and "spinning [them] around and around."

In November 1991, Jerry awoke, feeling a presence once again. The room was filled with an orange-red light that soon receded. The next day her mind seemed to be "turned up full volume," flooded with thoughts. Jerry felt as if she were filled with information of a "universal" sort, "soul stuff, unusual for me." After this, as mentioned above, she wrote intensely. Her writings included a hundred poems over the next one and a half months, whereas before this "I never wrote a poem in my life." She found the pressure of these thoughts and writing quite overwhelming and said, "I don't know where it's coming from."

In the months prior to our first hypnosis session on August 11, 1992, Jerry continued to have abduction experiences, including one episode just three weeks before, in which she consciously recalled seeing a UFO close-up and being taken by humanoid beings—whose attitude she felt was loving and benevolent—into the ship. There she saw shelves with instruments and vials, was seated on a chair or table, and had a complex dialogue with aliens she felt were "beyond what we would think is intelligent or even genius." One of them explained that they came from so "far into the future" that she would not be able to comprehend. Jerry remembers saying to herself, "This is great. I can see everything and I am so aware." In her journal she concluded, "I was convinced beyond a shadow of a doubt that what I was experiencing was real. They looked at me with their loving and all-knowing smile and simply said, 'yes.' I then said, well, if this is real then I am somehow living a double life . . . I had a feeling there was a definite reason that I and others like me were not aware of this other reality, at least not as aware as we are about this reality we have here and now."

Bob accompanied Jerry to our first hypnosis session. He came as a skeptic, but said that either "she's lying to me" or "it's really happening," and "She is not a liar at all . . . She's the most honest person I've ever met in my entire life." Nevertheless, some measure of Bob's resistance was suggested when he said that he "fell asleep" through much of the *Intruders* miniseries, which contained some blood-curdling abduction scenes.

Before the regression began we reviewed several of Jerry's abduction experiences, after which she spoke of her search for a church that would feel more compatible than the Catholicism in which she had been raised. "I kind of floated in and out through religions with friends," she said. A local Protestant church seemed the most comfortable to Jerry and Bob at first, but "they wanted to fully change our lives," and, in his words, "we took a break [from attending] and are enjoying it so far." Jerry was especially troubled to find that she could not speak with anyone in the church about her abduction experiences, for they regarded the phenomenon as "totally evil, of the devil." God, they said, "would never, ever make beings that looked like that," which turned Jerry against the church, for in her view, the aliens are "another intelligence, or another being, another reality . . . I don't feel they're necessarily bad or good." Once when Jerry told several churchmen about an out-of-body experience she had had, "they went straight to a high official who said, 'Well, don't ever do that.'"

After this we discussed Jerry's curiosity about the episode in Missouri when she was seven years old, and we decided to focus on that. We reviewed in detail the location of her home, which was on a hillside by a little creek and cow pastures, and the arrangement of the rooms in the house. Jerry shared a room with bunk beds with her younger sister.

In the trance state, Jerry's first images were of her pink-walled bedroom and of standing on the floor in her long flannel nightgown. The house was very quiet, and she remembered feeling anxious and pulled to go out of her room and down the hall. A strange bright light filled the room. Jerry had the thought, "I shouldn't be afraid 'cause I know them." Despite her mounting fear she felt compelled to come out of her room into the hall and then into the living room. Outside of the house, in the direction from which the light seemed to be coming, Jerry saw perhaps twenty or thirty small beings and backed up in terror. She could not move as several beings passed through the window frame into the room. "I wouldn't go out, so they had to come in," she said. Jerry felt that they became impatient with her and "just picked me up" from her crouched position. "I don't want to go out the window," she said, as the pressure became more intense.

To Jerry's amazement, the beings took her through the window, "and then I went up real fast." As if "stopped in the sky," Jerry could see the top of her house, the trees, and the ground below. "It kind of took my breath away to go up so fast." There was a "big thing above me" into which she was taken. Despite the coercion, Jerry feels that she was somehow a participant in this process, but "I don't know how, though." Weeping at this point, Jerry saw that two of the beings were also "floating up" her baby brother, Mark, and she worried that "he's probably afraid," although he appeared to be asleep.

Jerry felt like running away, but realized she was "paralyzed kind of up to the waist." Breathing heavily in the session, with her voice quivering, Jerry described the paralysis as being like a painful vibration. Then "a tremendous vibration" extended into her hands, and "I'm afraid it'll go through my whole body." I reassured her that reliving the experience would not harm her. "The vibration is so strong. I don't understand it," she moaned, afraid that she would not be able to breathe. "I can't do anything," and "I'm worried about Mark," she cried. The powerful vibrations seemed to shake Jerry's whole body. "Okay, do it to me, but it's not fair to do it to him," she said. "He's a baby! I just hate them for it . . . I thought at first they were all right." Crying as she recalled an earlier abduction experience, Jerry said, "I just thought they were cute and they wanted to come out and play."

The round room inside the ship into which Jerry and Mark were first taken was dark in the beginning. Then she saw that it was "dome-shaped. It's real white . . . It's got railings. It's got different levels, and they're way up high. The people and the machines are way up high."

There were two curved tables in the room, one for Mark and one for her, into which each of them was placed. "He seems small" in his, she observed. "I just look over at Mark and tell him to be good and not to move around," Jerry said through her tears. "He might fall or something," she worried.

Jerry then saw a small, "real dark," being standing at the railing on the next level "just watching" her, and noticed behind her a taller, lighter one she called "the leader." He looked older, "wrinkly and tight," with "a nice face" and a "permanent type of smile," wore a one-piece, goldish-yellow suit, and had a little bit of stringy yellowish-white hair. His hands were "long and skinny." This being communicated her name, "Jerry," to her, as if he knew her, which she found frightening, especially as she realized he seemed familiar to her as well. All this made the whole recollection more real to her. Breathing hard and her body shaking, Jerry cried out, "Oh, I don't know if I can do it! . . . Up till now it was just a dream," she said, but if she admitted that this being is or was real, then "everything else would be real." Her mother "was wrong" to have insisted that this experience was a dream, Jerry said. "I've got to stop thinking about what everybody else says . . . I have to live my own life. I can't keep pleasing my mother," she added. "I have to stand up for" what happened, Jerry said, "no matter if people will think I'm crazy or not."

The leader asked Jerry "if the medication has been okay up till now," which Jerry did not understand. She then relived an extremely agonizing procedure, involving the insertion of "something sharp" like a "needle" into the side of her head, which evidently had taken some time and against which there was "no medication." She cried out desperately and sweated profusely with her body writhing in pain as she tried to hold off the memories. "I think it would kill me," she said. I assured her that this recall would not kill her and encouraged her to scream as she described the instrument being driven from "a high angle" down into the side of her neck. "Stop hurting me," she screamed loudly and complained now of spasms and other uncontrollable movements in her legs (which I could see). Panting, Jerry then screamed out in terror, "I can't stop it! Ahhhh! Ahhhh! I hate doing this! Stop it! Stop this!"

Jerry's loud screaming and thrashing continued, and she cried out, "They're turning it! They're turning it! Ohhhh! It's inside of me. That's what he stuck inside of me. Ahhhh! That thing! They stuck that thing inside me!" I reassured Jerry as best I could that she would be better when this was over. "It's coming out," she said. "There's a leak. Something's dripping I feel like, in my throat." She was not sure whether this had been blood, saliva, or what. "They're letting me relax. They're awful. They're cruel. I thought they did something else. Oh [in a whisper], I didn't anticipate this." Jerry recalled that she was told that some sort of tiny object was left inside of her at that time "to monitor me" with no explanation other than, "We just have to do what we have to do."

"I think it's still there," she said. "I don't remember them taking it out." After this Jerry felt limp and tired. She did not know what was done to Mark, but said, "If they did the same thing, I'd kill them." She recalled little of what followed. The leader left while she stayed for a few minutes on the table. Then she saw flashes and dots of red and yellow light. She did not remember how she got back to her home, nor did she recall being able to tell her mother or anyone else any of the details of the trauma at the time. As the regression came to an end, Jerry and I speculated on the protective mechanisms that might until now have prevented her from recalling this harrowing experience.

"I'm drenched," she said, as she came out of the trance. "I didn't think you could do that," she said, "to put me out." Her arms and legs continued to "feel funny" for several minutes after the session was over, and she observed that she had also felt "like somebody turning you into a vibrating machine . . . like somebody put you inside a machine and you were part of that machine." Their "medication" evidently referred to this vibratory process, but she did not feel it in her neck or head and therefore the vibrations did not mute the pain. Jerry then recalled that when the anesthesiologist tried to give her a spinal injection to reduce pain during the delivery of Colin, she screamed loudly, for it seemed now to be "a similar thing" to what she had gone through on the ship. This session upheld Jerry's childhood conviction that the alleged "nightmare" at age seven was actually "a memory" of something that "must have just happened, then and there." Bob, himself somewhat in shock over what he had seen his wife go through, said at the end of the session, "It's a lot, a lot to take. At first I wasn't sure, and I saw it bugging her as much as it's been, the pain and everything. I started getting a little worried there." But then he found a way to "just kinda sit back. I was glued to the chair at first, but then I had to start getting up . . . "

The day after the session, Jerry and I had a follow-up telephone conversation. She expressed shock over how vividly she had relived the abduction experience. "I thought I would just have a few memories," she said. Jerry noted how reluctant she had been to look at or recall seeing the alien beings themselves. This was just too scary, for "what we see we know." She was confused about the intensity of her pain and wondered if some elements of her 1990 experience had become confused with the childhood experience just explored. In her journal she tried to sort out the elements of the experience from when she was seven and the 1990 one, but was at a disadvantage because she had not explored the adult incident under hypnosis.

In the weeks that followed Jerry described in her journal various other current and past abductions, related dreams, "vivid dreams of UFOs," visions, and out-of-body experiences, stimulated by the regression. It was also during this time that Colin began to tell of his experiences, and Jerry recorded these in her journal. Six days after the regression she wrote, "I feel that part of a long and heavy burden has been lifted from me since the regression." On August 27, she saw a beam or pillar of light appear in front of the stairs in her house and thought of calling me but did not do so. On September 21, Jerry recorded for the first time a frequent recurrent dream about a horse that she wished to take home in order to care for properly. At the end of the dream she felt "cheated of the opportunity to be with and take care of my horse." In September, Jerry began to ponder her sexual fears, and related them for the first time to the traumas of her abductions. It was Jerry's desire to overcome these fears and "have a normal sexual relationship" that led her to request a second hypnosis session.

This meeting occurred on October 5, and Jerry explicitly stated her wish to find out why she had tried to avoid sex "at all costs." Bob's sister, Anna, was present during the session. Jerry elected to explore the episode in which, as a thirteen-year-old, she had been terrified during the night by pressure in the abdomen and genital area. Before beginning the regression, we recalled the circumstances surrounding the episode, which probably occurred in the fall of 1975 when Jerry was starting the eighth grade. Although the episode was frightening, she told no one, including her mother, about it.

Under hypnosis Jerry's first recollection was of awakening to discover bright, white light illuminating her room. She felt a presence that frightened her and thought, "If I'm just real quiet they won't get me." The beings tried to reassure her by telling her not to be afraid, but this did not work because "they're so full of lies." Although "I don't want to see them," Jerry noted two beings, "one behind me and one over here [to her side]." They told her she must go with them and ignored her protests. "They grab me by my arms," Jerry said, and she felt a "gentle, soft, like velvet, cold" contact. This touch seemed to relax her, and next Jerry found herself "just kind of going with them. Slow. Slow going up. It's weird. I don't know how they can do that."

With one being on each side of her they floated Jerry "out the window, like the wall. It's like it's not there." She got "that paralyzed feeling again" as she was pulled up to a large craft. "This pulling on scares me," Jerry said, as her breathing became more rapid and shallow. She was taken through an opening into "this same stupid room" where "bad things happen." Two beings were doing something to a table, "like getting it ready or something. I just have no control." A taller being she has seen "a lot" since age five but does not like knowing told her not to be scared, but her fear builds nevertheless. "Don't they understand what they're doing?" Jerry protested. As a small child she trusted this being, but now felt betrayed by him. "He just doesn't have any patience," and despite her protests "they just put me on the table anyway. You can't argue with them."

Jerry felt "embarrassed" before the aliens as they took off her pajamas. "It's like they think they're doctors or something. I don't think they're doctors." Lying on her back now, Jerry felt somehow more "relaxed" and less afraid. One of the beings put his hands over her eyes and pressed something that "looks like a tube" through the wall of her abdomen above the umbilicus. With the instrument still deep within her abdomen, the being took his hands away from her eyes and Jerry felt more relaxed and also sleepy and tired. Next she noted that one of the beings was holding a shiny, horseshoe-shaped object with a handle on it as others bent her knees upward and apart. Crying now she said, "He's going to cover my eyes again. Why is he doing this? I don't think I want to know. I don't want to know what they did." She told the beings she will tell her mother, but they said that she won't. "She wouldn't let them do this if she was here," Jerry lamented poignantly. The beings insisted that she would not tell because she would not remember.

At this point I asked her if it was "Okay for you to remember now?" She said that it was, but exclaimed, "It's not fair!" which I, of course, agreed with. "They tried to make me think it was just a nightmare," she complained. "What do they think, I'm just an animal or something?" Jerry's fear mounted once again as she felt "pressure" inside her vagina. She objected that although the beings kept reassuring her that she would not remember, she has, in fact, remembered what happened. Then she cried and moaned, sobbing, "I just want my mom," as she felt "something round inside," a cramped and "pinched feeling . . . Why are they doing this?" Jerry cried out. "I'm not going to let them do this again!" I encouraged her to express her emotions. "Why won't this end? Stop it!!" she cried.

Finally, this part of the ordeal was over. It felt to Jerry as if something had been placed deep within her body, beyond the vagina, perhaps through the cervix. As an adult, she had had an abortion and this felt something like the D & C procedure. She saw the horseshoe-shaped instrument being removed from her body and struggled with the memory of what it held. "I don't want to go any further," she protested. I reassured her of my presence and left the choice to her. "It can't be," she said in a whisper. "Oh, I can't believe it. I'm too young for that. I'm only thirteen." I said it should not be done, but that probably from a strictly biological standpoint she was old enough. "I don't know," she wailed. "Oh. Oh. This can't be. It can't be. I just must be . . . I don't know."

What Jerry saw was a "baby" that was "real tiny, skinny." The beings seemed very pleased with their efforts and showed her this creature, which was perhaps ten inches long. She could not make out many other details except that there were tiny hands and the head seemed big compared to the body. The baby was placed in a clear plastic "cylinder-looking thing" where it floated in some sort of fluid. "Why would they do this?" Jerry exclaimed. "I don't understand this. I'm too little to have a baby. They just told me not to worry. I don't have to take care of it." I asked Jerry what feeling of connection she had with this fetus and she replied, "I think they made me feel it's not mine. It's theirs. It's a part of them." As Jerry lay on the table for what seemed like half an hour, the aliens seemed to be "working on the little baby." Then they brought it up close for her to look at. The beings wanted Jerry to feel proud of the accomplishment of producing this creature. But she felt angry, confused, used, and betrayed.

Jerry continued to express her intense feeling of shock and disbelief. "I didn't even know I had anything like that in me!" she said with a weak laugh. "If they're gonna do this, they should at least tell you," she added. I asked if she had been given any further information about what this was all about. The "leader person," she said, "told me that it was beautiful, and that one day I would understand, but it was about creation."

"Creation of what?" I asked.

"I guess, like a new being. A new race, or a new—I don't know. He didn't really say specifically." He just said that at "a point in time of their own" she would know. "They said it was beautiful. It was wonderful," and "just to trust that it had to do with creation."

At this point Jerry began a debate with herself as to whether to tell me the leader's name, which she apparently knew. To speak it, she said, would make him more real, give him a stronger identity. Once when she was writing he seemed to speak to her and asked, "Do you deny?—I think he was wanting me to admit him. I think he wanted me to remember." The name that came to her was something like "Moolana."

As the session neared the end, Jerry recalled being helped to get dressed, but had only "real fuzzy" recall of how she returned to her house and came back into her bed. She awoke on her back and at the time recalled only that a being had stood over her, "put pressure on my stomach," and was "doing something down on my private area." Her conscious memory had thus condensed the beginning of the abduction and what occurred on the ship. Reviewing her experience, Jerry noticed that she had had no control whatsoever over what occurred. As in the earlier childhood abduction explored in the first regression, Jerry felt that she "wouldn't have been able to handle it" were she to have recalled consciously the traumas of this episode. But now she felt that "I was supposed to tell somebody" and "they initiated" that process.

As she came out of the trance state, Jerry felt relaxed, but continued to express her indignation toward the aliens. "They have no right to do what they do and they're pretty arrogant about it . . . Just to take—don't they know, don't they know us well enough to know that a thirteen-year-old doesn't do that? . . . It's for their purposes. They're being pretty selfish." But at the same time that she felt she was "a tool for their design," Jerry also felt that she was participating in a plan that came from a "higher" place. "My feeling is it's not just them."

Jerry spoke then of Bob's increased resistance to accepting the reality of her experiences, largely, she believed, because of the implications for him of having his little son, Colin, involved. She recalled then that she and Colin have a similar inherited deformity of one of their toes and that Colin complains that "the owl bit my toe" while Jerry has also had the experience of having her deformed toe examined. She associated then to a silvery-blue flash she had noticed one night recently as she was going to sleep and saw DNA in large bold letters and heard the phrase "the marker trait," which has meaning in genetics research which Jerry was completely unfamiliar with.

We talked about the impact of this abduction experience on her sexual life. Jerry was raised with an accepting attitude about "getting married, having sex, having babies," she said. "Sex means getting married, having babies, caring, loving, and sharing." But obviously "they [the aliens] don't do anything like that." They have "no respect for feelings or love or relationships . . . " When she has sex now it revives traumatic abduction intrusions like the one we had just recalled. "When I have sex, that's how it feels. It feels like they're doing that. It feels like I have to just grin and bear it. When I'm having sex it's just like that. It's like I relive it every time . . . I transfer, I know I do . . . I have no control over this . . . I never knew where that feeling came from." At the end of the session Jerry had me look at a small, circular, indented scar on her abdomen which she associated with the procedures just recalled. Until this session she had not known "where it came from," but seemed to feel confident that it was the result of one of her abductions. I was unable to discover from Anna, who seemed a bit stunned, what her reaction was to this session.

In the days that followed this regression Jerry had a difficult time. She had trouble sleeping, cried a lot, and searched unsuccessfully for an alternate explanation. Anna, she said, was torn by what she had experienced in the session. She could not "believe" it, but had said to Jerry, "I know you're not lying." Anna's skepticism made Jerry's attempt to integrate the experience more difficult. Meanwhile, she had consulted another therapist who lived nearer than I to where she lived, in order to pursue her memories more actively. But the brief hypnosis sessions with him, which continued through the fall and winter, appeared to compound her trauma rather than help her. "I have more memories coming to me than I can handle," she wrote in her journal in January. The therapist would push her to move ahead faster than she felt ready, urged her to do weekly regressions, and threatened negative consequences if she refused to comply. She felt overwhelmed, and found it helpful to attend my monthly support group. I also urged her to "slow down" and then to break off the sessions with the other therapist as little integrative work seemed to be accompanying the uncovering that was occurring.

Jerry faithfully recorded the experiences of the fall and winter in her journal. Her fears for Colin, as recorded above, resulted in the evaluations already described. She herself had several dreams of nuclear war in which there was general panic and she heard herself say "it must be Armageddon." In one of the dreams she "looked out into this vast nothingness and saw a UFO, and it was slowly moving along with a beam of some sort shooting down onto the land." Flooded by memories of abductions that were coming up in the regressions with her therapist, Jerry wrote of the "shattering" of "defense mechanisms" and the extraordinary nature of a threatening experience that can recur unpredictably at any time. "I wonder if anyone who is not an abductee," she wrote in early January, "could possibly comprehend what it means to not have any idea when the next abduction could be" and "I would like to know how the mind works when a person is subjected to continual trauma and knows that it may not end."

At the end of January, Jerry had what she called a "horse dream" in which she was looking for "my horse" in a laboratory-like room with pools of water in little rectangular cubicles or tanks. When she looked at these it made her feel sad. "It was almost pathetic." She noticed one of the horses closest to her. When it turned its head toward her, it looked at her with "big dark eyes."

"These little horses were all hooked up to some wires in the water. They all had long arms and legs and they were all very skinny. Their heads look as if they couldn't hold them up even if they wanted to. But this one turned its head and looked directly into my eyes. I am not sure how I felt when it did that, but I felt that in its eyes was an awareness that reached beyond what I would imagine it would be capable of."

On March 4, 1993, Jerry came to see me, accompanied by a close friend who was also an abductee. The purpose of the session was to review what was happening in her life and plan for the future. Jerry talked of her feelings of ostracism and isolation, especially the pulling away of Bob's family, and the need for a community of understanding around the abduction phenomenon. She recalled now unexplained losses of fetuses that both her mother and sister had experienced. Further discussion of the dream of the little horses led Jerry to link them with another dream of hybrid baby girls with which she felt a strong bond. "You're our mother," the girls had said, and she felt the same bond with one of the little horses which she now thinks represented a human or hybrid infant.

Jerry sought a third hypnosis session because there were still, as she put it, "a few" of her abductions "that keep bothering me." She particularly had in mind the painful 1990 episode in which "I screamed and screamed and screamed" and the 1991 encounter from which she could still recall the smell of the sea and hearing the waves on the beach. As it turned out, her psyche "chose" an incident in September 1992, which has had an especially profound impact on her intimate life.

We met on May 27, and Jerry came alone. Before beginning the regression, Jerry talked of the increasing estrangement she was feeling from her husband's family, which made her decide to stop talking to them about her abductions unless they asked. Her mother-in-law, she said, "can't accept it at all. She thinks if I'm a good little girl and I say my prayers, it'll all go away." But Jerry wondered what did, in fact, affect her abduction experiences, for she felt as if she were "just fair game." She had a sense, however, that her fearfulness makes the experiences worse. One time she recalled being more relaxed and "not fighting" and "that one was painless . . . They were doing things. They did something to my arm that made my arm swell up. They were showing me things . . . I guess I felt more of the communication, more able to talk to them and ask questions, and I don't remember any real answers."

Jerry does not feel that they want "to cause me fear and pain and agony," and "deep down inside I think that what they're doing is somehow necessary." It has to do, she said, with "races, beings or whatever, coming together to make another creation." This "was very important," she said, and "as a single person, compared to this big huge thing going on, I should look beyond myself and know that it's for the greater good." At the same time Jerry noted that over the past year she had learned to think more independently. "I feel I shed some of my old beliefs," she said, and she no longer "blindly" follows "someone else or some organization." As a Catholic, she had been raised to feel that she was "being disobedient to God" when she followed "my own instincts" or asked questions that challenged church belief. As we prepared to begin the regression, Jerry lay on her side in a curled up position rather than supine as is more usual. She explained then that this is her habit at night as well—"I'm thinking I can shrink up and cover up and pretend I'm not here."

To her surprise, Jerry's attention lighted not on the abductions we had talked about, but on an episode in September 1992 in which a golden light, so bright it hurt her eyes, had filled her room. The alien beings seemed to float down and through her screen door and into the room. "They're just really odd-looking. Their eyes. I just hate 'em. I hate 'em," she said. "It's like they just look right through you . . . They go inside you," which gave her "a really weird unnerving feeling." She avoided looking at them, because "it's hard to put in words. It's as if I'd lose my self, and don't feel like I have any control." Once again the beings reassured her telepathically. "I don't think I'll ever get used to their way of doing things," Jerry said. "I don't ever get to where I feel comfortable going through the window." She does not like the sensation, and is curious "how they can manipulate matter, solid matter."

Jerry was frightened again as she was taken through the window into a familiar enclosure. "I know this room," she said. She had "mixed feelings" toward the leader, whom she knows. "He talks to me. The other ones don't," but she had foreboding feelings about what would happen when he was there. As her fear mounted I suggested a device for reducing her anxiety. She would split her consciousness so that Jerry One, allied with me, would observe Jerry Two in the room. Using this approach Jerry One "observed" that Jerry Two was naked on a table, unable to move her arms and legs, in a room lined with "lots and lots and lots and lots" of rectangular-shaped containers, "like drawers, in a cabinet," with "hardly any space in between." Inside of these drawers, or "incubators" as she called them later, were hundreds of "I don't know if you can call it babies or not, but little just I guess fetuses."

"To the far right" and "towards the bottom" was a little fetus or baby Jerry believed was "mine." Our device was becoming more difficult now, for Jerry found "I can't be emotionally detached . . . This has been going on since I was thirteen," she said, and estimated that perhaps fifty "procedures" involving implanting or removing something vaginally had occurred over the years. "It goes in waves," she said. "I go for a while" and "nothing happens," and then "they come and it seems like it's all the time." Quite a few times—she does not know how many—she recalled being taken to see what appeared to be hybrid beings. "That's the part I hate the worst," she said. "I think of them, the little little ones, as horses."

"Do they look like horses?" I asked.

"Just their eyes do," and "skinny, you know, long limbed," like colts, she said. "That's how I think about them." She recalled particularly the twin girls she felt were her offspring, but does not remember being brought to see them until they were as tall as Colin, who was now three.

Jerry believed that during the particular episode we were exploring, an embryo was placed inside her body. She thought this because the episode was relatively brief—"those are the quick ones." She has been given information from the aliens that they take DNA from a human male—"the sperm could be from my husband" or someone else—and combine it with an egg. After combining the male and female germ substance, the aliens alter the embryo in some way, perhaps adding a genetic principle of their own. This altered embryo is then reinserted into the female body, as in Jerry on this occasion, for "gestation."

Returning to her memories of this abduction Jerry described how the beings separated her legs "like in a regular gynecologist's office" but because she was paralyzed no stirrups were needed. Then a long tube was inserted in her vagina and she felt "a pinch." She knew this was one of the times that an embryo was inserted into her "because I've been through this before and I recognize the routine." The leader had taken an embryo out of one of the drawers and brought it over to her. "The other way" (when they remove a fetus from her body) is "worse than putting it in," for then she feels painful cramping.

Jerry related the violation she experienced during the abduction to not wanting her husband to touch her. She was beginning, she said, to be aware of "how I came to associate both of those, you know . . ." of how the rapelike abduction experiences had interfered with her intimate life. "They've ruined that," she said. "I don't know what making love is because I'm always still too wound up in being tense and fearful of pain, and I associate sex with pain." Then she added, "I don't think that that's their intention to ruin my sex life." Before concluding our exploration of this episode I asked Jerry if there was anything else she recalled. "I wish you wouldn't ask that," she said and added, "I don't like it when they touch me. They touch me all over . . . Sometimes they do their little feely touchies, and I just like block that out too . . . I'm just connecting from when my husband touches me, just anywhere, I just push him away . . . It's not because I don't love him. I love him, and I just never knew what it was. I never understood it . . . I just feel so bad," she said, "He's a very, very loving person, and he loves to be held and touched and I can't do that. I'm afraid to be touched. I just want to know what it feels like to be comfortable with that."

We worked further to distinguish the seemingly coldly analytic procedures of the alien beings from her husband's loving, caring attention. "They're just doing what they have to do with no respect for my feelings, and he has total respect for my feelings," she said, but "I kinda just react the same way." Until now it has sometimes felt "wonderful" to say no to her husband because it felt as if she was somehow stopping the aliens through him. "I'd probably cry for days and days and days," Jerry said, if she could enjoy being held lovingly by her husband. We talked further of her loneliness and her hunger to be held and touched, "safe, warm, kinda like when my mom used to hold me. Boy, did I used to love that. I haven't let that happen in a long time . . . They're coming between us," she said.

An "alienation of affection," I observed, which delighted her.

I asked Jerry if she recalled anything further from this episode. She remembered being taken down a dark corridor to another tiny room with a table she was pulled onto by the alien hands despite her resistance. After several beings stared at her "midsection" she became paralyzed again and they did some painful poking of her feet, right arm, and right hand. The "tall one" came over to her and she felt gratitude toward him, for he helps her by reassuring and touching her, "sometimes on the shoulder . . . I don't mind him touching me," Jerry said, but "I don't like it when he looks in my eyes because it goes like inside me. It's just too . . ."

"Too what?"

"I don't know. It's like someone just crawled right inside you and knew everything about you . . . I just kinda lose myself, and he kinda just gets in there and I just don't like it."

"Is there a way that you do like it?" I asked.

"Yeah, sometimes I guess I do," Jerry replied. "It's just kind of like I get ashamed about it because it's a sexual thing . . . It's not me. It's him," she added, "and there's no controlling it."

As Jerry's feet were being poked with needles, she was asked to look at a screen close up to her face that seemed like a TV monitor. She was angered to see that on the screen were home movies showing her dancing with Colin. One of the beings was staring at her, observing her reactions to witnessing this intimate family scene. She became angry over this blatant invasion of privacy. At this point Jerry noted that there was some sort of machine on one of her toes, making it numb. This was the one that had a mild nonalignment like one of Colin's toes, "squished up" and folded under. "I passed on" this "defect from my toes to him," she noted, and the beings seemed "curious or studying" that.

Next she was shown a picture or painting of Jesus in a white robe. Again the beings wanted to study her reaction to seeing this image, but then Jerry became sleepy and does not remember anything else. She returned to the room where her clothes were and she slipped them on with the beings' help because she was so sleepy. She then had a "picture" of "floating through the big tree in the backyard and right through the window into bed," still feeling pain in her hand. Bob was asleep, and evidently had been through the entire experience.

Before concluding the session, we reviewed once more the ways that Jerry had condensed her abduction experiences with human intimacy. Her ex-husband had insisted that she must have been sexually abused, and "we went through marriage counselors" trying unsuccessfully to discover a human perpetrator. Jerry believed that if they had been able to discover "someone else like my father or my stepfather" she would have had far less difficulty dealing with her sexual conflicts. Jerry's conviction about the reality of these experiences was increased by the fact that her younger brother has the same problem as she does with physical intimacy. "His poor wife" cries and cries, Jerry said, "'He won't touch me. He won't let me touch him' . . . Their little girl had something happen, I think." Her older brother's daughter told a story of "a light in the window and in the living room is where the monsters come in."

At the end of the session Jerry said, "I really do think that they do exist. They are real, and they are interacting with us, obviously not in any form that we're used to . . . There's a reason they're doing this," she added. She feels that they are "making—whatever you want to label it—another whole civilization." She does not know "whether they're going to take it and place it somewhere else, or it's going to be introduced here." Jerry, like many abductees, has dreams of the world as we know it coming to an end and relates her breeding role to this eventuality.

We scheduled a fourth hypnosis session for five weeks later, July 1, to continue Jerry's effort to "separate" affection and sexual intimacy with Bob from the memories of her abduction experiences. Her awareness of the source of the problem had helped "a little," but she still had to "keep telling myself over and over again" that "he's not them." Before beginning the regression Jerry and I worked out a strategy whereby we agreed that during the session I would specifically reinforce the distinction between traumata we might uncover that had been perpetrated by the aliens from the memories of her intimate contact with Bob. This therapeutic strategy appealed to Jerry, especially as family members were telling her that she should just "accept that's the way you are" and "Bob had started to be supportive," accepting that "maybe I'm just going to have to not want sex." We began the hypnosis session by focusing on the sexual encounter that Bob and Jerry had had on Saturday afternoon ("I seem to be more comfortable during the day," she explained), five days before. Colin had been asleep, and the older children were told not to disturb them. I asked Jerry to remember that afternoon and to report any intrusive thoughts that came up at the time or might occur now.

"Flashes of memory" came to her from an incident at age eight when she and several family members were returning during the night from visiting her aunt. Jerry had fallen asleep and awoke to discover that the car was stopped in the road. She became frightened as she saw "a face in the window, it's right there, right close," and a grayish, metallic-looking craft hovering nearby just above the ground with "lights coming out of the bottom of it." Her mother, who had been driving the car, one of her brothers, and her sister seemed to be asleep.

The being's face had "like devil eyes or something." Soon Jerry found that her feet and legs felt "like when they fall asleep" and she could not move them while she kept hearing in her mind "It's all right." Next she felt "like a bee bite," which caused "a funny feeling" to run down her left shoulder and upper arm and caused her to go to sleep.

When she woke up Jerry found herself lying on her back, apparently alone, in a dark place she did not recognize. She was so afraid that her teeth were chattering. The being she had seen outside the car was "in my face again, just looking at me . . . He must be the devil," Jerry said, "because I just think he's so ugly." The being told her that "he's just going to do a few things and then I can go home." Then she felt a squeezing sensation at her throat, as if from the alien's hands, and she was afraid that "he might kill me." After this another being pushed her over onto her side and seemed to be staring at her back. Although her fear of what the beings might do was intense, Jerry felt "like I know them . . . I don't trust them," she complained, because "you just never know what they're going to do."

"I don't like them touching me," she said as she remembered being touched repeatedly all over the back. It felt like many "little needles" and "a little pinch." Her terror derived from the fact that unlike "going to the doctor's office when they tell you and you know what's wrong and your mom's with you," in this situation "I don't know what's going on, and I feel that any moment they're going to hurt me." Then Jerry recalled that "they roll me back over" onto her back and did a lot more "just looking." She felt temporarily "better 'cause he's not doing anything," but "I kinda got a feeling there's something going on behind me."

Jerry felt intense embarrassment as she told me that the leader pushed her legs into a pulled up and apart position to "check me, and he brings a light, this really bright light, and I'm thinking what are they going to do. I don't like this. It's just, really, it's okay to, you know, poke at me and everything, but I don't like that. That's my private area, and I don't think they should be doing anything there, and I don't think my mom would like that." At first they were "just looking" inside her vagina, but then "they put something in there" that was "kinda like when I get older, the gynecologist, you know, kinda like that." This procedure was painful and Jerry screamed in terror for her mother. But "that doesn't stop 'em." It was "over real quick," and after all this looking and "checking," the being who performed it looked up to the leader and said "no" or gave some sort of negative response to a question of his. Jerry interpreted this to mean she was not yet ready for their reproductive procedures. This vaginal examination seemed to Jerry to be part of an overall "checking" the beings did.

This experience was mortifying for Jerry. She felt "like a rag doll" afterwards, "like I can't, I don't have any control over my body." In this instance, Jerry could describe the strange, hooked light that was put inside of her, but she was not able to recall the actual sensations that she felt at the time. "I wouldn't know the words like 'rape' [at age eight]," she said, "or something, but it's like that."

I asked Jerry if she had ever known whether she had a hymen, but she said that as a child and adolescent she did not touch or examine her genital area, perhaps, in part, because of the trauma of this experience. Jerry was "always very modest" about her body, and her mother would sometimes comment on this. Then I asked her explicitly whether there had been an anal examination. She had not intended to tell me, because "that's worse than the other" and occurred before the vaginal exam. "I just skipped past it. I just skipped past it," she said.

I asked what made it "worse."

"Just doing that," she said, was "totally completely disgusting" and more uncomfortable.

At this point in the session Jerry was struck powerfully by the ways in which her reactions to Bob's advances were patterned by the "scenarios" of her abduction experiences. She made a circular motion with her hand to describe the way in which her abduction memories were triggered by being touched by him and how intensely the alien and human experiences had become interwoven. For example, when Bob touches her on the back it brings back the touching/probing of the aliens. "When he starts with the touchy-feely stuff the tape starts running in my head," she said. When he spreads her legs prior to intercourse it brings back the memories of the aliens pushing her legs apart on the table. The act of intercourse is equivalent in her mind to the assaultive genital probing on the ships that she cannot stop, and when she says no and stops Bob, it is as if she is stopping the aliens, "even though I know I can never really stop them."

We then derived a strategy for their intimate interaction that would maximize the distinctions between Jerry's abduction experiences and her relationship with Bob. First, they would discuss and agree in advance that she would initiate the forms of foreplay and intercourse after a good deal of affectionate talk and she would have the option of stopping their encounter at any point without guilt. She would direct Bob's touch, which would be slow and smooth, focusing on her breasts (which the aliens do not touch), and would contrast the rapid, needlelike touching of the beings. She would initiate genital touching, and when Bob's penis was erect she would mount him, take him into herself, and be the more actively moving partner, which she assured me he would like. At every step she would be in charge and in control.

At the end of the session I summarized for Jerry the two parts of the strategy we had discussed—a psychological emphasis on distinguishing in her mind between the alien and human experiences and an action strategy that would reinforce this distinction. She looked forward to initiating the plan. I also stressed how especially traumatic the anal and genital penetration of her body at age eight must have been. For at this age, even more than at thirteen, a child has no way to understand or consciously record such experiences, for the psyche is simply too immature. Thus the memory becomes deeply buried in the unconscious, affecting later feelings and behavior in ways that the person has no way to comprehend.

Five days after her last regression, Jerry stopped by my office after another appointment she had in the hospital. She looked well and happy and said the strategy was working. Two days later Pam called her and she said that "whatever we did was totally successful." She tried what we had suggested with Bob, expecting to be anxious. "But it didn't show up at all." He did everything she asked of him, relaxed, and enjoyed her initiatives. She is certain now that what has been upsetting her has nothing to do with him, for otherwise their sexual activity would have bothered her. She was elated that none of her old, disturbing feelings returned. Bob is very happy about the change. "I can't believe it," he said. Several months later, the change was still sustained.

Discussion

Jerry's case demonstrates a broad range of abduction phenomena. She has experienced complex, intrusive reproductive procedures on the alien craft, including insertion and removal of what appear to be fetuses of some sort and has had encounters with hybrid entities. At the same time she has undergone the intense personal growth and philosophical and spiritual opening that often seem to accompany abduction experiences. Interestingly, her philosophical and poetic writings antedated her work with me and cannot, therefore, be attributed to our relationship or interaction. All three of her children, the products of two marriages, seem to be involved in the abduction phenomena.

The representation in her psyche of the hybrid babies as horses or colts, unconscious at first, reminds us of the variety of animal forms, including deer and different birds, in which the alien beings may appear to abduction experiencers. For Colin, the aliens were owls from the sky. This complex symbolization may be a product of the psyche's unconscious power of disguising threatening elements or may be induced by the mind-altering powers of the aliens themselves. Another possibility would be that some sort of deeper connection of the alien beings with the animal spirits themselves can occur, similar to the human-animal connections that are familiar in shamanic practices.

Jerry's abduction encounters, beginning in childhood, have been deeply traumatic for her, and their intrusive, rapelike, reproductive and sexual elements were buried deeply in her psyche. Because the memories of these experiences remained unconscious, she was unable to distinguish the physical aspects of human intimacy and sexuality from her alien traumata. As a result, Jerry was unable to enjoy or even tolerate physical contact with her husband, with whom she had a mutually loving relationship. The uncovering of core memories of her abduction-related traumatic experiences, through four long hypnosis sessions, allowed Jerry to psychologically separate human from alien reproductive and sexual activity and permitted us to devise strategies for reinforcing the distinction between them. Jerry and Bob were then able to enjoy a satisfying sexual relationship.

It is difficult for someone who has not been present during these hypnotic regressions to appreciate the emotional intensity of the traumatic experiences an abductee like Jerry has undergone. Her verbal expressions of rage and outrage and bodily writhings are something to behold. But over and above these abreactive expressions, which permit the integration of the traumatic memories, Jerry's case also illustrates well the other dimensions of abduction trauma—the lifelong personal isolation, the philosophical disbelief, and the fact that new episodes may befall the experiencer and his or her children at any time. Jerry is particularly eloquent in speaking of this last element. She concluded a poem titled "Regression," written during the winter of 1992–93, with the lines, "This wonderful technique gives relief to traumas past,/with one catch although, will it end with the last?/Because, unlike other victims of rape, incest, even war trauma,/we're not relieved of our continuing unrelenting other world melodrama."

Despite the great suffering Jerry's abduction experiences caused her, she, like many other abductees, holds to the view that there was something of great value, a creative dimension, "a definite reason," for the abduction process—perhaps the creation of a new race of beings in which she was participating. It is difficult to know whether such ideas represent an authentic realization by the abductee herself, are implanted by the alien beings, or result from a kind of identification with the aggressor.

Jerry, like many abductees, has opened her mind and heart to important philosophical and spiritual concerns. These are expressed most fully in her writings. In a November 1992 journal entry she told of a dream of the future destruction of the earth through nuclear war. That such concerns for the earth and its fate predated our work is made clear by entries from December 1991, six months before she called me. Writing as if she were receiving information from another troubled and perplexed source or voice that was speaking to her, Jerry described the beauty of the Brazilian rain forest, but followed this with the concern that "it was dying a very slow death ... Why was this most beautiful place being destroyed?" she continued. "You started to investigate further and you discovered that it was not the place that was dying. It was the inhabitants of the place that were killing it. You then were very concerned about this and continued exploring the rest of the earth and its inhabitants. The very same thing happened everywhere you went. You then decided something must be done. But what? How? . . . Does humankind have that much hatred for his future that he would destroy it?" she wondered.

Jerry's writings include consideration of a vast range of existential matters, including the nature of time, space, and the universe itself; the great cycles of birth, death, and creation; the mysteries of truth, spirit, and soul; and the limitations of material science. She believed that she was to write a book about Universe, Soul, God, and Eternity based on the ideas that were coming to her and organized her communications into chapters. Sometimes she seemed to pause in awe before the power of the information she was receiving and the implied responsibility that accompanied it. In a December 1991 entry she wrote, "But why would you pick an ordinary housewife to do such important work as this? And who is going to understand and even buy the book?" The answer she received was that she had chosen this role herself.

In an entry dated November 22, 1991, Jerry wrote from the standpoint of the archetypal creative force of the universe. "Imagine that your essence, your soul, was part of a whole, and as part of a whole you decided to give birth, to create. You then gave birth to your thought to create and made your thought into matter. As this birth came to be solid, you then decided you would continue to create, and after some time you decided you would like to be whole again. But in order to be whole again you had to gather up all of the fragments or pieces of your whole being. In order to become whole again you must be able to then understand that you have to then create and give birth to that thought. And in order to go back to your original form you must again reverse the process." She then likened this process to "the union of a man and a woman. The two give thought to create a baby. Their thought then becomes matter in the form of a baby."

A number of Jerry's writings are concerned with the relationship of the material world to the spirit world and the limitations of a purely technical or technological way of knowing. For example, in a November 1991 entry she wrote, "Technical data does not lead to the discovery of other beings. Spiritual data does." A month later she wrote:

> Science: manifested travel into space and time
> Spirituality: unmanifested travel into space and time
> Science: limited travel
> Spirituality: unlimited travel
> Both valid
> Which ticket will you buy?

Jerry has shown a great deal of courage and determination in confronting the disturbing power of her abduction experiences in the face of the community of disbelief that has surrounded her. She has also developed a marked increase in her ability to know her own mind, and think for herself despite or because of her isolation. These qualities were captured in a poem she called "Decision," written in the winter of 1992–93. She wrote of her battle to overcome her fears and the secrecy and silence that had always oppressed her. She had chosen, she said, to "no longer" let her abduction experiences "take all of me . . . At least I will have the dignity," she concluded, "of knowing and owning my own memory."

“All beings in the tank were identical. The frames of the tank were flush with the wall,” according to Catherine, who in this drawing depicts herself accompanied by two beings.

CHAPTER SEVEN

If They Would Ever Ask Me

Catherine was a twenty-two-year-old music student and nightclub receptionist when she called me for help in March 1991 after an episode a few weeks earlier that puzzled her. One night in late February she had completed work at about midnight and started for home. But oddly, instead of stopping at her home in Somerville, near Boston, she kept driving in a northerly direction, explaining to herself, "I guess I'll go for a drive" and that she would "put some highway miles" on her new car. When she returned home there was a forty-five-minute period for which she could not account.

The next day Catherine awoke at about noon, "flipped on the news," and saw "something about the UFO seen last night." Some of the news commentators tried to explain the object seen over Boston as a comet or a meteor, but the object traveled horizontally in relation to the treetops and Catherine said to herself, "That's an unusual meteor kind of thing," and "a comet comes down out of the sky and smashes and that's it." Also, a policeman and his wife reported that the object had stopped overhead and shone a light down on them. One of the TV channels showed a chart of the object's path from southern Massachusetts to the northeastern part of the state (Barron 1991, Chandler 1991). Catherine then realized with shock that although she did not recall seeing the UFO, "I was traveling the same direction as it was." Ironically she had recently been reading about UFOs and "halfway hoping to see one and halfway hoping I don't." Also troubling to Catherine and contributing to her contacting me was an unexplained nosebleed—the first she recalled in her life—that occurred shortly after the above episode, and the fact that she had found herself answering positively to most of the questions indicative of possible UFO encounters in a book about abductions.

In our first session Catherine was apologetic, fearing that she was wasting my time. She recalled a dream from age nine in which she was paralyzed and terrified as "some kind of creature" with long fingers, larger at the ends, came up behind her and grabbed her. The creature's hand felt cool. In her terror Catherine wanted to scream and "call for my mother, but I can't. I can't say anything." She also remembered another dream from the previous Christmas, 1990, when she was home visiting her mother in Alaska, in which she was in a spaceship with curved walls and there was something in the room she was in that was "like a big fish tank." She was not certain that this really was a dream. I did a light relaxation exercise with her to help her recall details of the nighttime drive. Catherine remembered the roads she had traveled and felt fear as she recalled driving twice through a wooded area in Saugus, about ten miles north of Boston. She also described a considerable fear of needles. Finally, she noted that she was in somewhat of a career crisis, feeling that "I'm not using all of the skills that I have."

Catherine and I both felt that this first meeting—which in retrospect was highly suggestive of UFO abduction experiences—was equivocal, and I suggested that she see what other memories would surface in the days to follow and asked that she call me in about a week. When she did not call, I called her and she said that she felt foolish calling me back, that nothing more had surfaced, and that she was busy filling out résumés to move on in her career. I did not hear from Catherine for nine months after which she wrote me a letter saying that she now had "impressions (memories is too strong a word)" from Christmas 1990 "of a ship in the field" behind her mother's house in Alaska. Also, she had become panicky watching the movie *Communion*, based on Whitley Strieber's book; had seen an odd light in a cloud moving across the horizon six months before; and discovered a small straight scar under her chin for which she could provide no explanation. All in all she was "wondering about it too much to let it all go" and had decided that "I would like to see if anything can be brought to light simply for my own peace of mind."

Over the next eight months I did five hypnosis sessions with Catherine and talked with her frequently. In our sessions several of her abduction experiences were explored in detail and extremely powerful emotions emerged. Catherine has attended our monthly support group regularly and has become an important support person for other experiencers. She has, indeed, changed careers and is currently attending graduate school in psychology. Catherine's case is significant for the clarity of observation that she brings to many of the UFO abduction phenomena. But in addition she demonstrates the ways in which personal growth and transformation of the phenomenon itself may occur as a result of a shift in the attitude and approach of the experiencer to the encounters, especially in relation to the terror associated with them.

Catherine grew up in Oregon in towns in and around Portland, moving frequently because of her father's work as a surveyor. He became disabled with back problems when she was a child, and was reduced to doing repair work and carpentry around the house and odd jobs for other people. He also had a drinking problem; would disappear frequently when drunk; and was given to impulsive, angry outbursts. Once when Catherine refused to clean her room he put all of her belongings in a dumpster and burned them. Her parents were divorced when Catherine was in college. She has virtually no contact with her father now.

Catherine's mother, Susan, is a teacher and works with handicapped children. When Susan was in college she saw a UFO ("lights in the sky that do things that planes don't do") that was witnessed by about three dozen other people. Susan, concerned about her daughter and curious about my work with her, called me from her home in rural Alaska, where the family had moved during Catherine's teenage years. I was impressed with her sensitivity and openness to Catherine's experiences. She expressed a belief in the possibility of life beyond this planet that might take unexpected forms. Catherine's only sibling, her brother Alex, is eight years younger than she. Catherine believes he may have had abduction experiences but doesn't know it. He had an unexplained mark on the side of his left hand that has the same horseshoe shape as two scars Catherine has on her left hand that she believes are abduction related. This mark has since disappeared.

Susan describes Catherine as having been "a free spirit, a little different" growing up. Searching for other sources of trauma, I asked Catherine about childhood sexual abuse, rape, and other possible violations. She told me that when she was about four a childhood friend of the family put his hand between her legs and touched her genitals. This was a disturbing experience—"here's this old guy that I thought was absolutely wonderful and I trusted so much, that my parents loved too, and it was like—I was shattered." Neither Catherine nor her mother believes that she was sexually or physically abused by her father or other family members.

The first abduction experience that Catherine recalls occurred when she was three years old. The memories were consciously triggered—i.e., remembered without hypnosis—by a disturbing nightmare scene in the first episode of the CBS miniseries *Intruders* in which one of the women abductees sees a barking dog at her bedroom window that "turns into" or masks the memory of an alien being. Catherine remembered waking up in the middle of the night and seeing a being, at her bedroom window with blue light coming into the room from behind it. The family's home was a one-story mobile home, and Catherine calculated that "this funny-looking guy outside the window" would have to be quite tall or floating because the window began several feet above the ground and the entity's thin torso could be seen in the window.

She described the being as having "huge black eyes, a pointed chin—his entire head is like a teardrop inverted. He's got a line for a mouth, nose I can't see totally well from where I am, but it's not like a human nose. It's just a bump. I can see nostrils, but not as large as ours are. He doesn't seem to be wearing any clothes. He doesn't really seem to have any color to him. He's got a bluish cast to him caused by the light coming from behind him. It's like he's backlit somehow."

The being in the window appeared to come through it and "materializes at the end of the [blue light] shaft." When "the beam came in and hit the floor," Catherine says, "I've got this impression of floating above my bed, like kind of being levitated out the doorway to the hall." Catherine experienced terror at that time and in our interview. "It's like monsters are coming to get me. But they're real. There's nothing I can do about it . . . I was crying out for my mom right then, and I was trying to scream for her to come but I couldn't move. The words wouldn't come."

Catherine sensed that when she tried to scream for her mother the beings did something to reduce her terror, which seemed to diminish after she started floating. After this Catherine saw five or six beings similar to the one in the window "in the living room moving around really fast, and I'm not sure what they're doing . . . They're like picking things up and looking at them, and putting them down again." After the busy rushing around "it seems suddenly [the beings] all got organized again, and they're all in a line. All the activity has stopped. The living room was very light [". . . it's definitely got a bluish tint to it . . ."], but it was in the middle of the night and there wouldn't have been any lights on." Catherine recalled that she was floated "literally through" the front door, face first, and saw that "outside it's light too. It's the middle of the night, but it's *light*." In the field outside her home "there's like a ship," and "it seems like a lot of light is coming from it, but there's more light than it would put out." Catherine's mother later told her that the trailer park had large, blue flood lights. The ship has a "disc shape. It seems like it's got a lot of lights all over it, but I'm not sure if that's right."

Our discussion of this episode ended at that point, but, as is characteristic of abduction memories, Catherine recalled many more details in the weeks that followed. Three and half months after the above interview she told me more about this episode in a letter. She had a clear image of one of the beings in the living room "picking up a teacup, holding it a few inches from his face, peering at it intently, and putting it back again." She remembered being floated "in front of him [the being that had been in the window] and out the door of my room to the hall, to the living room," and then "out the front door."

Catherine then described what had occurred on the ship, which we had not explored in our previous discussion:

> They take me into a round room in the ship with a long bench around the perimeter, except for the doorway. The bench had a red cushion all along it. There are other kids in there, maybe five or six, all under ten years of age. A taller, female being enters and says to me, "Do you want to play?" I get the feeling from her of a nursery school teacher, or a day care center leader. I'm sleepy and confused, but I say, "Okay." She seems pleased with this answer. I look at the other kids—they're older and taller. The room seems very bright. She goes to the other side of the room, where she came in, and brings back something. I think it's a metallic ball, and it floats. She flies it around the room doing loop-the-loops, etc., and some of the other kids are trying to fly it, but not as gracefully. They hit the walls, and it makes a metallic "clang" when it hits. There's a feeling of amusement from her when this happens. It gets to be my turn, and she says, "Would you like to try it?" and I say "Yeah!" because I want to show up all the bigger kids. She gives me a metallic rod, about a foot long, or maybe a little bit longer. It's about an inch in diameter, and there's a thick, short antenna coming out of the top. It's silver/gray, and smooth. The antenna is about four inches long, with a small ball at the end. The rod is like a remote control, and you point it at the ball to guide it, but you have to concentrate at the same time to control it. So I make it stop, hover, then go and come to a dead stop after moving very fast, and am doing it much better than the older ones were. The older kids who did badly are giving me glaring looks, and I get a sense of frustration from them. The female being comes to take the rod away after a minute, since my turn is over, and when she does, she tells me that I did very well, but I have to stop because I'm making the other kids feel bad because I'm younger and they didn't do as well. There's a feeling of specialness—of my being special to her because I did so well, and am so little, and she's proud of me because I did better than I was supposed to. There are a couple more kids who get their turn, and then she takes the ball and rod back to where she got

them. She comes back and says to us all, "You've done very well. We're very pleased with all your individual progress." I feel proud of myself.

Yes, there is more to this story, but this is all I have, so far.

The next encounter Catherine has recalled occurred at age seven and was recovered unexpectedly in our third hypnosis session in which the regression was open ended—i.e., we were not seeking the memories of a specific episode. The session began with her experiencing herself walking with two friends to their house carrying a big box of candy. She was wearing a Camp Fire girls uniform. A finger on her left hand hurt and had a funny-shaped blister on it, which made carrying the box more painful. As we talked about how the three girls and the friend's mother went from house to house selling the candy, an image of a woman who kept peacocks, and of the alley where they were kept, came into Catherine's mind.

She then recalled an event a week earlier when she was at one of the friend's houses and felt strangely drawn to leave and go down this alley to see the peacocks. It was rainy and the alley was muddy. Catherine was afraid the lady would come out of her house and yell at her, "because I know I'm not supposed to be there." She was throwing rocks at them to make the peacocks' beautiful feathers come up when she saw "a little white thing." This turned out to be "a little man standing there. He looks startled. He has a big head, big eyes, and he doesn't have any hair." He told her that he wanted to take her somewhere, but she felt she should not go because her mother had told her "I'm not supposed to go with people I don't know and I don't know him." The figure told her it was okay, but she felt afraid and angry, "because I told him I don't want to go and he's going to take me anyway." She tried to run away, but the being had "his hand on my arm" and she could not get away. Catherine began to cry in the session like a helpless child, and repeated plaintively, "I don't know him and he's going to take me anyway!!"

Still crying Catherine said, "He's taking me up. We're flying up . . . I can see everything down. It's scary. It's not supposed to happen. He's still got his hand on me. I can see everything below me and it's not supposed to be like that." After this she passed through a "hole" into "the middle of this room." Catherine thought of hitting the little figure, which was "as tall as I am," but she "couldn't move at all." He seemed to laugh. "He thought it was funny that I wanted to hit him," which she could tell "because I kind of heard him in my head." Inside the room the "little man" went to another room to get something and bring it back. "I said, 'What are you going to do with that?' And he tells me, 'I'm just going to make a little cut.' I say, 'Why?!!' And he said, 'Because we need a sample.' I said, 'NO!! NO!! You can't cut me!!' and he said, 'We have to.' I said, 'No, you don't have to!! That's mean!! You don't have to do it to me!!' He said, 'It's for scientific research.' I said, 'Well, why can't you cut something else?' He said, 'Because we need blood.'" He made a little cut on the fourth finger of her left hand, which hurt less than Catherine expected. With an instrument like an "eyedropper kind of thing" made entirely of metal, he drew in a small amount of blood.

Insisting that "we had to get the sample," the being said he would take her back. "But you didn't tell me why," Catherine insisted. "He said, 'I'm researching your planet.' I said, 'What's wrong with my planet?' He says, 'We're trying to stop the damage.' I said, 'What damage?' He said, 'The damage from pollution.' I said, 'I don't know about that.' He said, 'You'll learn.' And then we're going down again. I'm getting closer to the ground, getting closer, getting closer, and I'm on the ground, and I want to run away but I can't move. He says, 'We'll be back for you.'"

Once again Catherine found herself in the alley where the peacocks were. "I'm running and running, past the peacocks, running up the street," until she ended up "where I'm supposed to be now." Catherine calculated that perhaps fifteen minutes had passed, and no one seemed to have missed her when she rejoined a group of children watching cartoons on television at another friend's house. The memory of what had just happened to her seemed to fade quickly. Perhaps "I blocked it out," Catherine suggested. By the time she got to her friend's house "it was already fading." When she went back in the house "I thought I was just outside." It appears that the pain to her finger caused by carrying the large candy boxes was the entry point to the memory of the abduction episode which had occurred a week earlier. A small horseshoe-shaped scar remains on her ring finger to this day for which Catherine has no other explanation than the incident above.

The next episode that Catherine relates to the UFO phenomenon occurred when she was fifteen or sixteen. It involves unexplained lights on the hillside behind the mobile home where she was living with her mother, father, and brother. No abduction seems to have occurred. When Catherine and Susan were driving back to their home they saw "little lights" moving parallel to each other, close to the ground. Her mother pulled the car over and they watched for a few minutes as the lights, according to Catherine, did "strange things" that a plane would not do. Although once in the house Susan seemed to lose interest, she did suggest, "Maybe it was a UFO," and recalled her own sighting from her college days. Still fascinated, Catherine continued to watch through the window a complex movement of three or four lights on the hillside "all dancing around up there." At one point they all swooped down at once, but Catherine could not get her mother to take an interest in the phenomenon. "I said, 'Mom!! What's coming down off the hill?!'"

Although inconclusive, this experience seemed to epitomize Catherine's feeling of aloneness and isolation in relation to the UFO/abduction phenomenon. This feeling was reinforced for her by her sense that her mother "thinks it's my imagination." She longed to leave her "total hick town in the middle of nowhere," hoping that "maybe stuff like this won't happen if I'm in a city." Without the support of anyone else, Catherine wondered, "These lights. It could have been . . . who knows! It could be my imagination."

The Christmas "dream" of 1990 turned out to be the first adult abduction experience that Catherine was able to recall. The story unfolded in our first two hypnotic regressions. Her mother's mobile home is in a deserted area six to eight miles from a small town in south central Alaska. There are large fields behind the house. Christmas day fell on Tuesday and Catherine remembers that the "dream" occurred a day or two later. Before the hypnosis she recalled awaking the next morning with an "image in my head of being in a room in a ship . . . I spent about ten minutes just lying in bed trying to remember everything about it that I could and burn it into my memory as much as possible. I know it was very important to remember it. I didn't know why. Part of me was saying it was just a dream; it's no big deal. But part of me was saying, no, this is very important. You need to remember as much as you can about this." Catherine thought about the dream the entire day to see if she could remember any more about it but could not at that time. Later she recalled the details mentioned at the beginning of this chapter.

Under hypnosis we reviewed in great detail the layout of the home, Catherine's arrival before Christmas, her father's visit on Christmas Eve, and the quite uneventful activities of Christmas Day and the day following. Now as she recalled awaking with the impression of having seen "a ship" she felt that "it didn't really seem like a dream." She also felt "a little bit nervous" in the session. I encouraged her to "stay with the nervousness" and reassured her of her present safety. Then she said, "I can remember walking down the hall in the middle of the night and looking out the window in the living room and seeing a big ship out there, back in the field." Catherine believes she was in a "more than half asleep" state at this point. The field was like a frozen swamp in the winter, and the craft was "sitting on the ground" between the trailer and several large trees. "It's like a discus, but wider in the middle than a discus is and silvery metal. It's bigger than the trailer." She also noticed individual lights all around the rim of the craft ("white lights, kind of in a groove," she observed after the session). Her first reaction was, "It shouldn't be there."

Catherine's anxiety in the session began to mount. Wearing only "a big stretched out T-shirt" and with bare feet, "I can see myself going through the living room to the front door and putting on my mom's big heavy boots and one of her big heavy coats and going outside." Gasping now Catherine said, "I feel like I know I open the door and go outside. I don't want to open the door." I acknowledged her anxiety, offered support, and allowed her to choose whether to continue to report her recollections. Bravely she decided, "I'll do it. I see snow and it's dark . . . I'm just standing there with the door open, looking out at the snow. It's dark . . . That's what I see. My mom's car is out there. It's to the left." In the session Catherine began to recall (actually relived) feeling numb, with pressure in the chest. Beginning to sob and pant Catherine said, "I'm starting to feel numbness in my face now. My arms are starting to feel really heavy. Numbness is moving down to my hands. I'm feeling a very heavy weight on my chest and stomach. My knees are starting to feel numb now too . . . like Novocain numbness," she said later.

After standing for a short while in the doorway Catherine said that she started to go out of the house "to the ship." But she had difficulty doing so as "my entire body feels totally numb." She noticed that "there are creatures out there" by the ship. I asked her to describe them. "There's five of them, and it doesn't look like they have any clothes on. They should have clothes on because it's Alaska. It's the middle of the winter and it's cold." The creatures were "exactly the same size. They're standing in a row . . . They're kind of glowing, golden glow. They light up a little bit the snow around them . . . They have very big heads." She felt "in my head" that the beings were "waiting for me," and despite numbness in her arms and knees, she walked haltingly most of the way to the ship, describing these moments with panic-filled sobbing that required a great deal of comforting from me. As she got closer to the ship the beings "come around me in a semicircle. I'm trying to look at them and I can't. I can't see the faces. Their arms are very long. They don't seem to have any body features like we have. No nipples, no belly-button, nothing." They had no hair or evident teeth and the faces were expressionless.

Then Catherine declared, "I know I went in there but I can't go in there," meaning she could not face in the session what happened next. She described a metal ramp, angled at forty-five degrees, "sectioned to make one big ramp." At this point in the session it was clear that Catherine's terror had mounted to the point that she could not continue her story. I spoke to these feelings, encouraged a breathing and relaxation process, and gave her the choice of going no further that day. She said, "I don't feel that I can. I'm feeling a huge weight on my chest. Everything is going. I'm totally afraid to even think about going inside." After further acknowledgment of her fear, I suggested a trick or game we might play in which she would stand at the base of the ramp and send an imaginary puppet-spy with his eyes closed up the ramp into the ship with instructions to open his eyes upon our command and report back to her what he saw. She agreed, and the "spy" reported "a small, oval entryway and the walls come down, curved on the side—like being on the inside of a big egg. Everything is metallic." The puppet could tell there were other rooms, but could not "see any entrances or anything."

Catherine was then willing to go into the ship "on her own." She thought that she kind of "glided up the ramp." She noted further details of the curved walls and shape of the first room, which she called "just an entry hall." There was light but "not a specific lamp or anything." She saw an oval opening to another room and said, "I'm going to go in that room but I haven't yet," and added, "You can send a spy into that room. I'm not going to go into that room. I'll look at the room but that's it." I agreed to this, but encouraged her to place the spy more under her agency or control than before. She said, "He's invisible and nothing can happen to him and they don't know he's there." He was also "a kid," a boy.

In the room the spy saw, "lots of panels and instruments and scientific things, but they don't look like things we have here. There's kind of like a platform thing in the middle of the room. It's not that huge. It's maybe half the size of your living room downstairs, and you can still see the curve of the outside of the ship like in the other room . . . There's just calm. There's a thing on the ceiling above the people there in the middle of the room. It looks like it's long on hinges like desk lamps like we have here that you clamp on and you can move them around. And there's another being in here. He's waiting, and I think he's like the doctor or the medical examiner kind of guy. And there's all kinds of instruments and buttons and panels everywhere along the walls. There's almost kind of counters along the walls except for the entryway . . . The table in the middle is like solid, not like a table with space underneath it, but attached to the floor and it's like one big solid kind of block."

She seemed to feel calmer now, which appeared to correspond to a shift at the time of the experience. Wearing "just the T-shirt and boots. I don't think I have the coat," Catherine was "floated" into this room and "I know I'm supposed to get on the table." Feeling confused, Catherine noted that this "doesn't now feel that surprising," suggesting that the experience might be familiar to her. "They make me lie down, but I don't want to lie down. I feel something coming . . . The doctor guy comes over and looks at me . . . medical curiosity. I'm a specimen, not like a friend, not like someone you know . . . I think something bad is going to happen."

At this point Catherine wanted to stop the narrative and we agreed to do so. Before ending the regression we talked further of her recollections. She thought she might have seen her mother on the ship in a previous abduction. She described the small, very thin necks of the five beings she saw in addition to the doctor. "You wouldn't think that they could support their big heads. Their bodies are actually kind of frail." They seemed to be wearing no clothes, and the skin was "kind of whitish, like pale . . . When they came in with me they went to different stations in the room, like they had different things to do."

After the session Catherine debated the reality status of the experience. "I thought I was making it all up until I started crying. I still don't remember it like a real memory, as in I remember I went to work yesterday . . . Oh, God," she exclaimed. "The idea scares me! Obviously! . . . Well," she added, "I'm willing to admit maybe it's really not all my imagination. My reactions to other things make more sense if this happened. I'm not, like, a totally irrational person, which is kind of a relief! It doesn't seem like a dream. It seems more real than a dream, but not as real as me talking to you." Then she added, "I can't see why it [i.e., making up such a story] would benefit me emotionally or psychologically. I can't see any reasons why it would," an observation which expresses a central aspect of the debate over abduction experiences. Playing the devil's advocate I suggested that such experiences might make her a more interesting, dramatic, or exciting person. She objected that "if I had had this wonderful experience, *who* am I going to *tell* about this that's not going to look at me like I'm a total kook?" In order to "calm myself down," Catherine noted, "I tell myself, 'it's your imagination.'"

A kind of decisive indication to Catherine that she was dealing with something real was the power and authenticity of her emotions for which we could find no other source. "I don't think I would be sobbing for no apparent reason if there wasn't something there. I'm not given to bursts of tears for no reasons." Also persuasive to her in these initial recollections was the sense that she was compelled against her will to get up in the middle of the night and go out in the Alaskan winter toward the ship. "I did not choose to get up and go do that." The experiences themselves gave her a feeling of being "totally violated. It's what I would imagine a rape victim would feel like." Finally, I noted another quality to the tears, a kind of sadness. She suggested self-pity, but I thought it was something deeper. Affirming then her feeling of what I have come to call ontological shock, Catherine said, "I get the point—of having to realize!! Oh, God."

For support at her next hypnosis session Catherine brought a young woman friend from the nightclub where she had been working. A sticking point in her mind, which has never been resolved, was whether or not she was wearing her contact lenses during the Christmas, 1990, episode. She does not remember putting them in, but, remarkably, could see adequately throughout the experience, whereas without her lenses "everything would basically have been a big blur." Catherine had remained frustrated since the last session about having no definite answer regarding the reality of her experience. Without concrete physical evidence, she, like so many experiencers, felt nonvalidated by science and society and in danger, therefore, of being looked at as "insane." If she had been raped, she remarked, "I could go to the police. There would be evidence there. They could take samples, and people would not look at me like I had lost my mind if I said to them, 'This happened to me.'"

She began the second regression by briefly reviewing the steps leading to being forced to lie on the table. She noted the muted lighting in the room and again felt the loss of any will of her own. The leader or "examiner," though taller than the others, was still not as tall as she is. His skin seemed "very smooth—whitish, gray," and he seemed not to be wearing clothes. "He's looking at me like you look at a frog before you dissect it." Looking around the room Catherine observed that everything is metallic, "like brushed aluminum but darker." Other beings were moving about, seeming to perform various "specific jobs" like "pushing levers and buttons and checking things and getting things ready." Their movements were "very light, kind of like when you think of a cat. It's very graceful and lithe."

Catherine became increasingly distressed, panting and crying, as she described how one of the beings spread her legs apart on the table and the examiner stared at her face and genitals. She noted that she had no clothes on. The examiner "says something to the one on my right, and the one on the right goes off to the right side of the room to get something, and the examiner puts his hand on my leg, on my thigh, and it feels cold—not like human cold hands. I mean, it's even colder. I don't like it, and the other one comes back and he hands the instrument to the examiner." With a lot of support from me Catherine spoke of how she was unable to resist. "He's doing this to me and I can't do anything about it," she sobbed pitifully. "He's got his hand on my left thigh, and he's taking this with his left hand. It kind of looks like a cone, but with something else on top, and he's going to put it in me," she sobbed loudly. Her voice breaking, Catherine continued. "He puts it in. It feels cold. It's even colder than his hand. I can feel something going up inside me farther and farther. It feels like something's going up in the intestines, up so far. That's how far it feels like it's going." Although the instrument seemed to be pushed up "way farther" than the vagina, "It doesn't hurt. It just feels like it shouldn't be there. They didn't even ask me!!"

As the instrument was moved about on her right side in the region of the ovary for what she estimated to be ten or fifteen seconds, Catherine had the sense that "it's taking samples." After the instrument was retracted, the examiner gave it to an "assistant" who "takes it away to where he went to get it before." Although she did not see anything definite, Catherine had the strong impression that "tissue samples" were taken from "the uterus lining," the cervix, and perhaps the fallopian tubes.

After this I asked Catherine if anything else was done to her body on this occasion and took a kind of "inventory." She described a metal instrument, "maybe a foot long," that was inserted perhaps "six inches" into one nostril. Somewhat shocked, I said that would have gone into her brain. "That's what it was supposed to do," she responded. "The examiner came around, and he had this thing in his hand and it had a like a handle on the end. It was this long, flexible thing, and he kind of leaned over my right shoulder and he wasn't looking at me. He was looking at my nostril, and he put it in as far as he could. I didn't like it because I couldn't breathe very well, and then he hit something in the back and he just kind of pushed it, and he pushed it through whatever it was."

Sobbing, whimpering, with her voice cracking again repeatedly, Catherine said, "I could feel something breaking in my head. When he pushed it through, he broke whatever it was and he pushed it all the way through, up even farther." The procedure was uncomfortable, but it did not actually hurt. "I'm wondering what they broke . . . I don't know the anatomy, and he broke something to get it through, to get it into my brain. I don't know what it was. I want to know if it's going to heal." Responding to her worry about having "heard something snap" in her head, I tried to reassure Catherine that I doubted any permanent damage was done to her brain. Later Catherine commented that "I was afraid of bone fragments in my brain." I asked Catherine what she could see after the probe was removed. She said there was a little blood on the instrument and in her nostril, but she could not see that anything else was removed. The examiner gave the instrument to the assistant who took it "to the far side of the room where he came in and he does something with the instrument. I can't see what it is."

It was at this point in response to a question of mine that Catherine observed that she seemed able to see as well as she ordinarily could with her contact lenses but did not think she was wearing them. In this context a memory came to mind of the examiner "looking in my face. He's scrutinizing me . . . It's like they're trying to figure out what else they need to do." I encouraged her to tell me about his eyes, which she could see only "very, very, very vaguely" and found unpleasant. Nevertheless, she was able to recall, "They're very, very big. They're much, much bigger than our eyes, and they don't blink, and they're kind of slanted on his head. And they're all black . . . I can't see any pupils. I can't see the retina, no whites, nothing. It's just all black." I asked what was so disturbing about the eyes. "I think maybe it was because they didn't care," she replied. "Just like, scientific. A kind of curiosity. It's not looking at me like as a person. It's looking at me as their experiment. I mean, as yes they would care if I died because it would ruin their experiment. Not because they care about me as a person or anyone else around me."

The feeling, Catherine said, was one of "total helplessness. I'm scared because I know they don't care about me and I have no control over what's happening, over what's going to happen. They think they're superior to us . . . That's another thing I got. Total superiority . . . It wasn't even told to me or anything or, get up on the table, like the other was. It was just the attitude." I asked if this attitude came mainly from the examiner or from all of the beings. "From the examiner more," she replied, "but all of them."

After this, several beings let Catherine get off the table and they took her into another room. Once again she became anxious and saw only blackness ahead in the room, although she feels certain that she did see something more in it. We then had the following exchange:

JM: What's causing the blackness now?
Catherine: [Very softly] I don't want to know.
JM: You don't want to know? You do want to know? I'm sorry, I couldn't hear that.
Catherine: I want to know, but I think it's too scary.
JM: What happened in there is too scary?
Catherine: What I saw in there is too scary.

Deciding that she would like help in remembering, Catherine agreed to try the game we had played last time and sent a "spy" into the room with a flashlight which he would turn on for two seconds during which he'd look around and report back. What the spy saw was shocking to both of us. Along the left side of the room there were "cases" stacked in rows reaching from the floor to the ceiling, perhaps eight feet high. There were four or five rows from top to bottom and eight or ten from left to right, making about forty cases in all. "I know each case has something in it. It's all the same thing," Catherine said, but there had not been time to tell just what. We gave the spy another two seconds. This time he saw "like creatures in there, but they're kind of deformed looking. Each one has one of those things in it." Catherine said that she passed through this room on the way to another place, and was now willing to say what she saw during those few seconds with the aid of our spy with a flashlight.

In the cases are "like baby versions of them." They are "all in a liquid" and "all facing out" and "the cases are lit from the back." There appears to be nothing else in the room. The creatures are naked and "like standing up . . . When you go by a doll and it's in a plastic case and it's stood up like that. That's how it is." After the regression she described the cases as "like a window display in a toy store. They cover the entire thing with Barbie dolls and you can see through the plastic things and they're all standing there." At the top of each case she could see "the edge of the water [or] whatever it is. But they're fully submerged in it. The heads are large and in the same proportion to the bodies as the alien figures themselves. They're just like miniatures."

After passing through the room with these cases Catherine was led by two of the beings along a walkway, "curving around to the right, like following the edge of the ship," and through a doorway to another room. She was still wearing only her T-shirt. Then she entered a room that was much larger than any she had been in before. A path went through it, also curving to the right, but the proportions confused her. "I can see where the edge of the ship would be on the left, but it's way far away—maybe fifty feet away. I don't understand how it could have been that far because the other room was only like ten feet wide and this one is so really far away."

Catherine found herself in "a forest . . . I'm confused but it's there. It's in the room, and there's like trees and rocks and dirt and things off to the left. I can see them from where I am. We're not going that way. We're going around to the right. How can I be in the forest?" Incredulous, Catherine exclaimed, "It doesn't make sense!" for although "there's forest all around" she could still see the curving walls of the ship. "It wouldn't have fit. It wouldn't work." After the regression she reflected that she "looked way off in the distance" and "could see walls, but it didn't make sense in context." She said that the forest even smelled like one and contained pine trees. She estimated it was "high school gym size."

Finally, she was taken back to the room she had originally entered and was given back her clothes. "They take me down the ramp, and they're walking me back over the field and up the little hill and to the door. They open the door and I go in and I take off the boots and the coat and I don't think they followed me in, so I went back and got into bed and went to sleep." Her mother seemed to have slept through the whole episode. The "tank" on the ship Catherine had previously remembered appeared to her to be a distorted version of the liquid-filled cases in the room that she saw on the ship.

Before bringing her completely out of the regression, I reviewed with Catherine the reality status of her recollections. "I don't think it's a dream," she said, "but I don't think I was supposed to remember. That's why it doesn't seem totally real." Also, she added, "I don't think I want to believe it . . . But I do remember it . . . It scares me to think it's real."

It was not until her third regression that Catherine seemed ready to speak of the episode of late February 1991, in which she had driven, as if under some sort of compulsion, to a wooded area in the town of

Saugus, north of Boston. It was this behavior, whose strangeness had troubled her, that had led Catherine to contact me. The Saugus encounter is in some ways the central abduction experience of Catherine's case. She had just completed talking of the unexplained lights behind the family trailer when she was fifteen, and I asked "where you would go if you were to explore further."

She recalled then the drive from Somerville to Saugus, about ten miles. "I go out on roads I've never gone out before. Just for the hell of it." As she headed north "all the time I keep looking out up at the sky and I keep thinking about UFOs. I've been thinking about them a lot for the past few weeks. But I've been reading a couple of books about them, so I figure that's just me, because I've been reading the books. I keep half hoping to see one and half way hoping I don't." She thought about what to do about her job and whether to move to New York to distract herself, "but my mind keeps returning to UFOs." She followed signs to the Saugus Iron Works and sat for five minutes in the parking lot and then realized that "doesn't make sense." Becoming increasingly lost she passed through a residential section and came to a wooded area. She felt anxious about driving through it, but thought that that was the way back to the highway and "I have to." Catherine assured herself that "if someone tries to jump the car" she could speed it up and "run them over."

Catherine drove through the wooded area, but realized that was not the way to the highway and she had to go back through. Even more anxious than the first time, she drove through again, remarking, "I think something happened but I don't know what it is." At this point I stopped her narrative and asked her to go back more slowly through the experience of driving through the woods for the second time and to tell me any feelings that come up. She said, "I don't want to be there . . . I've got to drive out . . . I'm starting to go numb again." Although "my foot is on the pedal like steadily" and the pavement was level, the car was slowing down. The car stopped and the numbness increased to the point where "it's like my entire body has gone to sleep."

Although Catherine could see streetlights from outside of the woods, it was lighter around her "than it should be." Unable even to move her hands by then, she sensed "something coming behind the car on the left side, the driver's side, like a light coming up there." Something came up to the door and opened it, but Catherine "can't look at it . . . There's something there. I think it's one of them. There's a hand reaching out to get me . . . It's long and thin and very light colored and it's only got three fingers." The figure "presses me with its hand to guide me and I get out." She felt there was no choice. "If I had my choice I'd be speeding the hell away from there." Then "I'm going around behind the car and this being is kind of behind me to the right side and the car is on my left side." The being had "huge, black, almond eyes," and was "glowing." She believes that the light that came up behind the car emanated from the glow of the being itself.

Sensing her fear the being did something, perhaps with its "huge hand," to calm her. Although she found this "somewhat comforting that it didn't want me to be scared," at the same time "I don't like it to have that power over me." The being walked with Catherine for a way along the road and then "it either took me into the woods and something was there, or we went up and I'm not sure which one happened." Noting her confusion and also that we were both tired, I suggested that we end the session and she readily agreed.

After I brought her out of the hypnotic state, Catherine sobbed softly over the helplessness she had felt and the increasing realization "that there's more truth to it than not . . . I don't just start crying." I asked Catherine to pinpoint more sharply what her deep sadness was about. She said, "because I feel powerless, and I feel like they can get me and do whatever they want to me practically any time, and I can't do anything about it. And that's a very terrifying thought."

I suggested she felt, "You're not master of your own life."

She replied, "Within a very small framework, but in the grand scheme of things I feel like I don't [i.e., have control] . . . It's all because of them," she said. "My basic idea before any of this was 'you are master of your own destiny. You are responsible for your own awakening, for your own realization of the nature of reality!' . . . I totally believed that."

The further exploration of the Saugus abduction took place in Catherine's fourth regression, more than five weeks later and after she had seen the first episode of the CBS *Intruders* miniseries. By this time she was highly motivated to know "everything," asserting emphatically "I think it's better to know than not know." She was also beginning to become preoccupied with "global concerns," by which she means how "we're messing up the planet." These ideas, Catherine says, may have come about from "impressions that I've overheard," or "things that were actually told to me." Recalling the abduction at age seven, recovered in the previous session, Catherine remembered the being saying to her, "'We need to find out about the effects of pollution on your planet.' That's made me think a lot. It's made me wonder if it's more because they're actually concerned about keeping the planet intact for whatever reason, keeping the planet intact so their specimens don't all die on them in the middle of the experiment."

At the same time Catherine was also troubled about the fate of the earth. "I think they're right. If we don't do something immediately, it's going to be suicide for us all. I'm more concerned about keeping myself alive and my friends alive and the rest of the people on the planet alive because it's a fine thing to do rather than I'm concerned about some little bastards coming and taking me away and not messing up their experiment. Our motivations are totally different even though the goal may be the same." She has come to the conclusion from the aliens' "attitude" that "they have lost all the genetic material" and are using us for their needs. But, she added, "all the genetic experimentation is a big part of it but it's not the entire story . . . It's hard to describe, but it's a much bigger plan than just that." But if they do not continue the reproductive/genetic activity "then they can't proceed. But this is just one step."

After the above discussion Catherine determined to explore further. "I need to know about Saugus, because it's a huge, huge thing," she said. In the beginning of the regression Catherine briefly reviewed the events leading up to being taken from her car in Saugus with the additional detail that when her car door was opened she thought to herself, "Oh, Christ! It's one of them!" She now realized that in some way her trip north was forced upon her and "they made me think I was doing it for other reasons." With the memory of being in a spot in the woods not far from the car, Catherine recalled what followed. "He's kind of taking me up, up the diagonal. We just kind of fly off. We're not going straight up. We're going across too. 'This is too fast! Why are you going so fast? I'm going to fall off the beam! I'm going to fall down there!!' And he just kind of says, 'No, you won't.' Everything is speeding by on the ground and we're going up together and we're coming to the ship."

The ship was "huge. Everyone else should be able to see it too and I don't know why they can't. It's got lights all over it. It's just like silver metallic, but it's got lights all over it. It's fucking huge . . . he's taking me inside. We're in a hallway. There's some more of them waiting. There's like four of them now. They're pulling at my clothes . . . *I'm pissed off! 'Stop it!! I'm perfectly capable of doing this myself, thank you,'* and they kind of get an annoyed attitude." Catherine thought (she could not speak) "some snotty comment like 'Why don't they just go rent a porn movie.' They don't get it. They don't know what a porn movie is. I don't think they understand the concept of voyeurism or anything like that."

Then naked, Catherine was led by the beings into an enormous room "the size of an airplane hangar." Catherine was amazed to see "hundreds of tables in here! There's hundreds of humans in here, and they're all having things done to them." The beings walked her through this room and she saw rows of tables on either side, separated by perhaps five feet, with many that were empty, and about a third to a half with human beings on them to whom procedures were being done. She estimated she saw between one hundred and two hundred humans in all in that room. Under the tables she saw drawers where she believed instruments were kept.

Catherine got onto one of the tables and noted a black man with a beard on her left. She was forced to sit up and an examination began. "They're running their little fingers down my spine like they're counting the vertebrae in my spinal column." Repulsed by the touch she exclaimed (mentally), "'What the hell's that for?' 'To make sure everything's okay,' he says. 'I could have told you that,'" she says angrily. "They're feeling my arms, my legs, my ankles, feeling my neck, my thighs." She asked more questions about the purpose of this examination. They told her "there may be things you don't know about." A taller being came over and stared at her and told her that she asks too many questions, that it is good to cooperate and that what they are doing "isn't bad. It's necessary. 'I won't try to hurt anyone.'" Although she felt these were "bullshit answers," they made Catherine feel calmer and "a little more accepting."

As the being stared into her eyes, Catherine felt she had no choice but to look back. I asked what the experience was like. "I think he knows everything about me. He knows exactly what I'm thinking. He's answering the questions before I even think of them," as when he told her this process was necessary "before I even asked." Looking into the being's eyes is "scary" for Catherine, "but then parts of me are kind of beaten down. I feel just calm and peaceful." In addition, she felt the figure "wants to know me personally. I'm trying to think, 'why, you don't care about me as an individual.' But it's hard to think it. It's hard. It's hard to think anything that's against what he wants me to think." The being insisted, "No, I want to know about you. I care about you." The struggle of wills continued. Her resistance was "making him work harder than he feels like he should have to." Her mind said "bullshit" as the being tried "to tell me that he loves me." Finally, she conceded, "Maybe he's right. Maybe I just don't understand . . . I think maybe I was wrong for thinking that he was lying . . . I just don't understand them. That's why I think that." She persisted with the thought that he does not know what care means, and he responded, "'No, we know. We just don't feel it as intensely as you do.'"

The debate was now apparently over. "He's won, so he walks away. He goes around to the foot of the table. He says to me, 'Are you ready now?' I say, 'Ready for what?' He says, 'It's time.' I say, 'I wish you would answer something.' And he says, 'You shouldn't wonder so much.' He tells me they're going to take it out and I'm thinking, 'take *what* out?' One brings around a cart with like a tank on it. It's like a cylinder and it's filled with clear liquid. He's putting my feet up and spreading my legs out, and I'm thinking, 'Oh, God, what are you going to do?'" I urged Catherine, who was clearly distressed, to breathe deeply and return to her center. I assured her of my presence and that the worst was "almost over."

What followed was the most disturbing experience of Catherine's recovered abduction history and the most difficult few minutes of my work with her. As she sobbed and panted, at times crying hysterically or expressing rage, I needed to assure her repeatedly of my presence and to express my sorrow over what she had been through as I asked for details. My sense was that she was determined to follow through to the end despite the fact that she was reliving a powerfully traumatic experience.

The tall being inserted "a big metal thing" in her vagina, which was intensely upsetting. Then he took a longer and thinner "version" of this "and put it up inside me!" She felt that he was trying to reach something inside her body in order to cut it off. Sobbing forcefully she said, "Oh, God, Oh, God. He's taking it. I can feel him cutting." With short, neutral questions interspersed by me to elicit what she was seeing or feeling together with expressions of support, Catherine reported, "He's cutting inside me. I can feel it . . . He's got it. He takes out this hunk . . . He takes out the thing he put in and there's something attached to the end of it. It looks like a fetus . . . I can see it." I asked how many months she estimated it was and she replied, "I'd say about three months but I don't know enough to know for sure. It's about the size of a fist."

I asked Catherine if the fetus looked human. She said, "It's kind of hard to tell. Eyes like theirs." The examiner "seems proud. I get this feeling [she pauses]. He takes that other metal part out, the part that was spreading me apart" and "gives it to the little one who has the cart with the tank on it and he wheels it away." The examiner was saying to

Catherine, "'You should be proud of yourself,' and I keep thinking, 'Why?'" She felt she was used like "a glorified incubator." She thought, "'You fucking lied to me, you bastard! You don't care about me,' and he said, 'No, I do.' He says it again! I say, 'You fucking bastard. How dare you? How could you do this to me?' He's coming over to my side of the table and he's looking into my eyes . . . I'm trying to fight him!"

Once again Catherine was angry, crying, and very upset. "He's trying to do the same thing to me as before. You will not do this to me again! You will not," she yelled. "He says, 'Why are you resisting? Why are you making this so difficult for everyone?' And I say, 'Why have you fucking ruined my life?' [sobbing] He says, 'We haven't ruined it. You won't even remember it. I said, 'Bullshit! I *will* remember it,' and he's putting his hand on my head and make me feel kind of . . . [she stops] 'You will not do this to me again. I will not let you.' I try to fight, but I feel like he's winning,'" by which she meant calming her down. Again Catherine was told that this is "necessary" and "all for the best in the end," and "I'm saying, 'Won't you even tell me what the end is? How do I know that?' and he says, 'We can't tell you.' I say, 'You won't tell me a goddamn thing.' I say, 'How many other humans have you done this to?' and he says, 'It's a very large number.'"

Catherine felt that despite her intense effort to fight the being's influence she was "losing it" and was positively affected by his assurances that they care about her, are "sorry this had to hurt" her, and "didn't mean for that [her suffering] to happen." After more communication of "meanings" concerning the aliens' experiments, plan, or project—no word seems quite right—and further assurances that "they're not going to hurt me," Catherine said simply, "You should have asked me." They also told her again that she would not remember. I asked her how it is then that she and I are able to recover these memories and she replied that since they did what they needed to do it does not matter anymore. After one last reassurance to "be calm," given as much with his eyes as with words, the examiner left, "the little guys" took Catherine off the table and walked her back through the room with all the tables.

Looking around at the people on the tables, Catherine felt sad "for all" and felt "like I should try to start a riot or something but I can't do it." They took her back to the first room she had entered where her clothes were left. "I put my clothes on, and they're trying to put them on me again. I'm just like, 'Please [very annoyed], they're my fucking clothes. Let me put them on.' They tried to help me, but they just ended up fumbling. They were kind of leery of me at that point. They really don't want to aggravate me more. They're kind of scared of me." Although in her paralyzed state, Catherine could give little expression to her feelings, "they could feel the emotions there and they get kind of scared because they don't feel that intensity so they really don't know how to handle it, especially now that the taller one's not there. I don't think they have the capability to calm me down that much if I get upset again."

Stepping out of the ship "into this empty hole, we should go falling straight down but we don't. We go back on diagonal." One of the beings floated Catherine back to the passenger side of her car and walked her around to the driver's side. The door was still open with the keys in the ignition, and she thought, "Someone could have stolen my car." The door closed as if on its own—she is not clear just how this happened—and Catherine drove out of the woods, noting that it was then 2:45 A.M. and about forty-five minutes have passed for which she cannot account. As she drove out of the woods, Catherine felt anxious and also "silly." In her fear she raced home, driving over a hundred mph ("I wanted to see how fast my car could go"), which also served "to get out some aggression." Once home she went right to bed and fell asleep.

The next morning on television she saw and heard the stories about a UFO or a "comet" that had followed the route she had felt compelled to travel, and called a friend, who was seeing a therapist to whom she told Catherine's story. The therapist knew me and my work and arranged for Catherine to contact me. After our initial interview Catherine did not call me as we had arranged. Nevertheless, I called her after about a week when I had not heard from her. "That kind of made me realize that you really were interested and I wasn't just totally crazy or making something up."

After the regression Catherine, Pam Kasey, and I considered the possibility that her impregnation with this fetus might have occurred during the Christmas abduction in Alaska. Arguing against this was the fact that the fetus seemed too fully formed for a two month pregnancy. Catherine recalled another episode in late October or early November 1990, "which made no sense at the time." She had found herself driving in the middle of the night on deserted roads and pulled off the highway at a rest stop. "I was really scared to be there because, again, I was waiting for something." She waited for what seemed about fifteen minutes, but does not recall anything else happening before she drove home. By Christmas Catherine had gained some weight, which she began to lose following the late February abduction.

She recalls no other pregnancy symptoms and we have not explored the late October episode further.

We reflected again upon the reality status of this experience. She had recently read David Jacobs's book *Secret Life,* which contains stories of reproductive traumas, and wondered if perhaps "I'm subconsciously picking up on stuff that's there," although she had never considered herself to be suggestible. Pam observed that Catherine had told her before she had read *Secret Life* that she thought the Saugus episode "had to do with a fetus." Then Catherine asked herself "why," if these memories were not authentic, "would I be coming up with these bizarre, traumatic stories?" In the end Catherine felt she was left with only two choices. Either she was "crazy" or "I don't know what else. I mean, other than it's actually happening."

Finally, we discussed the sincerity of the examiner's expressions of caring and affection. She could acknowledge that from the alien perspective and commitment to their enterprise they might feel affection as we might toward a pet animal that was being used for experiments. But to her this was "not an excuse because they know, they know that we have more of a consciousness than that. They know what they're doing! They know how traumatic it is for us and they don't give a fuck." Two days after this session Catherine wrote me a note of appreciation for the help she had been receiving "at a time when my foundations of reality have been shaken."

Over the next two months Catherine wrestled with many questions concerning the physical evidence that might corroborate her encounters, their reality status, and, above all, the shifts in consciousness that might best allow her to adapt to the phenomenon and even, possibly, have a more mutual dialogue with the aliens. We met on July 27, 1992, to review the ways in which her attitude had shifted, which she attributed in part to what she had learned from talking with other abductees. She continued to have visitations and possible abductions. With regard to physical phenomena, we noted a tiny lump in front of her right ear, for which she could not account. One night in mid-July she drew three circles on her leg to remind her to ask the aliens to allow her to see some of their writing, which, if they were to respond, would increase her confidence in the reality of her experiences and also related to her desire for a more mutual exchange of information. But the night she drew the circles (July 15, 1992) she was visited again ("they came down and paralyzed me") and she was too scared to pursue her questions. As it turned out she was abducted that night (see discussion of fifth regression below).

Although the reality of her experiences as we had recalled them in the sessions "is going up and up and up," she had concluded that they are "not part of the normal consciousness," i.e., they occur in, or reflect, a nonordinary state of consciousness. The implication of this was that "I've got to change my world view even more than I had already." She also had decided that "my reaction to the whole thing is going to determine the nature of that particular experience . . . If I am totally petrified and lashing out at them and basically turning into a hostile animal, then that's how they're going to have to treat me," she observed. "If I am calm and somewhat rational," she said, "I think a lot more can be accomplished as far as at least my enlightenment about the whole thing."

Toward the end of mastering her fear and making herself "less of a trapped animal," Catherine had been imagining "the scariest thing that could possibly happen." But instead of "letting it go crazy and having the same horrible, horrible experiences that I've been having" instead she would "go along with it and not totally struggle and fight because that just makes me out to have more control over me, which in turn makes me notice less and get more wrapped up in my fear and my fighting instead of being more aware and having less physical control, less mental control where I could actually have a dialogue with them and get a straight answer and have them show me something that would be helpful to me or whatever." Toward the end of controlling her "animalistic" reactions and moving "higher along the evolutionary scale," Catherine developed what she called her "no fear mantra . . . If I get afraid for any reason, I'm just sitting there saying, 'Don't be afraid. Don't be afraid. Don't be afraid.' And it works." After several meditation sessions in which she intentionally imagined aliens coming into her room she found that she had more control over her fear and could calm herself down.

Recently Catherine has come to the conclusion that the aliens are "more advanced spiritually and emotionally than we are" and therefore "they don't have the need to be as emotional as we do." This means that "if I'm going to get anything useful from them I've got to deal with them on their level." This also means to her building up a "core of inner strength." This is not "something that they can take. It's not something that anybody can take." She does not expect the invasive procedures to stop, but she can diminish their traumatic effect. "I'm not totally saying," she adds, "here's my body, do with it as you will. It's more a realization that it is going to happen." Inviting them to "show me some of your writing because I want to learn more about you and my part in your whole plan" is "a totally different concept" than "screaming at them, 'Why the hell are you doing this to me, you fucking bastard.'" Perhaps "that is something they will answer" and is "eventually going to help them because I will be more participatory in their plan." Catherine attributed the changes in her consciousness, "this spiritual growth, this psychic growth," to the "major, life-shattering" impact of the abduction experiences themselves.

By the time of the July 1992 meeting Catherine had already noted several changes in herself which were the direct result of her shifting attitude toward the abduction experiences and her generally greater psychological openness. The abductions themselves operated as a provocation. "There's got to be some experience that totally changes everything and how you perceive everything," she observed. Catherine attributes her capacity to take advantage of the impact of her abduction experiences to the exploratory work she has been doing in relation to them.

She has noted greater intuitive abilities in relation to other people. She can "feel people's auras," the energy fields around us that some especially sensitive people can see, and is more sharply attuned to the emotional states of others, which she finds "very useful . . . I can actually kind of check people if I'm meeting someone, and I want to find out if they're giving me a line or actually sincere and kindly in what they're saying. I can tell what their intentions are . . . This whole experience makes you open up to so many levels," she concludes, "so many other possibilities. Everyone has these kinds of abilities, but we shut them off because our society says, 'No, it doesn't exist,' and denies it all. And I'm opening back to this." One of the more difficult phenomena that Catherine and many other abduction experiencers have to deal with is a virtually constant flow of sensory experience, especially light flashes ("staticky kinds of things," she called these at one point), intrusions of patterned color images (while typing in one instance) and, to a lesser extent humming, buzzing, and other sounds. Gradually, these visual sensations have diminished and auditory ones have increased. The neurophysiological shifts that underlie these sensitivities are unknown.

It was October 26 before we were able to schedule a regression to explore the experiences of July 15. In our July 27 meeting Catherine reported that she knew "something" had happened around two o'clock in the morning, "because I had looked at the clock." Earlier that evening, in addition to the circles she had drawn on her left leg with a permanent marker, Catherine had also written, "Show me some of your writing." She experienced unusual light streaming through her window ("as if someone had a big spotlight outside the window"), and had the impression of beings in the room—"one of them coming towards me with a big wand thing with a light on the end of it and pointing it towards me, and that was a half-dream remembrance." She found that her right leg and then her entire body was becoming numb. "I tried to scream, 'NO!' but I couldn't. The words wouldn't come out. I couldn't get any sound out, and it came out choked like, 'AHHHHHH.'" She seemed to be too paralyzed with fear to completely follow her own new approach to the phenomenon.

At the beginning of the October 26 session Catherine began by wondering if abduction experiencers are chosen because of greater auras or energetic, vibratory fields that protect them from the consequences of "messed up childhoods." Catherine, Pam, and I speculated for a few minutes about the relationship of the abduction phenomenon to greater vulnerability and woundedness in the backgrounds of experiencers. In the case of at least some abductees the aliens seem to be entering the energy fields or responding "to certain vibrations of a quaking soul." We also talked about the possible dimensions of reality from which the abduction phenomenon might emanate and the various "sensory aggravations" that had occurred in Catherine's life since our last formal meeting. She expressed a desire to know "why all these things are happening," and we agreed to try to seek meanings in the regression as well as the narrative of her experiences.

Just before the regression was started, Catherine said that the search for meaning seemed like "the next logical step . . . I am past the I-am-not-insane, at least ninety percent of the time . . . and past the is-this-really-happening, and past the well-I'm-confabulating-things, and all the denial, and being absolutely terrified about it, I mean, it's the next logical progression."

In the hypnotic regression, our fifth, Catherine began by reexperiencing light streaming into her room, "like a huge searchlight," and she again heard what sounded like human voices outside her room. She tried to wake up from her half sleep state but felt "they're not letting me." Once again they gave her soothing messages and she was "furious because they always do this!" Crying she exclaimed, "They never let me remember anything for real and I've even asked them to." Two of the beings floated her out of the bed, "putting me into the beam." She told them not to hurt the cat who was "hiding" after "zipping upstairs" to get away. Catherine felt that the intense control the beings imposed upon her was to keep her from fighting back. She believes her opposition "makes them very nervous."

Catherine said, "If they would ever ask me about anything that they wanted to do, maybe I would be more willing, and even though I've been trying to be less scared, and try to get them to talk to me more, they still really don't want to. And I try to ask them questions that I think that they will answer, and I still get the same run around bullshit answers." Nevertheless, she said that in an abduction two weeks before, for which the memory "blocks" were still too strong to explore, the beings did give her some meaningful information in response to her request to "show me the end."

We returned to the beam of light and she described passing *through* her window, the porch, and a tree. She saw her apartment building getting smaller and smaller and the city receding below her. As she went up, she felt that although she was wearing only underwear the energy from the beam kept her warm. She was brought backwards up through "a hole in the floor" of a ship and found herself in a room with a wall that was more rounded than other rooms she had been in. "They want to talk to me about something," she felt. There were a lot of other beings "milling around" and a few other humans being taken to various places on the ship. The beings took her along a wide corridor that led around the ship and she could see stars "just hovering there" through a window.

I asked if she was still half asleep or fully awake by then and she replied, "It's not really either . . . once they put me in the beam, I kind of shift to this other consciousness that's not either, really." I asked what this consciousness is like and how it differs from "our ordinary waking consciousness." She replied, "It's like I have access to an entirely different part of me that I don't have access to in normal waking consciousness." In this altered state Catherine knows "more of them. I know more about them. It's not like kind of half-knowing maybe something happened like when I'm awake." The knowing, she said, is just as real as in our ordinary consciousness. "It's the same thing as knowing something here, but it's just like that, the door was shut in my mind and I don't have the key to it and they do." In the hypnotized state she was fully present to this other realm of information.

Catherine thought irritatedly that this long trip through the curving corridor would have been unnecessary if they had simply brought her in the other side of the ship, and she got the impression that the beings were, as ever, irritated with her questioning, oppositional attitude. They came to another room with a sliding door that slid open upwards. The room seemed to transform from a typically spare spaceship room with tables, curved walls, and perhaps a viewing screen into an ornate executive conference room complete with shag carpeting, mahogany paneling, and a large viewing screen. As Catherine recalled when she reviewed the tape of this session, she had had the impression that "the more I thought of a corporate executive conference room, the more it appeared," but when she realized that this was a kind of staging, the conference room "images just melted away to reveal the previous images, and finally the actual room."

During the regression she was aware of the simulation of a conference room and objected to their concocting this just for her benefit. But she was told, "We have to have a conference, so you have to think it's a conference, so we're taking you to a conference room so you can be in that kind of serious frame of mind instead of making your usual smart-ass remarks that you always do." "When this happened," Catherine observed, "I was just starting to not fight them. I was just at the very beginning. I'm not where I am now. So it was a very different situation. I was relating differently than I do now." The "goofy" tricks, she felt, were appropriate to her level of consciousness at the time. Once she was through the staging theatricals the room was returned to its original state and Catherine was told to sit on a small, cold metal chair.

She was then shown scenes of nature on the screen, "like a camera panning a forest—trees, and there's a deer, and you know, moss and dirt and needles on the ground, and I'm getting this impression like this is so beautiful, this is so beautiful." But she felt that her emotions were being manipulated and resisted, making "them have to work harder." Looking back she thought this was "okay because it gives me a little bit more control in the situation, and if they want me to listen to what they have to say they need to treat me as an equal being and not use all these manipulative tricks on me."

"Other nature things" appeared on the screen, "like Grand Canyon, and like okay, great, I've seen this on TV. It's going to the desert. It's going to be pyramids. I'm seeing more Egyptian, ancient things, like hieroglyphs and pictures, pictures of pharaohs and things, and I'm getting this thing, this was your life . . . I'm like, Oh! That's cool . . . It's kind of like a travelogue of my past lives." At that point she became intrigued, "because I like the ancient Egyptians a lot, so this is cool, if I was there." Then they showed her a current picture of tomb paintings with the paint flaking off, "but then it switched to me painting it." But in that incarnation she was a man and as she watched this scene "this makes sense to me . . . This is not a trick. This is like useful information. This is not them pulling a bunch of shit like everything else." Catherine now felt that her insistence upon a more reciprocal exchange of information had been affirmed.

I then asked Catherine to tell me more about this image of herself as a painter in the tomb of an Egyptian pyramid. In response to my questions she provided a great deal of information that she seemed to know as the painter, who was called something like "Akremenon." Some of this Catherine could have known from reading standard books about Egypt. Other details, such as the process of making the paint which she seemed to know well, and appears to be consistent with one textbook account, would not most likely have been known to her. She described the man's skin color, his clothing ("just a wrap-around loincloth") and headdress, which denoted the higher status of royal service when compared, for example, to a slave. What was striking at this juncture was the fact that the quality of Catherine's experience was totally transpersonal, i.e., she was not having a fantasy *about* the painter. Instead she *was* Akremenon and could "see things from totally his point of view instead of from me watching it."

Catherine described the light in the room in which she was painting (close to the outside, a kind of maze passage to get to the tomb); what Akremenon was painting (the blue headdress of a pharaoh's wife who was wearing a white dress and was holding a little jar as an offering in an act of atonement to the god of the dead, Anubis, in order to be buried and obtain eternal life with the pharaoh); another artist working on something "down lower"; her pleasure in doing this work ("I could do much worse for myself. I could be having to cut out limestone blocks for the outside"); and the rare blue stones "from one of the conquered countries" he ground to make the paint. Akremenon learned his craft from an older painter when he was a boy. After completing this scene "I've got to paint a bunch of dire warnings for grave robbers."

The wife's name was Tybitserat and Catherine called the pharaoh, who she said was of the Middle Kingdom and of "average" importance, Amen Ra [this is confusing, since Amun Ra is the name of an important Egyptian deity, not a pharaoh], but added, "To be perfectly honest it didn't matter to me that much as long as I had a secure position, I didn't really care." Later, reviewing my manuscript, Catherine wrote that it was difficult to remember the name of the pharaoh as this "was not central to what they were trying to show me. It wasn't the meaning or purpose of that life. I could be confusing several past lives even!" She believes that the pharaoh changed his name and "got rid of a lot of the gods." (Perhaps she is referring here to Ikhnaton, the New Kingdom pharaoh who abandoned polytheism and embraced monotheism.) Catherine also knew many details regarding the appropriate size of various figures on the panel she was painting ("regular people are small, royalty is bigger, and the gods are the biggest") and the complex problems of proportionality the painters faced as a result of the demotion of former gods by the pharaoh.

After showing her the scene in Egypt one of the beings asks her, "Do you understand?" What she realized then is that "everything's connected," canyons, deserts, and forests. "One cannot exist without the other, and they were showing me in a former life to show that I was connected with that, and I was connected with all these other things and I can't separate out as I've been trying to do." To Catherine this means that "I can't continue the way I've been going, and I can't continue to fight them the way I have been fighting them because I'm connected to them too. When I fight them I'm only fighting myself and I'm fighting my connections to all these things which you can't fight. It's there."

She asked the aliens why they needed to use such "theatrics" to show her this, and they reply "'to make you understand, to comprehend the implications. To put you in the right frame of mind.' And I'm kind of like, hey, now we're getting someplace!" She also seemed to learn from this episode that certain emotions, like "love, caring, helpfulness, compassion" are "the key," whereas others like anger, hatred, and fear are "not useful," especially fear. "Fear is like the worst one. They were trying to get me to get over fear, and that's why they were trying to scare me so badly, because I would eventually get sick of it and get over it and get on to the more important things."

I asked her to explain further how scaring her so badly would get her beyond fear. After a while the human body cannot handle it, she explained, "because it's just like you're on overload all the time, and number two, you just get sick of it because you can't really concentrate on anything else . . . When you get saturated," she added, "then you go beyond it . . . You have to make a conscious decision to get beyond it . . . It's when you finally say I can't continue like this then you get over it . . . You decide to rid yourself of that fear . . . I kind of explained to myself why it didn't help me to be immersed in the fearfulness anymore." I pressed her to capture more precisely the process of transforming her experience of fear. "I still get fearful sometimes," she said. "It's kind of like I had to push myself to the next level. I had to take the next step, get to the next plateau . . . You let it go. I mean, you feel it, and you let it pass through you, and it's gone . . . you don't hold on to it like you did before . . . I had gotten to that point where I had decided to go on to the next level." What she had learned during this abduction was "the next lesson," which followed naturally from the emotional shifts that Catherine had achieved "the week before" this abduction.

Before we ended the regression, Catherine said that her next goal or "role" was to "show others the way beyond fear . . . It's kind of like, you have mastered this lesson now, so you are fit to go show others the way." She also said that she had asserted to the beings once again that the "cheap theatrics to make your point" in creating the illusion of a "big global conference room" were unnecessary. But they insisted that "It's going to have more of an impact on you than if we just tell you this is a conference room." She seemed finally to accept that "they have their reasons and I shouldn't question them." We also came to the conclusion that her stubborn self-assertiveness and questioning though "making it a hell of a lot harder than it needs to be," might have been a productive element in her developing process.

Remaining true to herself to the end, Catherine thought again, as they led her back along the corridor, "We could just go out the other side." Then "We go and stand on the same place" and the floor "kind of disintegrates beneath us, and we go down the beam, and geez, through things again, which is very disconcerting, and they put me back . . . I get in bed, and I lie down, and I kind of pull up the covers halfway, and one of them pulls up the covers the rest of the way." Finally, she makes the passing observation that when the beam of light comes down it is blue, but that it is white at the end of her abductions when she returns on or in it. As we reviewed the session Catherine, like other abductees, suggested the things she had experienced "are like not from our space/time," which, to her was "just another example" of how "all these things are connected and you have a connection to them."

In the next abductee support group, which was two weeks after this session, Catherine shared her ideas about how to deal with the terror associated with the abduction experiences as several members of the group seemed trapped in their fear reactions. "I think it depends on how you're interacting with them," she said. "If they come to you and your first reaction is to act like a scared lab rat and curl up in a corner of the bed and try to hide, like a rat in a corner of a cage, and they have to come get you for whatever reason, they're going to treat you in that manner. But if you react like, 'Okay, let's deal. Let's try to have some kind of meaningful interaction,' I think their reaction is going to be a lot more respectful and a lot more on a peer level than if you react in an immediately fearful way." Later she shared the process of dealing with fear (getting "to this saturation point, so sick of it that you get beyond it") that she had learned in the last regression. Then, she said, "you go on to the next level and learn whatever thing you have to learn at that level. But the fear is the barrier to getting anywhere else."

Discussion

The unfolding of Catherine's case has followed her sense of need and desire to know more about her experiences. As a result, a number of areas remain unexplored at the time of this writing. For example, Catherine told Pam Kasey in a conversation in October 1992 that she had had "a flashback" that went "through my head over and over again" of being in a nursery with many bassinets. A female nurse brings one of the babies over to her and tells Catherine that she must hold it. She feels repulsed and disgusted and tells the nurse that she does not want to. "It was very hard not to start crying during class when this flashback happened," Catherine said. We have not looked into these suggestive images further.

Nevertheless, Catherine's case illustrates many of the characteristic features of the alien abduction phenomenon. Her candor and courage, ability to recall details, straightforward articulation of her experiences, and, above all, her self-critical and hardheaded quality of mind give her story special value. Initially Catherine was ready to reject as inconsequential the suggestive experiences that she could already consciously recall. She sought help reluctantly and held as dreams abduction experiences that she later felt to be real, though undergone in another state of consciousness. As more and more details of her troubling experiences were remembered with intense emotions, Catherine clung to her doubts as to their actuality, looking with me for conventional explanations, until she acknowledged in a note after her fourth regression that "my foundations of reality have been shaken." Most instrumental in Catherine's emerging acknowledgment of the personal truth of what she had undergone, was her sense of herself as a person not given to the expression of strong feelings without a solid basis in actual experience.

The acceptance of the actuality of her experiences, whatever their source may ultimately prove to be, has permitted Catherine to deal more effectively with the powerful affects and bodily feelings that accompany them, especially terror, rage, and grief, and to reach a higher or more creative level of consciousness. Of special value to her personal transformation has been Catherine's decision to let her fears fully "saturate" her being when the encounters occur, rather than to aggressively fight the threatening energies embodied in the alien presence and activity. This has not meant a blind surrender to the aliens' purposes, but is, rather, the recognition of the need to let go of control in the face of mysterious forces that she cannot usefully oppose.

Catherine's shift in attitude from antagonistic battling—a stance that was initially useful for maintaining some sense of personal integrity and agency—to a kind of active acceptance has had several results. It has enabled her to undergo considerable personal growth that has been manifested by a desire, which she is already implementing, to help other abductees come to terms with their experiences and by a deepening sense of concern for the fate of the earth's environment. Information regarding the pollution of nature and the breakdown of the earth's interconnected living systems has been given to her during her abductions. Although Catherine distrusts the aliens' motives in apparently trying to arrest these processes (perhaps they just want to protect their experimental arena and subjects), she can see the purpose we share with them in preserving the earth.

Catherine's accepting and open attitude seems to have begun to bring some response to her desire for a more reciprocal relationship with the aliens, or at least to bring answers to some of her questions. Instead of being told, as so many abductees are, that they are "not ready to know," recently Catherine has had less traumatic experiences and has had the interconnectedness of the earth's living systems powerfully revealed to her during her alien encounters. Also meaningfully demonstrated to her but more difficult for a Western mind to accept, was the interconnectedness of all consciousness as manifested in the experience (convincing to her) of a previous incarnation as an Egyptian court painter.

As is true in all abduction narratives, Catherine's case raises more questions than it answers. For example, what is the technology or process—we hardly know the right words—whereby our minds can be so deliberately tricked as to see a forest on a spaceship or an executive conference room instead of the more stark or "typical" room on the craft? And, finally, we are faced in Catherine's story, as in so many abduction cases, with questions about the real purpose or meaning of the hybrid reproductive project or experiment, which are disturbingly manifested in Catherine's experiences. She describes rows of hybrid babies in some sort of incubatorium, and tells of a huge room with hundreds of tables on which human beings are undergoing procedures in which they have not agreed to participate.

CHAPTER EIGHT

DELIVERANCE FROM THE INSANE ASYLUM

Joe, a thirty-four-year-old psychotherapist with a professional development consulting firm, wrote to me in August 1992 that he had had "a variety of ET experiences going back to early childhood" and felt the urgent need, "as scared as I feel," to "air out these closets." As a designer and leader of adventures in nature, Joe helps people to overcome their fears, including fear of the dark. At the same time he recognized that at the time he contacted me he was struggling with "my own fear around the dark." About three months before he contacted me, while a massage therapist was working on his neck, Joe suddenly had an image of lying on a table, surrounded by small beings with large heads, one of whom was putting a needle in his neck. He screamed in terror and could no longer deny the disturbing power of his experiences. He learned of my interest in the abduction phenomenon through another abductee and also from the roommate of a woman who was assisting me in my work and he wrote a letter summarizing his experiences.

When I first met Joe his wife was expecting their first child in one month. The exploration of Joe's alien encounters in the context of his wife's pregnancy, labor and delivery, and his own evolving role as a father, has given us a rich opportunity to examine the relationship of the abduction phenomenon to Joe's consciousness of the cycles of birth and death over time, including the recall of a dramatic past life experience. In four hypnosis sessions between October 1992 and March 1993, one before and three after his son Mark's birth, we explored the complex dimensions of Joe and Mark's relationships with the alien beings, and Joe has struggled to integrate the alien-related elements in his own identity. The personal liberation and growth that this integration has permitted has been a remarkable aspect of Joe's case.

Joe, the seventh of eight children, was born and grew up in a small town in Maine. His father, who sold leather and threads to shoe factories, died of Alzheimer's disease a year before I met Joe. Joe calls his Irish family "pretty typical Roman Catholic," and said that they were manifestly happy but actually dysfunctional and that his parents tended to be cold and "emotionally tight . . . I wasn't hugged or kissed much" and "spent a lot of time outdoors, where I felt safe and accepted."

"We grew up playing in the outdoors," he told a seminar group I was teaching at The Cambridge Hospital, "hunting, trapping, fishing." Joe, like other abductees, has felt that his relationship with what he calls "the ETs" has provided emotional nourishment, support, and love "when nobody else would." He describes his mother, Julie, as a fearful person who initially did not want to hear about his ET experiences, becoming frightened and flustered when he would try to tell her about them. He remembers waking his parents one night as a child to tell them of a frightening experience and being told "you just had a bad dream." Joe does not believe that any of his siblings have had abduction encounters. "I've talked with them since," he says, and received a "mixed reaction from denial to acceptance."

From the time he was eight until perhaps fifteen, Joe liked to spend summer nights sleeping outside on a porch with his younger brother, who does not necessarily believe in the abduction phenomenon but told him recently "you were always afraid of UFOs." As a teenager Joe became aware of how separate and alone he felt, "basically, you know, social-puberty blues." Recalling that time in one of his regressions, he said, "I just don't connect with anyone. I don't fit in."

Joe's wife, Maria, is a psychotherapist, five years older than he. They met at an alternative therapy teaching center and had been married for five and a half years when I met him. They had been trying to conceive for two years, but Maria had two miscarriages. Joe has seen two babies, toward whom he felt very loving, in "a very lucid dream" and wonders if they might not be fetuses of his and Maria's who were taken by the ETs. Joe identified intensely with his wife's pregnancy and in his August letter to me wrote, "Not unlike the labor my wife will be facing, I am experiencing 'contractions' that are increasing with intensity and are painful if I resist." He, in turn, found that often after Mark's birth the old feeling of not fitting in came back and he felt like "the fifth wheel." Joe finds Maria, who he does not believe is an abductee, basically supportive in relation to his experiences and says that they look to each other for love and support.

From the time he was a teenager, Joe pursued spiritual understanding, and has participated and taught in a number of personal growth–related activities, including various mind/body healing workshops, psychosynthesis, different forms of meditation, and membership in a spiritualist church. When he was twenty he spent a year living alone in the forest in northern Maine. Joe has worked therapeutically with people in alcohol recovery, with sexual abuse cases, and incest survivors. For several years he has been leading what one of his brochures describes as "team-building workshops, staff trainings and customized retreats" which are offered for individuals and organizations. He is also a valued consultant to other professional development groups, especially those that lead excursions into nature, including Outward Bound schools. Yet behind his personal quest and professional competence Joe has always sensed fears of darker forces, outside of his control, that were associated with his ET experiences. About ten years before we met, Joe was told by one of his own spiritual counselors that he would some day be "working with people from other planets."

Joe has had dreams of contact with alien beings as far back as he can remember. Sometimes he would wake up with his penis feeling sore. In his regressions he recalled experiences in which sperm was "extracted mechanically," and "I've also seen infants that I felt they were showing me because they were, in part, my own." Joe believes that the ETs were interacting with him "even in the womb," and in our last regression he recalled seeing the beings around the hospital bed when he was only two days old. Like many people with a history of abduction experiences, Joe had many unexplained childhood nosebleeds. When he was a small boy he had a recurrent nightmare that a witch, like the one in the *Wizard of Oz,* would fly up to his bedroom window, force him to look into her "huge eyes," and make him "climb out onto the broom" by "hypnotizing me . . . Once I looked at her in the eyes I was all hers and she would whisk me away." Throughout his childhood Joe was fascinated with but afraid of UFOs. He would sleep outside, but have difficulty falling asleep because he was "afraid that as soon as I did somebody'd come and take me away." Additional childhood and adolescent experiences emerged in the hypnosis sessions, including an experience from the age thirteen to fifteen period which will be described shortly in detail.

Joe continued to have fears of UFOs and alien beings through his adolescence. Once at sixteen or seventeen when he was experimenting with LSD, he became panicky to see a "small ship" about two hundred yards away "and there's somebody in it looking at me." He thought "it was checking me out" and "went just a little bit out of the way and dropped into the trees." During this period he had another experience in which he looked in the mirror in a bathroom in his home and felt himself "starting to sink, sink, sink . . . All of a sudden," he said, "I was looking through a window and it wasn't me. It was an alien who was face-to-face looking at me, and he was round-headed and it was very knobby skin[ned] and kind of just warty, bumpy, and I think it was like gray or green" with "maybe a small mouth, thin neck, and then, WHAM! The reality of it hit me. I just freaked because I felt incredibly vulnerable. It wasn't like 'Hi, my name is whatever and how do you do.' I felt like here's somebody from another dimension who's going to go, Whoosh, pop, I'm gone. I'm history. They find Joe's shoes in the bathroom."

The intense pain and fear that Joe felt in association with the memory that came to him during the May 1992 bodywork session drove him to seek my help. He told the seminar group, "As a therapist is working on my neck, I had a memory which totally blew me out of the water. And it was of being on a table, about this high, not quite so wide, being surrounded by little people with big heads, and they're putting a needle in my neck. I'm terrified. I'm screaming, and I almost broke out in hives in that moment." He was also anxious about the impending birth of his baby and his own imminent fatherhood.

Our first hypnosis session was on October 9, about ten days before Maria's official due date. Joe spoke of anxiety-filled dreams of UFOs, alien beings, and procedures on a table in a subterranean room hewn from rock. Maria herself was having dreams of "the baby coming out and talking" about travels of his own on a spaceship. Joe told of other complex, dark dreams involving mythic snakes, fish, dark birds, sexually threatening women, mythic gods, and winged horses. Vast, wind-swept landscapes and scenes that seemed both epic and fairy tale–like, captured his feeling of loss of control, helplessness, and fear as the time of his baby's birth approached. Images of a needle being put in his neck during the spring bodywork session returned and Joe said, "I want to release some of this fear" in the session. We reviewed the various abduction experiences of his life, but in the end opted for an open-ended regression in which his unconscious mind would suggest its own direction and focus.

Joe's first image under hypnosis was of a nonhuman being with a triangular face and a large forehead, narrow chin, and large, black, elliptical eyes. "It's inviting me back," he said. "It's moving away and drawing me forward. It's got long, thin arms. God! It's got a long, thin back." Breathing heavily with mounting distress he said, "It can move gently. It can move fast. It wants me to lie on the table. It's looking me deep in the eye, telling me to relax." Whispering now, Joe said he was "scared" because "I know they're going to do something . . . my whole spine aches. My groin is on fire."

Sensing that his fear was running ahead of his ability to handle it, I asked Joe to locate this experience in time and place. It occurred in his home when he was a "young teenager," fourteen or fifteen perhaps. It began in the late evening when he was feeling separate and "very alone." Feeling restless and needing to go "outdoors and connect," Joe walked behind the barn in the rear of the house as if "drawn" there by a "real subtle" force. He walked through the barn ("sometimes the barn feels a little spooky at night") and looked up at the stars. "That's when the ship came down. It came right down. Bam! There it was. Small." The ship was "kind of round, but oblong. It's kind of like an egg," a "standing-up egg." The craft was "real symmetrical . . . more oblong on the top half" and "about four feet off the ground," with some sort of "feet" holding it up.

Joe felt frightened as a thin figure whose face is "all lights," dressed in a one-piece, tight-fitting, black outfit, approached him. He felt he had gone with this figure, whom he calls "Tanoun," many times before and his greatest fear in this moment is that he will not want to come back to the earth at all. Joe felt impelled to go—there is no choice in that sense—but "I am more aware of the choice that I can go and not have to come back." He felt a tightness in the neck as the fear he feels with respect to the torn allegiances between the alien and Earth realms mounted "in my heart." Sobbing, Joe said, "I'm not alone with him. I know that. That's okay. But he's not there every day." The being communicated to Joe that he must return to "work with them [human beings]" and that "I've got to have my foot in both worlds."

With his "very round face . . . right next to mine," Tanoun put his hand on Joe's shoulder—"he's very comforting"—and "we kind of walked, kind of floated" into the bottom of the ship, which seemed "much bigger inside than outside." Tanoun took Joe down a hallway to a large room with a table that he has often been placed upon. With one hand on his head and one on his hip the being reassured Joe, who felt "this guy really loves me, and in a way I don't feel anywhere else, and that's kind of scary" because it made him feel so "different from everybody." Joe lay on his back on the table with his arms at his sides and noted he is wearing a "white, metallic" robe. He avoided looking into the being's eyes to reduce the intensity of the connection for which he also yearned. To "open to them in my stomach" would make the relationship with the beings "more real, and it makes me more like them, which is very hard to integrate."

In addition to Tanoun, who was "in charge," eight to ten slightly smaller beings surrounded the table. "The one on my left" held a large needle, about a foot long, with a kind of hilt, as Joe anticipated intense pain. Tanoun told him to look "inside his eyes" and relax by losing himself there, but he feared "I would disappear" and "not come back" if he let go completely. "He's inviting me forward, and I'm scared. It's like I'm inside him—inside his head, inside his eyes." The needle penetrated the left side of Joe's neck below the ear, "almost like against my skull." It was very painful, but became less so as "I look more into his eyes." Once in, the needle was moved around, and the pain ceased. Joe felt both that "they're taking a little bit of something out, and they're putting something in" that will make it "easier to follow me." He said, "They're putting a picture in my mind" of a "small, silver, pill-shaped thing that they're leaving there" which has "four tiny, tiny little wires coming off it." After the needle was removed Joe was told, "We are close. We are with you. We're here to help you. We're here to guide you, to make it through your difficult times."

After this Joe was taken further down the hall to "another ET" who seemed to him "like a head honcho." He was sitting in a chair surrounded by light which appeared to come from him. This being was taller than the others, with a face that appeared more human. "He's putting his hands on my head. It's like he's baptizing me. He likes me—he's energizing me. He's blessing me. He's giving me something to help me hang in here . . . He's giving me strength and knowing, just knowing that I'm not alone. I'm loved, and I'm connectable" with both them and on Earth. Sobbing, Joe said that he was told "they will be closer if I let them. You don't have to be so far away . . . It's just a matter of time when it will be okay."

Experiencing the struggle to be both alien and human, Joe felt confused as he returned to the earth. He perceived "the part of me that's waiting for me behind the barn looks like them." This was "the part of me that didn't want to go." Joe "pulled back from that part of me because he looks a little like them." Soon he was "by myself behind the barn" feeling "kind of mixed up." His body felt tight and uncomfortable as he returned. "I don't know where I belong," he said.

During this abduction event Joe received an "incredible" glimpse by "going in their eyes, by just connecting with them and leaving any separation," a sense of what it would be like to go completely into the ET or alien world. "I would just go out of myself" and go "anywhere"—to "world, space, planets, distance."

"Your body or your consciousness or both?" I asked.

"Without my body, sometimes in my body. I become wind. I become space. I become matter. I spin, swirl, slow, fall . . . " In the alien form Joe could experience different energies, "dancing drops, orchestras and music, crash, bang, hard places, dark places, vast, vast, vast . . . I feel bliss. I feel love. I feel connected. I feel unsafe. I just am in whatever I want. I dance. It's dance everywhere . . . dance with different beings, different lights, different energies." It is "so different" from "everyday life, being Joe" that it is "just so difficult to integrate" when he comes back to Earth.

Joe did not recall exactly how he was returned to the starting point behind the barn. He recalls walking into the barn, then to his house and upstairs to sleep. But just before I brought him completely out of the trance state Joe remembered being told by Tanoun, "your baby is one of us," the "us" including Joe himself in his alien identity.

Joe and Maria's baby, Mark, was born on November 10, about three weeks after the expected due date. About a week after Mark's birth, Joe wrote in a note to me, "As I write, mother and child are napping together, home finally after five days in the hospital. Mark Joseph was born by C-section [necessitated by breach position and infection] last Tuesday, and watching the surgery brought me full circle with my own experiences aboard UFOs. It was interesting, enlightening, and reassuring, and it gave me greater trust and surrender to this whole process."

A second hypnosis session was scheduled for November 30. We talked first about events connected with the hospitalization and birth, which had been especially stressful for both parents. Mark had seemed to talk to his mother in utero, and she had several dreams in which Mark told her the name that they were to give him. Joe spoke of an abduction experience that occurred two nights after he brought Mark and Maria back from the hospital in which two beings were beside him and sent "a shot, like a burst of energy" into his head using a blunt instrument and left him feeling "foggy" and disoriented. Maria was with him, but he does not recall seeing the baby. This was "the most conscious experience I've had to date."

In the beginning of his second regression, Joe talked of his care not to burden Maria with his ET experiences, despite her receptivity; of his own resistance to accepting their power in his life; and the feelings of vulnerability that they created. "I'm afraid to face a part of myself," he said. With strong emotion, verging on tears, Joe spoke of his distress at the discovery of how "intimately involved" with the alien beings he was, "in cahoots," a kind of "double agent . . . working with them" and thereby betraying his Earth partners. "I'm split," he said, "I lead this secret life, and the secret life is that I've spent a lot of time with them."

But more than this, what troubled Joe was his sense that he was a willing partner with the alien beings in using unwilling humans for a breeding project. Even "last night" he had had an experience in which he saw a large alien infant head with huge, dark, beautiful eyes with whom he felt a special connection. Although he felt that the hybridization project might be worthwhile, "an evolutionary thing" ("every time you hybridize you get more vigor"), he was not sure what species would benefit most. Almost crying in his distress, Joe said he had the sense that he had had a recent sexual experience on a ship "with a woman who didn't want to." We decided to try to get more information about this in the regression in order for him to expand his awareness of his complex dual role and to increase his power to choose freely.

At the beginning of the regression Joe saw a "parade of images" on a spaceship, a variety of "people" who seemed to be derived from a chaotic gene pool. Some of the beings were ugly, even evil looking. It seemed like a kind of interplanetary "United Nations." The total impact of this was to give Joe a "harmonious" feeling, as if he were being shown "they're all there in good company." His own form kept changing, "like a chameleon." He felt "more comfortable in a shape like them . . . somewhat translucent," with a large head and big elliptical eyes, long and thin in the torso, light grayish in color, the hands a little webbed with long arms and fingers—three and a thumb.

"Wow! I feel like I'm inside me. It feels very elegant, very graceful. I feel like I'm no longer walking—just kind of moving, most like swimming. It's opening up to a part of the ship that feels like a heart valve, or something. It just feels very, very spacious."

"Etherical," "fluid," and a sense of "vastness" were other ways Joe described what it felt like to be in the alien form. He felt "incredulous," doubting his experience and wondering how he could hide from himself that "I also exist here, on the ship . . . I'm just so much more comfortable." He felt then an intense struggle between his "humanness" and the humanoid identity, which he had maintained as separate. Yet he also felt that he was "more fully integrated in being human" than most alien beings.

Joe called the race of beings to which he belonged in his humanoid identity the "brotherhood" or the "Obasai" people. Thought processes for these forms are intuitive and "not linear . . . I feel like my thoughts are available to everyone, and there's nothing to hide. There's no shame. There's a sense of oneness, and we can have different ideas and opinions, but yet, there's still a blending harmony . . . This part of the ship," he said, "is for integration . . . it spends a lot of time around Earth." Other projects are not Earth related and involve "other dimensions, other galaxies," but "time and space is not an issue." Travel occurs when "you just think yourself there."

Joe told then of an experience he had had just a few days earlier as "Orion," his humanoid identity. He felt himself to be seven and a half to eight feet tall, although he sensed that he could make his body taller or shorter. A blond woman of about thirty-five he called "Adriana" was brought to him so that he could "make love with her" and "give her my seed." Although Joe felt that "she's been involved a long time," it troubled him, at least in his human self, that a part of her was frightened. Adriana, Joe said, was walking her dog at night when she was abducted and was "in a sleepy state" when the beings floated her into the ship. "Part of her freaked out when she first saw the ship," Joe said. He felt loving and gentle toward Adriana, stroked her head, reassured her that "we care for you" and encouraged her to relax. "I wouldn't copulate with her without on some level her cooperation, her agreement."

Adriana was placed on a slightly tipped platform with her head higher than her feet. She was kept in a sleepy or dreamy state ("mentally they create, like, this web . . . they just drape her in this soft, gentle energy") as she was undressed by small beings. "There is that fearful part of her that just totally doesn't want any of this to be happening," and when this resistance "bubbles up to the surface" the beings controlled her through a kind of energy "massaging."

The sexual or reproductive act itself was quite brief. Three or four of the beings watched as Orion inserted his small "almost hollow" penis (erect, "but not real hard, not like, you know, unbending . . . it just works its way in"), perhaps half the thickness of a human one, into Adriana's vagina. The testicles were "just some bumps" on his body. "They don't hang out or anything." Although there was fondness and love on Joe's part, "it's not horny-passionate . . . It's not a rhythmical in and out intercourse. It feels more like just a rocking embrace . . . very soft and very gentle," quite "natural . . . like a kinship." Joe wondered "if I changed my shape, because when I lay down on her I'm not much bigger than her, and just kind of rock a little side to side . . . It doesn't take long. It's more like intent. It's not like I have to work myself to a place where my body will release it. I can just put it in and release it." A clear fluid "just oozes out." Although Joe or Orion caressed Adriana lovingly, she appeared to be split. Part of her "is totally present" and "the interaction feels beautiful," but her "stormy" fearful part felt violated.

Reproductive acts like this, Joe said, are "necessary" so that "humans aren't lost in their race and their seed and their knowledge," for "human beings are in trouble . . . A storm is brewing," an "electromagnetic" catastrophe resulting from the "negative" technology human beings have created. Adriana's fertilized seed, for example, might be taken "back out of her" and "then we'll grow a baby that's got a lot of human in it" and "raise it" as "one of our own . . . If the humans totally die—we have their children." The purpose of this hybridization program, Joe said, was evolutionary, to perpetuate the human seed and "crossbreed" with other species on the ships and elsewhere in the cosmos. Joe spoke sadly of the inevitable further deterioration of the earth. Many humans will die, but the species will not be eradicated.

Joe felt conflicted about the information he was uncovering. On the one hand as "a father" with "a business" he feared ridicule should he make his knowledge public. On the other hand, he felt a sense of urgency in relation to his fellow humans. But the "defiant, fearful, egocentric" part of him stands in the way of taking full responsibility for what he knows. His "human side" has doubts and at times fears that the dark-eyed beings are "sinister" or "malicious," with "renegades" from other ships that toy with us "to make us a good breeding cow." Yet in his Orion self he does not sense anything of this sort.

After emerging from the trance state, Joe felt shocked at what we had uncovered and anticipated that he would need a great deal of support in order to come to terms with the complex and disturbing dimensions of his identity. He felt "a little incredulous" to discover that he was living "a double existence," but the emotional power of the session, together with the objective clarity with which Joe could experience being Orion, convinced him of the authenticity of what he had just been through. As a man brought up in an Irish Catholic family in which emotion was "squashed," Joe was astounded to see how his abduction experiences had become a "conduit," opening him to a wide range of strong feelings. He was particularly concerned for the distress that was caused for Adriana and other human beings by acts such as Orion's. In the end he came back to the affirmation the session had given to his lifelong feeling that he "should have been born at a different time or on a different planet."

The session was both validating and confusing for Joe. Over the next few weeks he struggled with the task of reconciling his ET and human identities, his stress as a new father, and, perhaps most important of all, his sense that Mark too was connected with the alien world. A third hypnosis session was scheduled for January 4, 1993, to explore these areas.

At the beginning of the session Joe spoke of his own feelings of neediness that Mark's presence was stirring and the lack of nurturing he had experienced as a child. His relationship with Maria seemed to be "on hold, emotionally, sexually," as she was so occupied with Mark. Furthermore, Joe had not "felt them," the ETs, much recently, missing the "support and love from them," which was adding to his feelings of sadness and loneliness. At the same time as "my reality has been rocked" by the affirmation of the ET encounters, "my heart's just blown away by this beautiful little being . . . How do I balance my needs, the baby's needs, Maria's needs" and the "deeper levels" being stirred by the ET experiences, Joe asked. Sometimes he feels he is "being shuffled" like "a deck of cards." Joe wanted to explore under hypnosis "my connection with them," but at the same time he did not want to "abandon my kid" as he now felt "abandoned by them."

Before the regression began, Joe told of a recent dream in which a lifeless baby was taken out of a plane crash. He picked up the baby and cleaned it off with muddy water and saw that there was something odd about its back. Then he gave it to somebody else to take care of. He associated this with his sense that Mark is "from there . . . incarnated" as "part ET" and he feared that they might come and "take him out of my arms . . . I cannot handle you guys coming and taking him," he said. He felt vulnerable and unsafe himself and worried further that he could not "protect" Mark, and wondered if he should.

Joe's first image under hypnosis was of an ET showing him "a tray that they put babies in to weigh them." He also saw babies in high seats who looked human except for big eyes and bony eye sockets. "The ETs are gentle with the babies," and three grays, one of whom was "the same one that works with me a lot," were feeding them a "green, clear liquid" by putting the end of a cylindrical silver and glass tube in their mouths and letting them suck it in. One of the babies was Mark, who was fat as he was in "real life" at the time. Mark was looking up into the ETs' eyes and seemed relaxed. The beings sponged the babies with a green liquid, as if to put energy into their bodies. The liquid seemed to be the same substance they gave the babies to drink. The ETs seemed to have "a primary relationship" with Mark and the other babies, "and they're not going to let me interfere with their relationship with him."

Joe had the feeling that he once went through a similar experience, and felt sad for Mark because he knew "there's a painful part" that lies ahead for him. He described that "we [Joe, the ETs, and even Maria in part] worked out this relationship with him before he was born to us [Joe and Maria]." Mark himself was once a gray ET, but his consciousness or soul "went from being an ET to being born as our baby . . . this wasn't any light thing for him." There are risks for Mark's soul, Joe noted, in "being human, going into this body . . . He's making a big commitment." I asked him to explain. "It's kind of like putting on a wet suit and scuba gear, and it's putting on a denser existence and you can get trapped in it. It can get stuck on you . . . you begin to believe what your body tells you and you forget how to energetically disconnect from it . . . that you're vaster than it." Just maintaining a physical existence can be "all-engrossing." Fearfulness and the preoccupation with the care and survival of his physical body could make Mark forget that "he's more than his body" and that "it isn't a life or death matter for him if his body gets hurt or the body dies, or he isn't socially accepted." If Mark's energy were to become totally focused on his body "then the fear will be overwhelming and he'll just get stuck."

I took Joe back to what he had witnessed as the beings worked with Mark. They were inside a ship in the dark, "probably so that Mark isn't distracted by other things." Joe was wearing just a T-shirt and Mark was in his diaper. The ETs were doing a kind of "remodeling" of Mark so that "more of him," more of his energy, could manifest. "He connects with them . . . On one level he knows what's going on . . . his soul [is] totally mature as us, but he also has forgotten . . . and it will be a process of bringing that level of awareness back up to speed." Joe saw the beings holding "crystals against his head," moving his hands, and "shining a light in his eyes and on his hand . . . It was like a light came out of his eye to look at his hands. They're helping him make that connection . . . They've got him on his back and they're stretching him, moving his arms and legs." These interventions would enable Mark to be more connected and less fearful and to experience "more soul, more energy, more heart." Because our physicality is so "dense," "a lot of awareness" is required to expand our knowledge beyond the technical level to wisdom. To liberate latent powers, we have to "levitate our body" and live without eating.

Joe felt the weight of his responsibility as a kind of "donor," giving Mark over to this evolutionary process while at the same time making sure "he doesn't forget . . . He's counting on me to help him remember." In his role on Earth, Joe felt like a man he read about who went "undercover," admitting himself to an insane asylum to discover the abuses there but became stuck inside after the people who knew him outside the walls died off. Joe likened his lonely struggle to sell himself and manipulate his way materially, indeed "human existence" itself, to living in an insane asylum. "Let's pretend, pretend everything's great. Let's pretend we're not all so fuckin' tight, so tense that we can't even walk straight. Let's pretend. You know, you scratch my back and I'll scratch yours." How, he wondered, was he to raise Mark in the insane asylum so that his spirit could be preserved.

At this point Joe's attention shifted to his own pain and loneliness and his relationship with the beings. He felt intense isolation and aloneness, as if he had contracted "into this hard shell" like an "egg that's just hard and dark." He remembered the ETs hands on him and felt he was "pupating or something" as blocked energies were released. He was seven or eight years old and in a vast space, as if underground. He experienced himself as if split between his ET and human selves. The ET part has got "his hands on my kidneys, my lower back," and his human self is "trying to relax and open and to connect to him. Oh, God! It's almost sexual." Joe was experiencing intense emotions at this point, expressed in sounds like "Ohhhhh" and "Ahhhhhh." These feelings, a kind of combination of excitement and pleasurable release of tension, grew stronger as Joe spoke of energy moving through his body. "The ET part of me is the most grounded and the least changing, and he's kind of like orchestrating. He has the most information of all of us. He's facilitating. Oh, but he's getting healed too." Intense energy moved up his spine and throughout his body. At first he felt "fragmented," but the "light. I feel it like light" brought his parts together. Joe seemed to continue to absorb energy in "slow pulses," which gave him deep satisfaction.

As he experienced his ET and human parts becoming integrated, Joe felt less alone. He could also connect with Mark. "It's like I'm more on his wavelength." I asked him what was the source of the energy he seemed to be taking in. "It's me, you know, it's our soul or essence" or "my ET self." It had always been there, but "I just wasn't energized to it. It was locked [from] me." I asked Joe what had happened during the seven- or eight-year-old period, but he could remember only that it was "a hard time." As he spoke of letting go of the blocks, he felt new waves of energy course through his body. More expressive sounds came forth as he felt "all these shivers. Chills . . . rolling, rolling, rolling through me. I feel like I'm expanding like a blown-up balloon." These pleasure feelings seemed to begin in the kidney region and radiate through his whole body.

After this, Joe's heart opened to Maria, and then he felt himself walking through and closing a patio glass door with the sight and sound of beautiful chimes. I asked about the relationship he had spoken of between his soul and his ET self. The ET self, he said, "was like another manifestation of my soul." Then he added, "I feel like I just integrated all parts of me towards oneness." He was struck with a powerful image of looking down and seeing his own body as if in "a hall of mirrors" and saw himself "on many different levels." The experience was intensely "beautiful" as Joe experienced "walking through these different membranes of myself." The levels were "getting together" in a "harmonious" order. "It felt real integrated." Joe felt that now he could be with Mark in the human insane asylum. "From a more grounded place" he could "walk him through the hall of mirrors."

At this point in the session Joe no longer felt himself to be in a ship. Rather, he was in "just space" or in "many dimensions"—there were no right words. A "gray" being seemed to smile at him and asked, "So how does it feel?" Joe described, "I'm not fragmented. It feels great. It's like oneness." He thought that although the ETs had participated in his process of integration that "it gets bigger than them, beyond them." Something more reciprocal seemed to be occurring, bringing them also closer to "oneness, closer to creation." The ET-human connection "enables them to become more than just ET and human . . . Working with us feels like it helps them go even higher." The ET that asked, "So how does it feel?" seemed genuinely curious on his own behalf.

After the regression, Joe and I talked of the implications of the session for his parenting of Mark. He felt he could more fully "be here for him," help him "stay strong and help him stay connected" with his "higher self." Joe felt that for himself his abduction experiences, especially as revealed in this session, were "like a rite of passage," a "step of growth" toward becoming "more human." He felt that as a human being he had been part of "an experiment that went sour," a kind of aberration of God's creation.

We returned to the image of this culture as an insane asylum. Joe pictured himself rocking Mark to sleep at two o'clock in the morning, "feeling the ETs" and "real comfortable with them coming and taking him." For now, he thought, they were helping Mark become connected with "his own soul." Joe called the whole alien-human project a "retooling . . . creating a different reality" where "there's the option of humanness." A necessary step in his transformation, he said, was "my humanness going into the pain . . . I'm more integrated. There's no doubt about that," he concluded. "I feel like I'll be a much better parent."

This session had powerful reverberations for Joe in the weeks that followed. He continued to feel the "fragmented parts" of himself coming together and an increase in "my soul's love and energy." Four days after the session he had a kind of energy crisis or "kriya," evoked during a massage. He sweated and shivered and felt intense pain moving from place to place in his body, starting in the kidney region and flooding his spine and head. "I was moaning and rolling around, overwhelmed by the experiential/emotional/physical pain." His "ET guides" were holding his hands and head, and he was flooded with scenes from his past, "a full circular tray of sixty to seventy slides containing sixty to seventy separate experiences . . . It was like the ETs held my eyes open and manipulated time so that I experienced each tray in one to two seconds . . . I felt they controlled the multidimensional shifts." Maria came into the room several times, but he did not want to speak to her out of fear of interrupting the process. He found himself simultaneously confronting his parents angrily as a child and feeling understanding, compassion, and acceptance toward them.

Joe spent the next two days recovering from this experience. On the third day he had a vision of a "giant vaginal hairball. It was gross, slimy, and dirty. I couldn't distinguish much, just two legs and a hairy crotch. At first I was revulsed, but stayed with it. It clarified into the hair of a goddess being born. She had long black and gray hair, now clean and brushed, flowing from the vaginal lips. I could 'see' inside and saw the beautiful, wise, young/ageless face of my goddess, my feminine self. I felt a flood of love, comfort, and warmth in connecting with her, and knew her birth was my integration of male-female. It was a beautiful vision.

"The outfalling of all this is I have felt much more centered and grounded. It is much easier and cleaner to discern what is best for me—how I can best love and honor myself, the one that is wild, outrageous, impolite, and divine; the one that is emerging more and more. I feel my ET guides have been playing a big role in this and they want me 'whole and healed,' i.e., up and running, ready to go, when the shit hits the fan geographically.

"One of the last things I did Monday morning was to call a business associate of mine and present him with the idea of him taking over a lot of work of my business. It just doesn't feel important any more—it's not what I'm truly being prepared to do."

During the weeks that followed Joe continued to have a number of ET experiences and to recall earlier contacts. He felt he was struggling to develop a more comfortable relationship with the beings, whom he regarded as his spiritual teachers, peers, and helpers. "When I feel vulnerable they are totally present with me. I feel their genuine compassion and understanding." He likened them to "sensitive psychotherapists" that "impel us to grow" but "don't handle us with kid gloves." They even seemed to "arrange contact" with him when he was in special emotional pain. Joe also began meeting with other abductees in our group for more connection and support.

In mid-February I arranged for Joe to speak about his abduction experiences to a psychiatry seminar group at The Cambridge Hospital. In this public "coming out" he was able to take this uninitiated group of largely skeptical psychiatrists and other mental health professionals through his story in a disarming and convincing fashion that left the group curious and more open to the expansion of their reality. At the end Joe spoke of the "incredible amount of terror" that he still faced, in particular around lack of control, but reiterated his belief that the purpose of the alien-human relationship is ultimately "a benevolent situation."

Joe requested a fourth hypnosis session to recover memories of his ET contacts during his own infancy so that he could further experience "myself coming together" and deepen his understanding of Mark and strengthen his role as a parent. We met on March 1.

Before the regression Joe noted that he was nervous, remarking that "every time I go down into a session I come back up and it's like the world is different." It has been difficult though "thrilling," he said, to see the world "as this cognizant, intellectually understandable place . . . I would not change my seat on the bus for anything," he said. "But it's also scary."

Joe told of a complex recent dream in which Mark turned into a very thin, white "ET baby right before my eyes!" In the dream he was given Mark to hold, but surrendered him with guilt to three women in an underground chamber. He said he wanted to go back in the session to "when I was an infant in this life, when I was a newborn and was still connected to the part of me that was an ET, more so than I am now."

I questioned whether Joe should begin the session with an expectation of where he wanted the regression to take him. Nevertheless, his first image was of being a two-day-old infant alone in a hospital bed and feeling vulnerable and unsafe. Sobbing and moaning he voiced "empty" feelings in his abdomen. "Oh, God!!! I've never felt so alone!!! Ohhhh. It feels so foreign. It feels so cold . . . just like isolation. Just, ohhh. Everything is so far away. Everything. It's like harsh! It's bright. It's loud. I don't feel nurtured at all." A nurse was there, who "helps, but it's like she doesn't really see me. She changes me; she dries me; she feeds me," but she did not connect with him. A familiar ET is there. He has black eyes with "a blue light to them." The nurse seemed not to notice the alien, but the baby felt trust and "his love for me . . . He [the alien] feels like a midwife . . . reassuring me, touching me, bringing me back, telling me it's okay." The eyes of the alien changed, "like clouds moving across the sky." Joe saw concern, sorrow, and compassion move across his face. The nurse left and Joe saw a female alien beside him. "They feel like parents," he said, "They're just really nurturing me. Really, really giving me love, really helping me feel okay."

The aliens assured Joe that they had been with him in the first two days, but "it was I who had left. It was I who didn't see them . . . I got so scared during birth that I shut everything out. I shut them out. I feel like they're holding me, and as I was born, this, it's like a river, being in this current. And this river just swept me away, I got scared." I had the sense that Joe was talking about the birth process itself and asked him to talk about being born. Then he began to writhe, breathe loudly, slowly at first, coughing and choking, arching his back and hunching his shoulders and grimacing. "I'm scared," he said. I asked where he was. "I'm moving."

"Where?"

"In the birth canal. It's tight! I'm just scared! I don't want to go!" Emitting more powerful groans, shuddering and choking sounds, and gasps with each intake of breath he said, "I'm coming out! Oh, God!" I asked him then to try to give words to his fear. "It's like I want to do this [leave the womb and be born]," he said, "and I know it means being alone, and I want to be alone, almost to get away from my mother. And I'm scared. And, God! I'm afraid I'm going to be lost. Ahhhh!" Reassuring him that he was safe now, I asked if he recalled the delivery, for example whether there was a doctor or a midwife. "It was a doctor," he said. "I got so scared, I like, shut out! I went internal. I, I pulled away from everything. God! I went deep inside myself. It's just a scary place to be. Oh, God! I can't believe [sobbing now] I came back! Ohhhhhh!"

I asked if he "came back" to somewhere he had been before.

"Yesss!" he said. I asked about this. "Oh, God! It's such a horrible place!" I asked him to tell me about the horrible place. With intense feeling and conviction Joe told of being a poet named Paul Desmonte in a village near London at the time of the industrial revolution. Desmonte was arrested, tortured, and he died in prison after blaspheming against the political and religious establishment. I took Joe through the details of his arrest, prison experience, and confrontations with the authorities. In prison he was starved, kicked, and beaten with sticks and belts, which left him with broken fingers and ribs. Eventually they "tired of playing with me—till I no longer responded, no longer gave them the satisfaction of crying out." I asked how he had died. "Some would say of starvation. I'd say of hopelessness." After six to eight months in prison he stopped eating the little food they gave him, and the experience became "a kind of a healing." I asked him to explain. He said, "I had to face the truth of my own writing. Yes, I believed what I believed. I believed that man was greater, but I never went further than that. I always took my stand there and fought and fought and fought. But when I was alone I had to explore what that greater was."

"What did you discover?" I asked.

"My own fears. My own judgments. My own biases, and I began to experience them." Then "these ETs" came back. Joe attributed their return to his struggle to open to a greater sense of himself, to his letting go "of my bitterness" and of his head-to-head battle with the authorities. He discovered he was not "just a mere mortal stuck in this physical prison of a body and a physical prison of a cell, that I can travel and soar beyond these walls." As he "softened," he became aware of "them" [the beings] and no longer felt alone.

At this point Joe became quite overcome with feelings of awe and wonder. "Oh, God! You mean I'm not alone?" he asked, as if to the universe itself. Seeming to speak as Paul Desmonte he said that "in every fiber of my being" he had "feared the vastness. I have feared the nakedness. God, the vastness of it all. I cannot hide! I cannot hide from myself. I cannot hide from another. It is my shame. It is my sense of unworthiness that I hide, that I want to hide from, that I don't want others to see."

I asked Joe to tell what happened from when the ETs came back to the moment of Paul's death. He said he was afraid he would lose them again and become lost himself in "this transition" to death. I encouraged him to be in the moment of transition. "I'm afraid to feel it. Oh, God!" he said. I assured him we were there with him and he would be okay. Joe surrendered then to a state that is hard to describe. It was not unlike the birth process he had gone through earlier. He moaned and coughed, cried out to God and wanted to be held. He felt as if "I'm being squeezed out of my body . . . I'm contracting. Ohhh. I'm not totally present. I'm kind of scared and I'm just starting to space out. I don't want to space out." Moaning and coughing more intensely he cried, "Ahhhhhhhh! Oh! Ohhhh God. They're pulling me." I encouraged him to let himself go. Soon he seemed to lighten. A lot of the alien beings were around him, tickling, touching, and rubbing him. Laughing he said, "It's good to be here." It seemed "delightful." Joe still had some sort of body, as if on Earth, but simpler, lighter, "thinner."

"It's good to be back," he kept repeating. "This is much more real."

I asked Joe to explain to me, an "Earth stuck" person, what this other realm was like. "There is a golden thread that connects all life together," he said, "and that you, as all life, are connected to it. And it nourishes you, both as much as you would let it and yet no matter how much you negate it, enough to at least sustain your existence. It is a world of choice and this world, your world, is beginning to make choices that honor that connection, not that you have been lost from this connection, but you have journeyed far from it. Never lost, not without reason to explore, to explore what it is like to live without this connection." I asked if he knew why we had journeyed so far. "To explore its outer limits," he replied, to see "how far" from our source we could go. "Many are tired of" that journey and "are now working, flowing, and struggling back to their source."

I asked Joe why he made the "choice" to return, via his particular Earth mother. He said he returned to "the scariest place to go to face" fears of unworthiness, where he could no longer hide. The ETs, from whose love and nurturance he had turned away, had promised they would be with him. With the help of his spiritual guides he chose his mother precisely because "it would be tough" and "dense." Her fear, tightness, tension, hiding, and pretension were a "reflection of my own," he said. I suggested to Joe—one of the rare times I have ever made an interpretation to an abductee—that the sense of constriction in Julie's womb, and the terror of aloneness when he could not bond with anyone in the hours after his birth, had led to a disconnection from his source and had plagued his life.

"We got to know each other, me and the fetus," he said, "and it got tighter and denser and denser and denser . . . I wanted to be birthed," he continued, "to get out of this womb, to get distance from this woman," but in the fear he just "tightened and tightened and tightened, and I pulled away from everything. I shut out everything."

As the regression was nearing the end, Joe spoke further of the confusion, isolation, and despair he had felt in "this horrific world." His deepest fear is of cutting himself off once again and becoming "lost from the source" and "lost from them." Spreading his arms and breathing deeply he said, "Would you choose oneness, or would you choose insanity? The choice is definitely oneness." Before coming back from the trance state Joe spoke further of his feelings of nakedness and vulnerability and the difficulty of integrating his spiritual being while residing in the density of a "physical body." As he "returned" to the room, he felt a rush of energy and "lightness." He embraced his vulnerability. "It's beautiful," he said, "It's like what I see in Mark. You know? He's incredible! There he is, you know. He's got nothing to hide. That's how I feel right now." He also felt as if he were "waking up in an alcoholic daze" and realizing he had been "living with a batterer." But he now felt "strong enough and grounded enough" that he would "not go back" [i.e., separate himself from his source].

In the discussion after the regression, Joe spoke of the aliens again as "midwives" helping him stay connected with his divinity. He had the image of a being in a rushing river and the current has become too strong. The aliens are as if on a rock on the shore, "and I'm holding on to them." They want him to "stay connected through this" instead of becoming lost in his fear. Until recently "part of me stayed shut" from them. Now he has become "aware of them" and "connected with them." He finds his familiar being "beautiful" and sees "emotions move across his face" like "clouds moving at high speed." He reflected again on Paul Desmonte's self-righteousness, and how, when he was beaten and had no resources left, he faced the truth of his antagonism and softened. It was at this point that the alien beings, familiar to him even then, returned and "I could see them." Though he was martyred before his potential could be realized, Paul Desmonte did succeed in getting "a village talking a little bit."

Discussion

Joe's case contains many of the familiar features of the abduction phenomenon, but also takes us to the edge of our knowledge and understanding. One of our fundamental ontological distinctions or categories—the separation of consciousness from the physical world—is challenged by his experiences. We wish to know whether UFOs and their occupants are or are not from, or of, our physical world. To Joe his experiences, like those of all abductees, have the quality of coming from outside, of occurring in the external world. Yet some of them clearly challenge this notion. For example, the panicky sighting of a UFO and its occupants while experimenting with LSD as a teenager seems purely a product of consciousness.

* * *

At the same time, Joe's abduction-related experiences are as real—more real, he said on one occasion—than those that occur purely on the physical plane of reality, and there is no indication that he is psychiatrically disturbed or that these experiences are the product of some sort of psychopathology. As in virtually all abduction cases, this leaves us with the choice of searching—vainly, I think—for ways of explaining the phenomena within our existing world view, or, instead, of collapsing our rigid separation of psyche and reality, of inner and outer, and opening ourselves to expanded ontological possibilities.

The exploration of Joe's case has occurred in the context of his wife's pregnancy, the birth of his son, Mark, and his emerging role as a father. Themes of birth, death, and rebirth are constant in his material. His own feelings of vulnerability were stirred by the helplessness and needs of his infant son. But, in addition, the advent of the baby into Joe's life opened his consciousness to the recollection of his lifelong relationship with the ETs, who were agents of love and nurturing as well as trauma. Mark, like Joe himself, has a dual human/alien existence but is closer to his alien connection or source than Joe. Through Mark, Joe discovered his own human/alien double identity.

At the core of Mark's ET identity is the separation of his body and self or soul. During the third regression Joe was witness to extensive feeding, massaging, and other procedures ministered by the aliens to Mark, the purpose being to maintain Mark's connection with his divine source and to prevent him from confusing or limiting his notion of himself to his body or human ego. It seemed to Joe that the ETs were agents of Mark's ensoulment, virtually remodeling him as an integrated human/alien being. Joe's responsibility as a parent is to keep Mark connected with his higher self. The danger to Mark in this world, which Joe likened to an insane asylum, would be to succumb to the restriction of consciousness that derives from competition, intrinsic financial pressures, and especially the pretenses of civility that are the hallmark of the business world. The alien beings appear to serve for both Mark and Joe as what he calls "midwives," delivering them from the madhouse of our culture to another state of consciousness more compatible with the viability of the planet's life.

The aliens are also agents of Joe's own integration and reensoulment. In the second regression he discovered he possessed both a human and alien identity, which many abductees are discovering about themselves, and that he is a kind of double agent, functioning as a bridge between the earth and the realms from which the beings derive. During this regression Joe also experienced the ships and the alien realm as the home where he was most comfortable and he felt the temptation never to return. But his human task has been to integrate the alien/human dimensions or parts of himself and become a being that is connected beyond his material or earthbound self.

In the third regression Joe had the intense experience of feeling his alien and human parts come together, a profound, ecstatic expansion, a kind of rite of passage, that contained both terrifying and joyous sequelae which extended and deepened the process in the weeks that followed. Joe's experiences, especially those related to Mark's birth, demonstrate dramatically the separation or discontinuity of his being or soul from his body. The lightness of the soul's experience in the "spirit" or "other" realm—our language fails us here—contrasts with the density of the physical body as experienced in the earth domain.

The aliens are experienced by Joe and many other abductees as much more closely connected with the divine source, or *anima mundi*, than are human beings, who are struggling to overcome their extreme separation. Therefore, the coming together of these dimensions of his self brings about, virtually by definition, Joe's own deeper connection with the divinity, a sense of oneness with all beings—essentially his reensoulment. Curiously, the alien beings seem, conversely, to long for a deeper connection with humans, as if the greater density of our embodiment or physicality contains some sort of appeal to them. For Joe, like many abductees, the deep connection that occurs through the large dark eyes of the aliens is a central part of the process of alien-human connection and evolution.

Joe, like virtually all male abductees, has had traumatic abduction experiences involving forced manipulation of his genitals and taking of sperm samples as part of a human/alien hybrid "project." But Joe has also experienced himself as a participant in his alien identity as the agent of this genetic or reproductive experiment, giving us a rare look into the alien side (although mixed with his human perception) of the hybridization process. He felt some guilt—perhaps there are human feelings mixed in here—for copulating with a human woman, during which he deposited his "seed" in her. He felt love for this woman during the act, but it is understandable that a part of her would feel a violation.

Through his own experience Joe seemed to have access to information about the nature of alien genitalia and the process whereby the beings deposit a seed or some sort of reproductive substance in a human body. Joe, like many abductees, was given information by the aliens that this hybridization process was for the purpose of creating a new species that represented a reinvigoration of life, a step in evolution. "Vigor" seems a strange word in this context when one thinks of the listless hybrid children that have been seen by so many abductees aboard the ships. The current direction of human activity on Earth, Joe knows, is leading to the extinction of our own and countless other species. The hybrid process was a way, he has learned, of preserving the human genetic substance, though in some other form. What we also cannot know, of course, is in what reality all of this actually occurs.

Finally, in the fourth regression, Joe opened to a profound past life experience. This material came forth as a result of a choice that I made to pick up on the phrase "I came back" when Joe was experiencing himself as a two-day-old infant. This required at least a degree of openness on my part to the possibility of past life experience and a "return" to Earth from another domain. Otherwise, I could have ignored the phrase and asked him, for example, to speak more of his experience as a two-day-old and subsequent events. The past life experience seemed not to be arbitrary. Rather, it reflected Joe (as Paul Desmonte) expressing his values and truth, but from an attitude of arrogance, of limited and polarized consciousness, evoking antagonism, and resulting in martyrdom. He embraces the same values now, the conviction of greater human possibilities, but his consciousness has evolved to the degree that he can communicate his truth in a fashion that also embraces those whom he would wish to persuade. The past life experience seemed to be important not for the fact of another discrete existence, but rather as reflecting a stage in the evolution of Joe's consciousness over a span of time greater than a single human biographical existence.

This session was also remarkable for the similarity of the intense reliving of Joe's experience of birth in this life and the death of Paul Desmonte. In both instances there was intense emotion, fear, and ultimately release, as a transition occurred from one state to another. The sense I had was of life as a cycle of birth and death, of transitions from one state to another, evolving over time, and, from a larger perspective, hardly distinguishable one from another. The alien beings—"ETs" as Joe calls them—appear to have been with, or at least available, to him as protectors and guides of his spiritual evolution over time, showing up when his consciousness would open and expand, as before Paul's death, and lost to him when his psyche would contract, as after his birth from Julie's womb.

This observation may prove to be important for increasing our knowledge of the psychological conditions under which human beings are or are not able to experience the alien presence in their lives. If, in fact, the alien beings are closer to the divine source or *anima mundi* than human beings generally seem to be, then it is possible that their presence among us, however cruel and traumatic in some instances, may be part of a larger process that is bringing us back to God, or whatever we choose to call the creative principle after, as Joe phrased it, "a journey that has been taken very far," a "journey that many are tired of and are now working, flowing, and struggling back to their source" from.

Joe's own journey has resulted in remarkable changes in his life. He has been able to turn over many of the daily tasks of his business to an assistant, which leaves him free to pursue his spiritual and therapeutic calling. He has been willing to "come out," to go public as an emerging leader in the teaching of consciousness evolution. He has been willing to acknowledge his ET experiences and identity and openly share his knowledge with others. Joe and I have presented together on several occasions. I am impressed with the matter-of-fact and non-threatening way he can take an audience through his own doubt and emerging awareness that he has, indeed, opened to intelligences and experiences that are profoundly changing him and perhaps millions of other Americans.

CHAPTER NINE

SARA: SPECIES MERGER AND HUMAN EVOLUTION

Sara was a twenty-eight-year-old graduate student when she wrote to me requesting a hypnosis session. She was planning to travel soon and wrote that she wanted to be hypnotized before she left "in order to release some emotions and information that feel close to the surface and to lessen some feelings of anxiety and confusion that have been increasing in intensity." Many details of Sara's file have been omitted in this narrative in order to protect her anonymity.

In the letter she said that a couple of years previously, in the course of massage treatment for pain at the base of her skull, "I had the experience of small beings communicating with me telepathically." She also found that she was spontaneously making drawings with a pen in each hand ("I never used my left hand before") of what she took to be alien beings, focusing especially on their eyes. Her drawings also included passageways and "some sort of subtle body field" like an "entity's subtle body."

Sara is one of an increasing group of abductees who bring a degree of spiritual interest to the understanding of their experiences. Her search for meaning, and the struggle to stretch the boundaries of her own consciousness, enabled her to achieve powerful insights in a short time. In her letter she also wrote that recently she had begun "receiving information linking other entities to issues of planetary preservation and ecological transitions, especially polar and geomagnetic reversals." The desire to serve, "to do something constructive for the world," is vitally important for Sara, although she does not yet know the form that this will take.

Sara grew up outside an industrial city. She calls her Protestant upbringing "conventional" and describes herself as committed to experiencing reality as clearly as possible. Sara has never taken drugs and does not drink alcohol. She links this to her encounter experiences and she believes that since she has stopped consuming caffeine, chocolate, and almost all sugar, her experiences have become much more conscious and clear.

Sara's father has died. Although he was intelligent, Sara wonders whether he was dyslexic, and she suspects that that interfered with his ability to do the paperwork necessary to be more successful professionally. A frustrated man, he was physically and verbally abusive to Sara's mother and verbally abusive to Sara. She witnessed frequent arguments between her parents, and on occasions, she saw her father physically abuse her mother. Frightened by her father's temper, Sara would go into another room to avoid being hit. Sara recalls that her father was kind to her when she was small, but when she began to excel in school, he became quite distant. In contrast, Sara's mother is quite successful professionally.

Sara was especially close to her maternal grandfather, who died when she was in her teens. He was "very benevolent," and "we used to sit just for hours, sit there, and I would read [to] him . . . He was my source of support, a really good role model." For about ten years after he died, Sara would often have the feeling that her grandfather was in the room with her, especially when she was at her desk working. She recalls a "funny" room in her grandfather's house. As a child, she would frequently go into this room, shut the door, and sit there for a long time. In a "not quite awake" state, Sara would experience a kind of "hazy energy" in the room, but she recalls nothing else about it.

Sara was an intellectually precocious child, and she was reading on her own at a very early age. She was especially drawn to mysteries and books about ghosts and poltergeists. The family went to church almost every Sunday. "I didn't like the idea of original sin. It didn't make any sense to me . . . I liked the Holy Spirit a lot." She described the Spirit as "like the connective tissue that binds all of reality together." By age eleven or twelve, Sara was considering theological questions such as a resolution to the dichotomy of good and evil, and she was drawn to reading about other religions.

While Sara was an undergraduate, she participated in studies of extrasensory perception. Her interest in integrating the discoveries of physical science with explorations of spirituality and human consciousness have continued. On one occasion, she experienced electrical sensations in her body. On another occasion, "I felt like I got out of my body and I couldn't get back in, and I was gone for about two days." She was quite frightened by this experience.

After graduating from college, Sara married Thomas. She became increasingly unfulfilled by the conventionality of their life together. He would "blow holes in everything I said I felt," Sara said. She and Thomas remained married for several years due to a strong love between them. In addition, Sara desired "some sort of ordered, comfortable" existence.

About a year after she was married, Sara became very ill. Although there is no outward evidence to support this, Sara connects this illness and later intense pain in her neck and head to the otherworldly presence in her life ("They knocked me down," she said). While out walking with Thomas one afternoon, her legs suddenly gave way and she collapsed. She developed a fever almost immediately. Her condition was quite serious, and she was forced to go on disability from work. Her recovery was a long one, and during this period she and Thomas grew further apart and eventually divorced. The couple had no children, and to her knowledge, Sara has not been pregnant. Regarding her illness, Sara claims "It was for my own good," an intervention that seems to have moved her onto her present spiritual path.

About five months before she wrote to me, Sara met a young man named Miguel. When Sara and Miguel sat down to a meal at their second meeting, he immediately brought up the subject of UFOs and told Sara that he had seen a spaceship (this kind of synchronicity or serendipity is commonplace among abductees). Sara refers to Miguel as her "extraterrestrial friend." Miguel reported seeing alien beings in his dreams, and Sara felt that he may even be a "representative" of an alien species. He sometimes acted so listless that his behavior reminded Sara of the hybrid children abductees see on the ships. He was in an incubator as an infant and often showed "a huge neediness" according to Sara. At the same time, Sara valued the opportunity to discuss her encounter experiences with him.

Sara's abduction history is mixed with memories of various sorts of paranormal experiences. She has a very early memory—"six weeks old or less"—of "being picked up and moved and looked at." She believes that "someone was taking a picture . . . It was like the first moment of self-consciousness," she said. "I can shut my eyes and I recall it." Experiences related to ghosts "were a permanent fixture of my whole childhood," beginning at least as early as age four, Sara recalled. "I became a premier ghost story teller." Sometimes she would build her stories around embellishments of portraits and tell "past life stories" based on imaginative recreations of their lives. She would concentrate on the eyes in the portraits and become "mesmerized." The portrait would take on a "living vibrancy" and fill out into a "three-dimensional contour."

In addition to the ghost story sessions, Sara used to play what she called "seance games" with her childhood friends. Once at a slumber party, she asked her best friend, Annie, who was also the smallest, to lie down on the floor and said, "'We're going to try to levitate you.' I don't know where I knew about levitation either, and we went all in a circle. I think I was at her head, and I started saying something, and then it was like, now okay, and the girl went up, you know." Each of the children who were present had "a sense that something weird had happened," and afterwards no one spoke of the incident. "I remember that night very vividly," Sara recalled. "Oh, God! That night the whole room was very strange . . . There was a lot of electricity in that room. I think after that it was not even conscious for the kids." I asked her if they told anyone about it. "I don't think they even thought of the idea of telling." It seemed to Sara as if there were "a suggestion they don't tell." A couple of years ago, Sara asked the girl she had floated, "Did we lift you?" and the girl said yes and that everyone present was frightened by the experience.

Later, during the regression, Sara connected this knowledge and capacity to the floating experiences into, inside, and out of spaceships. "I feel like I'm levitating around the ship," she said, "like someone's giving me a demonstration on levitation. Like showing me, 'Oh, you can levitate!' And so they're letting me levitate, they're letting me play, basically. They're basically letting me levitate all the way around the ship and up and down."

Although the ghost story telling stopped when Sara was about nine, she continued to feel a presence in the house at times. "When I was thirteen I used to feel stuff in the house all the time," she recalled, "like things coming up the stairs . . . I didn't really look too hard. I'd duck under the covers pretty quickly. But I used to say, really loudly, like in my head—I'd never say it out loud. I'd say, 'I'm not ready yet! Excuse me, but I'm thirteen and just wait.' That happened a lot. A lot, a lot, a lot."

During our first meeting, Sara discussed the intense pain in her head and neck that she had mentioned in her initial letter to me. Expanding upon her letter, she said that during physical therapy a couple of years previously, she "started seeing a lot of figures in my head, and sometimes they would seem to be talking to me." She would shut her eyes and "see these little guys up here in this corner of my head, and they were kind of light, really yellow and light, kind of rounded . . . After I started seeing these guys the pain disappeared." The figures "looked yellow and round and sort of benevolent . . . The most overarching feeling I get is calmness. They're so calm." They had "very light" bodies with big heads. She recalls no prominent facial features of the beings, not even the eyes. Nevertheless, she felt (and feels) a lot of love from and to them. "It feels like home," she said, "like the ideal feeling of, uhm, like a warm family." After initially connecting with these beings, which she calls "light beings," Sara began to put her hand on the spot in the back of her head when the pain became uncomfortably intense and she'd "tune in" to the "light beings." She calls this "listening," and she found it to be helpful in reducing the pain.

Sara also mentioned two experiences that occurred about six months before I saw her. During one of these, "something" appeared to be looking at her from the bedroom door as she lay in bed, a presence which was confirmed by the man she was seeing at the time. "All I can describe is like an outline. It was skinny. It was skinny. That's all I can remember." During a separate incident, she experienced something in her bedroom next to the bed. This presence was also confirmed by the same man. Although it was emotionally difficult at the time, she sat up and tried to reach out with love and compassion to the entity. After that, the presence seemed to dissipate.

About a week before she was to come East to meet with me, Sara was in an automobile accident, the effects of which repeated the intense pain in her head and neck that had begun five years earlier. Because of this car accident, she was forced to delay her trip several days. Miguel was driving the car and became dizzy. He started to "space out" with distortion of vision, and they both felt as if some "magnetic" force were pulling the vehicle. The car went off the road, over an embankment, and "folded in on itself." Sara suffered cervical strain and wrenching of tendons and ligaments, and she was taken to the hospital in an ambulance.

When Sara would shut her eyes after this accident, in addition to seeing the "light beings," she could also differentiate a second type of entity. When "I shut my eyes I see them . . . I see these guys . . . down in a little row, like three or four little dark guys. Like gibbering." Later she said, "it seems to me like these guys are in my head." In contrast to the "light beings," she described the "other ones" as "frenetic." Shortly after the accident, she felt compelled to do her "listening" every day and to write down the information she obtained. She felt this would render additional accidents unnecessary.

A few days after the accident Sara and Miguel had an experience in which an unexplained green/yellow light penetrated their room. Miguel is ordinarily fearless, Sara said, but they were both terrified, and he appeared to have lost consciousness for a time during the incident. Sara felt as if she were "physically pinned down" and unable to move. She saw "three things hovering above me" like "three shrouded heads," and thought to herself "something like, listen, we're communicating. This is for real. Something like that, like, get your act together and start writing it down." Then "the whole thing kind of dissipated."

Sara has also observed unusual craft in the sky. On one occasion she was with a girlfriend and they both saw "a strange thing hovering above." Sara looked up, "and for a split second, I felt like . . . I felt like I was there and I was here. I felt like I'm in that spaceship, looking at myself. I felt like I was two places at one time, and then I started to think, 'Oh, wow! That's another whole possibility, you're coming back to see yourself.'" On another occasion she saw what looked like a star. "But it wasn't time for stars. It was like an afternoon. Really bright. Too low, but at a distance." After a while "I kind of got fed up with it. I'm like, if you're not going to do anything, then I'm going home. So I got in my car and started to drive away, and then it came at me, and then it came at me really quickly and flew over me . . . It looked like a flying star. It was just so bright." At the time she thought to herself, "God, I've got to tell Miguel," but she did not, and "it was like I forgot about it!"

Sara's wish to be hypnotized grew out of her desire to "know what's true . . . I don't want to know a story that I make up or anybody else makes up," she said. "I really want to know! I really want to know! It's the only thing that's important," even though "it may be really complicated and really overlaid and everything." She wanted to "get at" what "these little guys are." Finally, Sara wants to be responsible for her experiences. "To tell you the honest truth," she said, "I don't know if I believe myself . . . There's a part of me that really, really does. But there's a part of me that doesn't, and that part feels like it's destroying me."

Sara's first words after being brought into a trance state were, "I see my grandparents' house . . . I'm oscillating between that and my white canopy bed which was in my parents' house when I was little. I'm remembering a lot about falling dreams that were a series of dreams I had in that bed, where I'd wake up really suddenly and grab the bedposts to keep myself from falling any farther. I felt as if I'd been dropped or had fallen from something very high back into the bed. I had quite a few of those, and I used to wake up feeling as if I might be close to having died." I asked her to describe the sensations further, and she got "a real sensation of silver and like some sort of shaft, like an elevator shaft that I'd fallen through." There were further images of "white, shiny material" and a "place I've just fallen from." Then she shifted "to being in a field" and was "looking at what looks like a spaceship from a distance of maybe a hundred feet, and I'm outside and alone in the field."

The ship was a "white-domed thing" and had "a thing on the bottom and an entry that's vertical" and "there's light emanating from it . . . I see a lot of things that look a little bit like skeletons, but a cross between a skeleton and a walking insect. That is, they're walking up and down these inclined planes . . . There's light coming from—see one of the doors is folded down, and there's light coming out of it, illuminating the little creature that's walking up and down the inclined plane, looking a little bit like a thick skeleton. He has some sort of a bubble thing on his head, but I get a sense of filaments—then I just go right back to sliding down something into bed . . . Vertical. The descents were always vertical. So fast! So fast! Like almost rudely fast."

Sara recalls that she used to wake up terrified from these abrupt descents from the ship, "terrified that I could have died . . . That was not very careful . . . It's a good thing I caught the bed or I would have missed it," she said. Her next associations were to a long, shiny, white cylinder and the sense of her head hitting a "trapdoor." She felt as if she were going back in time to "a place where I was dead." Then she saw a being in what looked like a big, silver chair or throne made of metal. Although his head was "the most bizarre thing I've ever seen," nevertheless, she recognized him. There was an "outer orb around the head. It's translucent, and I'm seeing inside to a skeleton face. Inside the skeleton is not exactly like a human skeleton . . . There's this outer filamenty kind of orb around him, and the smile is kind of sickly, like a skeleton smile. But I don't feel, you know, scared. They're not mean at all, and they're nice. They're nice . . . No one's trying to scare me. It's not their fault they look like that."

Like many abductees Sara has had a name for this familiar entity. She calls him Mengus. "He's family, really, kind of benevolent," she said. Next she recalled herself first at age ten and then five, inside the ship ("I'm littler than he is"), "right up in front of" Mengus, "standing right next to him." She communicated with him "dreamy, like in my head . . . half telepathically" and "half verbally" in English. "He just kind of nods his head." She asked Mengus "what are you guys doing here on Earth?" and he replied, "Oh, we're just looking around."

Sara then saw what looked like a control panel on the ship, like a cockpit on an airplane but even more metallic. "I kind of float over to this stuff" and she asked Mengus what everything was. He told her "this is our transportation system." She pulled at various things, "but like nothing's on so I don't do any damage . . . He kind of lets me, you know. He's really benevolent . . . like here's this little girl, and she's just looking around, and isn't that fine." Although she sensed "a real warmth and benevolence . . . it is mixed with a very steely emotion. Serious. This guy is dead serious." Mengus said something like, "You're young now, but this is like preparation, and this is really important . . . We're leading you into this pretty easily, but this is not a joke, and this is not just to fly around, and this is serious business, so pay attention." It was "just like, 'Don't screw up.'" The great amount of love she felt from Mengus enabled Sara to really listen to him and appreciate that "there's no margin for error . . . I have a weird sense that he's dead now," she said, and "I kind of feel sad."

I asked Sara what made her feel that Mengus was dead. She replied, "I can just read his vibration, and when I go to find it now it's just like it's died and been recycled [see Paul's explanation of what happens when a being dies, chapter 10]. I can't access him anymore and he feels dead." Mengus "was really nice. I would say, maybe like my first real teacher." She has "the weirdest feeling that one of the little things I drew, the baby ones, was . . . Mengus's new incarnation."

Returning to her experience as a young child, Sara spoke of the floating/levitating phenomena described earlier and the sense that these capacities, although "really fun," had come to her from "past life." They were "not fun in just the conventional sense," but part of how we evolve. "I consciously understood that true fun can be a lot of work and processing." The vibrational energy of the translucent beings, Sara said, "was much more elevated than the conventional vibration you feel here . . . They're just so much more conscious! They don't keep everything suppressed in their unconscious. They're just awake. They're awake and they're responsible, and they're receptive and they're concise and precise and their eyes are open . . . Their hearts are open too. They're not afraid, and they're not stingy and selfish about their love, and it's just really nice. They're so, so, so, so nice . . . I get the sense there's a translucent thing on the back of their head . . . Our heads aren't translucent, you know are covered with hair and everything. We cover up all of our little things that we don't want people to see, and they just, it's like wide open. You can see right into it, and they're telepathic so they can't keep secrets that way. So as a result everyone's just a lot more together. They're not in denial the same way. I like that. God! I like that a lot. I wish I could be with them again."

Sara felt that to be with these beings, at least in the happy innocent way she had just spoken of, she would have to go backward in time, "before this life . . . I think I'll try it," she said. Next she found herself flying in a white spaceship with a number of little windows. It was flying over a desert area—"We're just whizzing around, and I can see down below and it's so beautiful . . . I don't know if I've ever been happy like that in this life, just like unreservedly, all the time, happy. Wow! We come over this ridge, and there's this big expanse of desert, and I see these reds and these yellows and oranges, and it's just like sensorially just scrumptious. It's just delicious." In this life her body was skeleton-like, "like Mengus's . . . It's creepy, and your bones are kind of little and brittle and it's kind of creaky. You walk in a very disjointed sort of way." Again Sara was struck by the joy of the maneuverability she felt within the space vehicle, how "just neat to zip around" it was.

From this alien/past life perspective Sara spoke of the "stupid" things that humans do and the temptation to confront them directly. But "it's much more useful to be subtle and make sure they thought of it themselves." Human beings are "so egocentric they won't change otherwise. They haven't. They've got this ego thing that they like to hold on to and they get really threatened . . . " At the same time there are "precious" things about human beings. "They can smell flowers, for instance. And that's like so incredible, and they get to feel the sun on their skin." As an alien being "I was operating out of less physicality, so you're lighter at one level . . . There are certain advantages. One is you don't get into these things like depression. But on the other hand it's a little disjointed and a little bit removed . . . The olfactory sense is not there the same way. You don't get the depth of smell, for instance," she observed. At the same time the aliens have seen "a bigger picture," and have more insight and patience. Also, "You have this thing in your head that [enables you] to access any kind of information telepathically. So you have this kind of informational pliability. I mean, you can get any information you need."

Sara felt that the purpose of her flight over the desert was to survey the planet for "planetary resources," in order "to see what is the survivability of an area like that" in case there were to be a "huge planetary shake-up." The desert area seemed to be a potentially "stable environment" in case of a major upheaval because it was high and flat. As she experienced herself flying in an alien incarnation she felt herself "going back and forth" between human and alien forms, as if trying to make a decision. The human body identity was aesthetically pleasing for its "flesh and things," while at the same time she was drawn to the greater perspective of the alien identity.

Sara returned then to the present and went on to describe a huge, ominous dark cloud covering the sky that seemed to exert a magnetic pull upon her, "like throwing dark, black tar over my head." The cloud seemed to Sara to embody the projected negative consciousness and vibrations of human beings. Its impact was debilitating and made her feel victimized. The cloud functioned as a kind of mask or shield to hide some sort of "hokey" craft of the sort human beings would design if they were to make a spaceship. This craft was the source of negative vibration and was piloted by a human being. It appeared utterly "stupid" from Sara's point of view. "I'm just loathing this whole thing," she said. The aircraft's "purpose," she said, was "ostensibly war," but not war to kill people. The war was "with people's heads . . . war to control people." She felt "this huge desire to shield myself from this thing."

Next Sara remembered childhood encounters of "levitating," "floating," and "bouncing" around her in the room with the white canopy bed. "I feel like someone's almost throwing me up and down." Two "Mengus-like guys" have been doing this. She felt as if there were a magnetic field between their fingers and her body. The bouncing about "was fun . . . I was laughing," and then the beings talked to each other, "not to me" and left headfirst through the window. These were friendly visits, "like coming over for tea," but the beings became "mad" after college because she was living such a "conventional, stupid life . . . a very short sighted existence," especially when she took a job in business.

Sara associated to another experience later in her life. She was alone, lying down on the deck to get a suntan, when "I felt something hovering on top of me." She saw a figure that "was like a cross between a Mengus being and a person." It was "human in shape, but lighter and free-floating." Sara received a communication from the being, "This is very important." The intention, she was told, was not aggressive, but some sort of test of "genetic compatibility or something," an "infiltration," "a feasibility test," "dimensional merging."

I asked Sara to tell me more of what she meant by "dimensional merging." She then described what I believe to be the central image of our first session. "It is like a plane," she said, "a sheet of translucent cellophane." There is "like a huge shattering of glass," and a "razor blade thin" slit opens between this Earth/physical dimension and the realm from which the beings come. In this context I asked her more details of the encounter. The being had "a light contour of a penis, but not like a physical penis" that entered her body. The experience was not like anything she had known in human sexual relations. "The being itself felt aggressive, and I did not like that part of it. There was not an emotional component to the whole thing on its part . . . It was more like a scientific explorer territory." I asked if there were orgiastic activity. "It was much, much, much subtler," she replied. "It was not entirely happening in this dimension," Sara said, "so you can't really evaluate it in the language and physical descriptive terms of this dimension because it wasn't really happening here. It was half happening here and half happening somewhere else." After this experience Sara "felt like I'd sort of been hoodwinked." The being "didn't give me the full story, and it just kind of said, 'Hey, trust me, it's important.'"

Then she said, "If a being were to project itself onto a sheet of cellophane, and [the] cellophane were to shatter through to this reality, and I could stand and watch, I'd do that." I asked if this had in fact happened ("come through") to her. "Yes," she said, about two weeks ago. She had gone on a ski trip. There was a large mirror in her hotel room. She arose in the middle of the night, and the place where the mirror had been appeared as a corridor. She attempted to walk down this corridor, but she bumped her head against the glass. Miguel had not gone on the ski trip with Sara, but "the minute I bumped the corridor Miguel was in the room, and I tried to scream out, 'Miguel,' but I couldn't scream. Nothing could come out." She was sharing the room with a skiing friend, who independently saw a silhouette in the room. Paradoxically she "just immediately went back to sleep."

The bump hurt a great deal, but the pain was compounded by the "interpenetration of the dimensions" as "the mirror opened up." It was as if "a being that looked like Miguel" or "a disguise of Miguel" came through. The being had "penetratingly dark" eyes, "dark, dark," and looked "insectlike" with "an overshaped head" and "a little, shrunken body . . . that's using the costume to look bigger . . . It hurt me," Sara said, but "the overall purpose wasn't to hurt me." It was rather "to explain something through demonstration," namely "this whole dimensional interpenetration exists." By "bopping me on the head," they "demonstrate, 'Hey! This is physically real.'" Otherwise, many humans are often too "dense" and/or too preoccupied to be reached.

"In a species sense" Sara has felt "compatible" with the Mengus type of beings, but the being in the hotel room seemed to be a representative of another species with which Miguel was connected, perhaps in a past life. In Sara's view these two species are trying to connect with each other as demonstrated in her association with Miguel. Each species, she said, has its own "vibrational plane," so that for two species to connect they must "create a new vibrational plane of interaction." This could be exemplified in a human relationship that, in effect, crosses the species barrier. This would be accomplishing an infinite number of things with "one beautifully concise stroke."

I asked Sara to say more about the being she saw in the hotel room. The head was the most prominent part of the body and was "shimmery," looking "reptilian," almost "snakelike, serpentlike" and quite elongated. "Red vein-things" made the head appear like "a body turned inside out." The creature was not "bad. It's nice enough." It was "almost like a sea creature, like a mollusk or a snail without the shell." It seemed vulnerable, in need of "understanding" and "cooperation" from her. For Sara to own that the creature truly exists "expands my borders of acceptability and tolerance . . . opening my heart to something that isn't the same as I am. That's good for me. I need to know that. I need to learn that and actively do that." It was "sweet," she said for the being to "put on" the Miguel costume in order to bridge the gap of unfamiliarity. When Sara looked into this being's eyes she saw "so much love" and felt love herself. She also perceived a "kind of sad" and "battle-weary" look, as if it were saying, "Give us a break!" "They're tired of everyone being scared of them . . . I feel bad for that guy," she concluded.

We ended the regression at this point, and Sara's mind began to doubt her experience and search for ways to "explain it [the session] away . . . It could be delusions and imagination," she said. But then she observed, "It's not imagination, either. I mean it is real. It's more real than imagination. But it's real in a hologram-like sense . . . like it's projected, but I don't know. I got bumped on the h . . . then you're right back to, 'My God! It hurt, didn't it?' . . . I went through something here, though, that was real," Sara concluded, "all this pain that felt like a searing, burning . . . " After returning to ordinary reality the two realities seemed "more on a par" or "much more equal."

The larger purpose of bringing these species together, Sara said later, was to bring about "personal evolution" in order to achieve "universal understanding." The intense pain was used to penetrate the density of human denial, to reach us when we are "asleep." Pain is the "extreme of physical tangibility." Each species brings something to the merger. The Mengus-like beings, for example, Sara said, are more spiritually advanced than humans, who need to become "a little more Mengusy." The Mengus-like creatures seek a greater physicality, "the ability to smell," for example. In the connection of species each retains some of its original elements.

This process of species connecting involves "tremendous, tremendous, tremendous love." Most ordinary human love, Sara said, is much more possessive, involving emotions like jealousy. This interspecies love is "more unconditional . . . I think that's everybody's sole reason for being here. Soul/sole, in both senses of the word sole." A few weeks later, Sara wrote to thank me for my help and said that "things seemed to calm down greatly" after the session.

Approximately six weeks after our session, Sara and I met for about one hour to integrate further the openings that had followed her regression and to discuss the possible forms that her life's calling might take. After some discussion of folkloric studies of UFOs, abductions and related matters, Sara suggested that the aliens may be assuming the forms of technology "in order to be more accessible to us," to appear, for example, in something that looks "kind of like an airplane to make it a little easier." She, like many of the other abductees with whom I have been working, spoke of the cataclysmic physical changes that may be ahead for the earth and wondered if somehow ecological and environmental concerns could unite humanity and help us transcend ethnic, cultural, and other boundaries.

Sara mentioned that she would sometimes sob because she missed "home," but for her this has "nothing to do with my Earth parents." It exists "in a different dimension." It was, rather, a deeper sense of connectedness that she missed. We talked further of what this other "home" is like and means to her. "Home is dimensional, not spatial," she said. But there is communication, nevertheless, between the dimensions. "You shut your eyes and there's always communication," she said. "The content is almost a hundred percent emotional," she added. It was difficult for her to describe this coherently. "It's all about . . . the emotion of love is the most . . . unconditional supportive life. I don't mean that in human life, but creativeness, . . . growth-affirming kind of love. It bowls you over. When you feel that, and when you feel that connection to that, the love feeling is so tremendous."

When Sara accesses this and other connected states, she says she feels "very happy." She says that "it feels like the magnetic field around me completely changes . . . like space or something is fluctuating, like, if you could see a thermal crack or something. It feels like that." She also feels that this state is somehow so familiar that she has always taken it for granted, and that if she focused her attention in this way more often, many additional things would become accessible.

Despite the joy she feels when she enters the other dimension, Sara feels it would not have been "ethically correct" for her to "jump" the chasm between the two planes totally or too readily. "In the past" she said it felt as if she "made a commitment, like an exchange student" spending a year abroad to be here on Earth. She was, in effect, in "an immersion program," has "taken resources" and has "a responsibility" to see it through.

One way or another, Sara expresses a desire to use "ecology as a way to help people make a . . . transition . . . People have to redefine philosophically what they mean by environment. People think, 'Oh, my environment.' But, it's like environment is [complete] . . . environment is . . . infinite. And it has an infinite number of characteristics, and they extend from physical to emotional-psychic to interplanor and cross-sectional . . . You are your environment . . . It's a much broader way than most people think," she noted. Sara spoke then of how difficult it has been for the human species to reach a "creativeness-affirming, life-affirming" place of unconditional love, which she related to all the ways "by which we differentiate ourselves," such as by creating gender, ethnic, and religious barriers. Ecology could be used to discover "commonalities" and "transform consciousness . . . If you truly, truly, truly do what's good for yourself, you're doing what's good for the world. The two things are synonymous."

Sara observed that she herself still experiences "emotional neediness." Using her metaphor of the exchange student here from another dimension she said, "I might be able to take a vacation back home, or be in two places at one time," but she says it may be more useful to reach a state of consciousness in which "it didn't really matter to me if I went home or not. Then I can go home because I don't *need* to go home." She talked further of how her spiritual path was her way of reaching a place in herself where she could "give love" both "there" (in the other dimension) and "here" on Earth.

Discussion

At one of our meetings, Sara asked me if I thought that the direction of her thinking and experience reflected something psychopathological—"like I'm making it all up." She was reassured to learn that other abductees had been struggling with the same philosophical questions. I then shared with her some of my speculations about the implications of the abduction phenomenon and where experiences like those she had described might be leading us.

SARA: SPECIES MERGER AND HUMAN EVOLUTION

Sara has been preoccupied since childhood with philosophical and spiritual questions, and apparently from an early age has exhibited certain paranormal powers, such as the ability at least to create the impression of levitating another child. These concerns and abilities seem to have been intimately connected with lifelong encounters with alien beings, beginning in early childhood with a mentor figure she calls Mengus who she describes as her first teacher. Sara's abduction experiences, fun and joyous as a child, but always at another level deeply serious, appear directly related to her personal and spiritual growth and her determination to find a calling that will give sufficient scope for her desire to serve the planet as fully as she can. Ultimately, however, Sara believes that at its core the abduction phenomenon emerges from a place beyond the physical plane and cannot be grasped through technology alone.

It appears as if from childhood Sara's encounters were a kind of preparation of consciousness for a life's work she strives to accomplish. This work appears connected to using an expanded notion of ecology or "environment" to bring about a paradigm shift from a consciousness of division and separation to one of openness, creativity, and unconditional love. Sara relates her own evolution in this direction to her encounters and her role as a kind of exchange student between the nonphysical universe from which the aliens or "light beings" emanate and the earth on which she has committed herself to live.

Sara tried repeatedly in our sessions to put into words the process by which the alien beings can enter our physical universe and she, in turn, can access theirs. One striking image was that of a powerful cellophane membrane that is shattered, creating a slit through which some connection with the other, nonphysical, dimension may become possible. She herself can access this other universe and she has longed to surrender herself over entirely to the other domain, which she, like so many abductees, considers to be "Home" and the place of her true parents. But she is constrained from going there altogether by the continuing earthbound challenges of overcoming her own egoistic needs, especially the desire to be loved. Sara, like other abductees, understands that as she transforms her own consciousness and shares this process, she contributes in a subtle way at a wider level. As she put it, "If you truly do what's good for yourself, you're doing what's good for the world."

Sara, like all abductees perhaps, is participating in some sort of project of species merger and evolution. The purpose of this project may be to create new life-forms that are more spiritually evolved and less aggressive, while retaining the acute sensory possibilities that accompany the dense embodiment of human physical existence. One part of our long hypnosis session involved Sara's memories of an encounter with an alien being that she experienced as occurring partly in our physical reality and partly in another, nonphysical, dimension. The most difficult aspect of the various kinds of interdimensional, interspecies connection that Sara described is the different vibrational frequencies by which beings from the other dimensions live and the radical adjustments that must occur for us to connect. Much of the intense bodily distress that Sara and other abductees experience during their regressions may relate to the bodily releasing of these vibrational incongruities that have been held in check, sometimes throughout the individual's life, by powerful repressing forces that may derive both from the human psyche and possibly from controls imposed by the aliens themselves.

Some of the most intense moments in Sara's first regression occurred when she recalled having been struck or "bumped" painfully when she misperceived a mirror in her hotel room for an open corridor, a "mistake" that might have been engineered by the beings themselves. Immediately following this shocking and painful impact she was able to recognize in her room a representative of another, more reptilian appearing, species of alien beings that was possibly connected with her friend Miguel as she has been linked with the Mengus-like species. The intense physicality of this sort of experience seems to be employed so that Sara and other human beings are forced to acknowledge the reality of entities and domains that our Western acculturation have taught us in recent centuries cannot, even must not, exist. Yet this kind of ontological, physical shock may be an essential initial step in the process of human consciousness evolution that seems to lie at the core of the alien abduction phenomenon.

CHAPTER TEN

PAUL: BRIDGING TWO WORLDS

Paul was twenty-six when he introduced himself to me at a UFO conference in New Hampshire. A sensitive, handsome young man, he is one of an increasing group of abductees that I have been encountering who have discovered that they have a dual identity as an alien (they do not use that word) and a human being. From the beginning of our work Paul believed that he had a mission to "be an example" of love and openness and to enable humans to overcome the fears that constrict us and prevent us from using the gifts we possess. The purpose of our work together has been to enable Paul to discover the depths of his complex identity and to take full responsibility for his transformative and healing powers. Paul and I have had two hypnosis sessions after our initial interview, and he has attended two support group meetings. In a small group, Paul, Pam Kasey, and several other abductees have been exploring the healing powers of their shared energies. When we first met he was living with his parents and administering his own advertising business. He was working to earn enough money to rent a separate apartment.

Paul, like many abductees I have seen, came to me after a disturbing series of meetings with a mental health professional who he continued to see until a few days before I met with him. He first consulted Ms. T., a psychologist, to explore "weird" experiences that had led him to question his sanity, including one, five hours after smoking marijuana, in which he saw "some type of being" on the stairs at home. He wished also to see if his experiences were related to difficulties growing up. He met with Ms T. somewhat irregularly over a year-and-a-half period. The therapy included four or five hypnosis sessions intended to recapture memories of possible sexual abuse by his paternal grandmother, which did not, in fact, emerge. What did come to the surface were additional encounters with unusual beings, going back to when Paul was age three, that were powerfully real for him and had a "shattering" impact on his worldview and "everything I've known."

As the sessions progressed, Paul found that he was becoming increasingly "awakened" to "a connection to like a foreign alien kind of thing" which Ms. T., perhaps understandably, could not deal with. One time in a session he asked for a sign of the reality of the beings, or the energies associated with them, which was followed by a loud bang near the office door. Ms. T. was frightened but willing to explore the noise. Paul was curious to check it out and felt an electric "crackle" in the room but found there was nothing visible behind the door. Ms. T. was wide-eyed with fear and Paul had to try to calm her down. He had the sense that something was "going to come" for Ms. T. over the next weekend and told her so. She volunteered no information at the beginning of the following session, so Paul asked her if anything had happened. Ms. T. reported that her bed had mysteriously bounced up and down. She revealed that she had been terrified, and, according to Paul, tried to ignore what had occurred except to clean the house "out of evil spirits." Her assumption, Paul said, was that anything "good or intelligent would greet you in a very comfortable way."

Despite Paul's feeling that Ms. T. was "suppressing" him because of how frightening she found the abduction material, a number of memories came out during the hypnosis sessions with her. He describes, for example, how in an early session he "expected to see my grandmother abusing me or something when all of a sudden . . . I see the ship, and I'm out in the back, the chimney, and there's these little people coming up and I'm like freaking out" (in our first regression we would explore this episode, which occurred when Paul was age six and a half, in much more detail).

In his last session with Ms. T. before he terminated the relationship, Paul recalled an abduction that he dated to age two or three, judging from the fact that he had on a one-piece, red pajama suit with buttons up the front that his mother confirmed he was put to bed in at that time. He found himself "on a table," and "it scared the hell out of me." He remembered then that the being had come into his room, "got my hand," communicated that Paul had to "just be strong here," and took him "through the door." Paul recalled only "flashes of a ship" and does not remember how he ended up inside. From the table he saw that the surrounding room seemed to be made of some sort of uniform metal alloy and that objects in the room did not seem movable. As he would try to get up, a being would press his head with two fingers and push him back down, seemingly effortlessly, and he would be temporarily calmed.

I asked Paul if he could see the being. "Not yet. I did in a minute," he replied. The creature was wearing a full "bodysuit" with "what looks like seams . . . All of a sudden he turned, like a flash. He moved so fast! And I'm looking at his head, and it was like classic. It was like big eyes, and I'm like they were dark, and I'm like 'okay,' [whispered], and I just lied back down." He thought, "Okay. Okay. Yeah. He doesn't want me to look at him obviously." Based on many other cases, I suggested that perhaps it was he who did not want to look, which Paul thought was "heavily possible for me too" as he was very frightened. Then he saw two other beings standing behind him, and thought "what's going on. Why won't somebody like talk to me here." Then Paul looked down and saw that "he's started doing something to my leg" with "long, long fingers. He or she—I don't really know—it looked like two fingers with a thumb, and he was just like feeling my calf, real lightly goin' up and down, and then all of a sudden I really felt my leg, my calf, it was like, it was pain, and I was like, 'Ow, my calf really hurts.'" Paul does not recall seeing an instrument but remembers that the leg felt "numb" and as if the being "injected something into it."

Paul further described the pain in his leg that he had recalled in the session with Ms. T., how "after that was done everything started settling back down, and then he started to get me up" and the being was "taking me out," when Ms. T. said "'All right, our time's up', and I thought 'Okay.'" As the session concluded, Paul still felt pain in his leg and Ms. T. asked him, "'Are you okay?' And I'm like 'Sure I'm okay.' You know what I mean? I don't know. Like, I guess so. I can walk. I can drive home, or whatever." There was talk about the possibility of a longer session, but the lack of feedback and Paul's feeling that Ms. T. had so much difficulty dealing with the impact of the abduction material on herself led him to terminate the therapy. Paul made a last attempt to get help from Ms. T. in a phone call a few days before he and I met. He expressed his difficulty in dealing with the memories that were coming up ("putting it all together"), but, uncertain herself all along as to what to make of the abduction stories, she seemed to have nothing to say except, in effect, "call me when you need me."

Paul's willingness to stay so long in this largely unproductive therapy relationship was not only related to how few people are qualified to deal with abduction issues. It was also connected with how alone and unaccustomed to receiving help he had felt throughout his life, and with his lifelong tendency, like children of alcoholics (both his parents did have drinking problems), to protect the adults around him from their own distresses. His decision to stop seeing Ms. T. was, in effect, also a decision to stop protecting her from the distress that his case was creating for her.

Before ending our first session Paul spoke of feeling "foreign" ("all my life to my mother I always said I was adopted. I'm not from here") and the difficulty that "people like me have to adjust and adapt to what is going on here if we're going to live and survive here." He related this to the negative, hostile emotional climate he encounters in mainstream society and expressed his desire to "be an instigator towards the positive" and become "an example of what someone can do" to free communication. "The model's there," he said but "this world has pretty much boxed it off. I've got to open that up."

Paul had suspected for many years that his mother's husband was not his biological father. According to Paul, his "father" was sterile and his mother had had a long-standing affair with another man whom she had hoped to marry, but who stayed with his wife when she developed leukemia. After he pressed her several times his mother finally broke down and confirmed Paul's suspicions; this was about a year before we met. The discovery that "my father was not my real father" added, of course, to Paul's feeling that he did not "belong here." Paul's father (his mother's husband) has had problems of impulse control. This may be related to the fact that *his* mother, according to what Paul's mother has told him, would "take their [him and his brother's] clothes off and do odd things that were abusive." Sometimes he would strap Paul for "anything that pissed him off." He also may have "exposed himself" to certain people and "made advances toward my sister [three years younger than Paul], but that's as far as it went. Nothing really developed from it."

Paul's mother seems, from his accounts, like a fearful person. He had the sense that she was intimidated by his intense curiosity and intelligence. "She would try to suppress me," he said, and make him doubt his own mental capacities. Naturally this uneasiness was aggravated by Paul's abduction-related encounters. "I would adjust," Paul said, "because I don't want to freak anyone out. You know what I mean, especially your mother . . . I'll keep this under wraps." The day after he saw the being on the stairs at home (described above), Paul told his mother about the experience. She admitted that at the top of the stairs, she had an "experience like an incredible feeling of fear that stopped her midpoint. She was coming down the stairs and she just stopped." To Paul this was "confirmation from someone else that something was in the same exact spot," but his mother denied that she had seen anything. His mother herself once had a two-hour period of missing time which she acknowledged when Paul asked her, but she had not given it much thought. "And I'm like, Ma! Two hours in the middle of the day! Didn't you question that?" Paul's sister has described a UFO that she and their mother saw together, but they are vague about the details.

When we met for Paul's first hypnosis session we began by reviewing his disciplined meditation program and the tensions he was experiencing during the process. He spoke of his wish to develop his psychic potentialities and of the altered state of consciousness and heightened awareness he would achieve during the meditations. He expressed the desire to "know more about my past," especially to find out "who were those people" from the encounter at age six and a half.

Before talking further about this time, Paul told me of a frightening incident that had occurred late at night or in the early morning a month before. He had been coming out of a dreaming sleep when he heard a loud bang like there was something "by the trash can out front." He looked out the window to check and, seeing nothing, went back to sleep. Going in and out of sleep he felt sure that there was "something in the room." But when he felt "something on top of me," and found he could not move at all, he became terrified. "I felt it further off, but then it came really close and, like, Whoa! That's too close. Too close! Back off! . . . As I'm freaking out, I'm trying to figure out where this thing is, and I was measuring it, like off the tip of my nose. . . . I could feel it right there . . . about four or five inches off my nose." He wanted to open his eyes but simply could not do it. He "dozed off again" (a common paradox in abduction-related terror) and then upon awakening was able to "break out of it" and look, but there was "nothin' there." Although fully conscious now, Paul found he was "gagging on my own breath because I couldn't speak" and was still unable to move for a few more moments. Although he saw no beings during this incident, Paul nevertheless felt invaded by "something [that] had definitely done something."

Although we were tempted to explore this incident further, we returned to the episode from age six and a half that occurred in the fall of 1972. As I characteristically do, I reviewed with Paul in great detail beforehand the context in which the encounter occurred so that I would already be oriented as the incident was explored during hypnosis and not have to impede his associations with distracting factual questions. Much of this information had emerged in Paul's meetings with Ms. T. We went over the layout of the house, the location of his room, and how his mother tucked him in bed that night. Paul does not remember falling asleep, but does recall looking at the bureau, getting up, and walking over to the sliding glass doors that separated his room from the back porch. A familiar voice in his head told him to go outside, as if "I was supposed to meet someone out there."

As he described his behavior that night, Paul spoke as if he were the adult Paul observing the actions of the little boy from "right behind his eyes . . . I don't feel like he feels," Paul said, "but I understand what he feels." The boy went out on the porch, looked up, and saw "a flash" of "a ship" that seemed to pass right over his house. "It was like the whole thing was lit," Paul said. It was round, "like a perfect circle . . . big . . . awesome," larger in the horizontal plane than the vertical, "long and smooth" on the bottom. "He [the boy] was deciding where we [the "we" refers to the boy-Paul and the adult-observing-Paul as if they were together] were going. So we went down the stairs [of the porch], and we went over to the chimney [a structure standing in the backyard] and then we sat down." The boy walked inside the chimney as if "we're hangin' out waiting for someone to come.

"I [Paul is now speaking from inside himself as the little boy] really had no idea who was going to be coming, and I was terrified when I started to realize that someone was there." Paul [adult] notes that "it had to have been brighter because inside the chimney I was able to see bricks, like charcoal bricks . . . I have to go with him [Paul as child] because I'm—we're—linked in a way, so we go out and that's when there were two groups coming up both sides of the house." The beings were small, about the six-year-old's height, with one that appeared to be somewhat taller. There were about four or five of the beings in each group. They "weren't human," Paul said, but nevertheless when "they started coming around" the boy felt more comfortable and "was thrilled" to be with them. They were touching and hugging the boy, who felt "a great calmness" and "really like home," as did the adult Paul in the session at that moment.

Pam Kasey's notes describe Paul's body movements during the regression. They provide a sense of the intensity of his experience. "Paul is having a lot of body reactions to this information, an involuntary tensing in resistance to each new piece, then often a nervous laugh, a deep breath, a sort of convulsion, and then a relaxing to the next piece. His face is grimacing, frowning, eyebrows furrowing, teeth clenching—then a laugh, head shaking emphatically, nervous, earnest, charged, overloaded. Further along it's getting harder—body twitching, tensing, head shaking, nodding, raising off the pillow, face contorting, expression changing every second or two—arms at his sides but hands gripping, opening, gesticulating."

At the start of the regression I took Paul back to that night and encouraged him to *be* the six-and-a-half-year-old boy rather than the adult observing him. "I get stopped," he said. "I guess it's fear . . . it gets more and more intense as I try to be him." As I encouraged Paul to be with the fear, Paul soon felt numbness in his face, spreading through his abdomen, chest, and hands. With his fear increasing he saw "a big eye in front of me . . . hands on me" and a sense that he was "shrinking . . . There are others there," he said, and "they won't let me feel anything." Then, feeling himself to be lying on his back naked in a room with a domed ceiling, he saw "tools and a bench." His whole body numb now Paul said, "I was able to look out a window . . . It's space . . . I just saw stars. I saw a lot of stars . . . It feels like it's moving."

Once again Paul had difficulty connecting his adult observing self with the little boy and his experience—"everything keeps me from connecting together"—but was able to say, "I can be me." Then he felt "plates, like grids, pushing down on my stomach." At my request he described one of the humanoid beings. He had no hair on his head and his eyes were large and black with no apparent irises. The nose was "flat, like an ape," and the mouth seemed to have "scales around it, like plates on his lips." The being let him up then and led Paul through a door. "I walk outside. We're looking at controls on the ship. It's a ship. It's a ship! It's a ship!"

The being seemed to Paul to be "a friend," showed him the controls, and told him "You're like me, on the ship." At first Paul did not understand, but the being explained further that "you're from here." Then, as other humanoid figures watched, the being led Paul to "a colander-like structure a little below, off to the side" of the center of the big ship. He said to Paul, "This is where we congregate." The figure that was escorting Paul showed him a bed that was rather like a human bed with sheets, but "it's floating." He told Paul, "They're your quarters. You're here. You're here when we go on these trips." In fact, the "quarters" felt familiar to Paul, for he estimated he had been there "seventy times." Paul felt confusion and disbelief then and now, but says, "I feel like I'm there. I feel like this is the room that he's showing me that I'm in when I go to where he takes me."

At this point I asked Paul about the timing and frequency of these many visits. Paul replied, "He's saying ther're all connected, that it's the same."

"What's the same?" I asked.

"The lives. They're all the same . . . It's close. It's close to my life. It's close to now . . . I was on the ship before I came here [i.e., to Earth]." At this point Paul was experiencing what might be described as if breaking through an information barrier, and he felt acutely the tightening and releasing of his body described in Pam Kasey's notes. "It keeps breaking and it's pretty confusing," Paul said. "It's plain to him [the being] and to me, but I can't grasp it here [i.e., in the context of the Western, rationalistic perspective that my questioning may have represented for him]."

At this point the being conveyed to Paul that he would "tell me anything I want to know." Our session became then an exploration of Paul's dual identity, human nature from an alien perspective, and alien-human relations over time. Paul himself was "kind of like a spy," put on Earth for a purpose. "He [the alien] says that your spirit is from here [i.e., the ship, not Earth]. He says what makes you up is here, and he says that the seeds of human being is how we integrated you into this, but you are from here.

"Home," Paul said, is "on that planet. They are very peaceful, very peaceful. They're not like here. They've been killed here before." I asked Paul where that planet was. "All right. All right. All right. You're not supposed to know. I'm not supposed to know. I can know. All right. All right. All right. I can see it. It's red, and it's—but it's blue. It's different. It swirls, like Jupiter swirls." The planet is "in this universe," but "farther away than you've ever known."

I asked how the beings get from one place to another. "It's like hopping," he said. "Energy, like folds into itself, and you're just somewhere else . . . everything folds, inverts into, and folds inside itself . . . Like you can move one at a time, or you can move like vast numbers of people . . . No one's supposed to know. People aren't supposed to know this yet." I asked him to say why. Speaking now as an alien, he replied, "We've been hurt here before . . . Your people hurt us." Paul said, "It's in your nature to be violent," and spoke of the human need "to control everything," and to isolate ourselves from other beings, including the aliens themselves. "Humans are just another form. You're another lifeform of energy. You think you're independent of life, and you can't be that way. You're causing death. You're causing a lot of death and it's your own. And we're trying to help you, but we came and we were killed by many of you."

People like himself, Paul continued, were "here to integrate, and it's slow . . . because if we came and tried to disrupt you people it wouldn't work. It didn't work before."

"Before when?" I asked.

Pushing ahead, as if ignoring my question, Paul continued, "You people are too violent already, you're too violent and you're too hostile. It's too imbedded in your nature and you have to come to grips on that. You have to understand that, and it's got to be a little bit of time . . . We can't come straight on. We can't come straight on. We have to integrate like with this."

I tried to take Paul back to what happened in his quarters on the ship, but he deflected this effort and persisted with his struggle to understand "the information that's been locked in me . . . It's more than we can understand." I felt I had no choice but to let him continue. He told further of the trouble the aliens have had in their encounters with human beings. "There's a lot of us [i.e., dual identity beings] here," he said. "When we come straight on it ends up in power. Everyone here is so wrapped up in power." He spoke of how difficult it is "for your species" to "truly open up to another." As a human being, but identified also as an alien, he has had a great deal of "trouble here." He is trying to help human beings but has felt attacked. "Any new thing coming in is attacked . . . I'm trying to do what I need to do to help you, but I'm under attack . . . Human beings think that they're it, that that's it. But there's so much else here . . . There's so much life, yet human beings want death. They're choosing destruction, and they keep choosing it over life, over connection, over creation. This is hell here . . . Everyone has tried to explain that to you. They've tried to tell you that this place needs to turn around. Human beings keep tripping over themselves."

Paul spoke further in prophetic tones of human stubbornness, unwillingness to accept what we have done or to receive help. "That's why people like us, coming here, get caught up in this and then we get sick like you." The alien beings can be "physical on your plane" but also be "connected with others not on your plane." That enables the aliens to "accept others" and to "communicate and integrate with them . . . Human beings," he said, "can't even integrate with something of your same plane, let alone something of a different plane. You can't even accept the life around you." Segregation, isolation, and fear characterize the human attitude toward life, Paul said.

Paul spoke of how difficult it is for him to exist in both human and alien identities. "It's harder for us to be fragmented like this," he said. In his human self he feels the great "pressures of your human society . . . It's just too much. It's so stressful. It's just so stressful." He spoke then of how we have expanded a kind of protective "shell" into "a whole separate thing . . . This shell is just supposed to be a minor thing. It's a minor thing and you people hold on to it. Like you've created this new thing.

It's a whole new thing. It's like a little layer. It's one little layer, and you've got that being your whole universe . . . It's like one layer of appearance."

As he tried to answer my questions, Paul found himself "bouncing" or "jumping" back and forth between his alien and human identity or perspective, which he found difficult. The flow of his thoughts seemed to have a direction of their own, almost independent of my questions. "What is the purpose of controlling something you don't even understand in the first place? What are you controlling? . . . I don't understand . . . You're controlling nothing," he continued. "If you look at frequency and energy, and the way it's structured itself around the form, and you start going deeper, and you start to understand evolution, the way that connects itself to molecular structure—it goes on for eons and eons! It's further than you can fathom, and it's tried to tell you that too, and you don't understand it."

"What's 'it'?" I asked.

"Consciousness," he replied, "Higher forms of consciousness . . . You're not going to understand infinity, but it's there!" The alien beings, Paul said, have access to this higher consciousness which "flows through you" and is an intelligence which exists or moves everywhere.

"Person to person, nation to nation, world to world?" I asked.

"Universe to universe," he said. "There's consciousness on every single level. It's infinite," he said.

I explored further with Paul what information he had received in his abductions about this higher consciousness. "It starts as an energy you can't perceive," he said, but then it evolves and starts "becoming intelligent." It can "bow into different dimensions . . . engulfing shapes just like a cell engulfs another cell and takes a new shape. The energy mirrors another shape; it engulfs a new shape and networks with that. It communicates with it and understands it. It learns, and it grows. It creates. That is creation in action, and it becomes more and more intelligent. It grows. It has more to choose, more choices." Paul spoke further of how matter and energy "flex back" or change one into the other in various ways. Relating this creative process to his perceptions of human recalcitrance Paul said, "You don't want to change and grow. Change to you is fear. Change means destruction. You've got so many backwards viewpoints on the way it really works! You hold on. You change for a second, and then you hold on, forever. You hold on forever."

Curiously, Paul revealed that the alien intelligence does not really understand why human beings are so destructive and resistant to change, and I asked if the intrusive physical procedures they perform on our bodies are to learn about this. "That's a part," he said, and acknowledged that the "poking" and looking are for the purpose of understanding, helping, and "adjusting," but added without explaining, "There's mistakes that have been made." In the last analysis "we [Paul as alien] don't understand why you are so stuck" and have not learned. "It poses some complex problems for us," he said. "An organism that gets to be at such a degree of destruction should flip back and learn upon itself. It should understand . . . like you stretched it out to its maximum, and you should understand that you're going to break . . . we don't understand why you choose destruction."

Intervention and change are possible even without understanding, Paul said. "We can turn it around," but "you're going to have to accept more changes that go on. The changes are going to get faster, and it's going to be harder for you to change . . . The intelligences are present now . . . it keeps opening levels." We have changed some already, he allowed, but our "human nature" would resist further change. I expressed my impatience and asked how Pam and I might participate in bringing about change. He said it was hard for him to be "like a spy" that has been shown "different levels . . . There's a lot of stuff around you that keeps knocking at its door and then it stops. It's happened before. It's happened so many times." In spite of the great distance human beings have come, they still isolate themselves. "That knowledge of evolution, that learning process, has grown all along with you. You have memories from the beginnings. The beginnings, if you can even fathom that. And I can't fathom that, because I'm part human and it's hard for me."

At this point in the session Paul seemed to be feeling a great sadness as we contemplated together how "out of balance" and "lost" we humans were. In a nostalgic tone Paul spoke of how "very comfortable" he felt "here on the ship . . . I want to go home," he said, and "It's on the ship. It's home. My home is there" (quite a number of abductees say this). I reminded him of his disbelief when first shown his "quarters." He said, "It's not hard for the real me to accept. I know that. It's hard for the shell-Paul to accept." The planet he had spoken of earlier was the home from which he originally came, and the "quarters" are "where I am when we're traveling, when we're exploring." Paul realizes now that he has been shown "a greater time distance" than he had previously understood, including past lives.

Struggling with his disbelief over what he was experiencing, expressed through nervous laughter, tense body movements, and interspersed comments like "this is bigger than I thought" and "it's so weird," Paul told how "we came across this planet" thousands of years ago. "We've made connections here before" with "primitive life-forms . . . I'm being shown dinosaurs, in a way . . . This is old. Reptiles—oh my God! That form we were able to make contact with."

"The reptilian form?" I asked.

"Yup. That form was—that was smarter than humans," he said, laughing. I asked how this had been communicated to him on the ship. "I don't know," he replied. "It's like a memory in a way. It's hard. This is hard because I know now that I can make contact again."

"With whom?" I asked.

"Contact with my brothers on that planet. In space. In that ship." I suggested we call the aliens "brothers" since to Paul the beings we call aliens he felt were, indeed, "my people" or "my brothers."

At this point images were coming to Paul faster than he could deal with them and I encouraged him to recenter himself and take his time. He said that "This reptilian form was very intelligent, a form of energy that spanned up to that point, and it did very well." I asked what happened to them. "It just evolved past this time. It allowed for new forms of life," he replied. "Oh, man! They can tell time. They can feel time. They understand what's coming in the future. They [the creatures in reptilian form] understand. They're so compassionate . . . They were able to have an understanding, a compassion towards the future of *your* existence."

Paul then felt waves of energy passing through his body, pushing on his abdomen, and causing his hands to feel like "needles." He said this felt "comfortable" and "integrated in a way." The memories he was having, he noted, were "not foreign" and "very clear." We were coming close to the end of the session, and I asked Paul how the memories he had recovered today could be useful to us as human beings struggling to understand our violent proclivities. His first response was that what he had recalled today was helping him to "understand more of who I really am . . . I am a cross between—this is hard to understand for me—between what I've known as me and the brothers who were with me, what the human beings would call an alien." He could not quite grasp the word, but it came to him that these were something like the "TA" people. The TA people have evolved over a long time but differently from humans, Paul said. They did not expect so many problems when they "integrated with you." I asked him why they had decided to integrate with us. "That's the way creation works," he said, but "humans aren't ready to do this, and we are ready . . . *we* want to learn."

I asked Paul why the evolving alien-human relationship was surfacing so much at the present time. He said that human evolution had reached the point where we were better able to accept the connection. "The human perspective or evolution has grown to a point that it's intelligent enough to accept more, but it's on the border. It's teetering back and forth." As we talked of these matters Paul felt intense heat energy passing through his body and concentrating in his hands. I wondered what it would take to push us "over" the border. "Accept *everything*," he said. There is "so much" to accept. For his own growth this meant more fully accepting his TA identity, "what I am." He spoke of how hard it is to be "in the middle."

"I'm in between. I'm more than a TA person. I'm more than a human being," he said. "This is hard! More people are finding out just how in between they are themselves."

It seemed as if with each expression of opening to the elements of his complex identity and the responsibility this seemed to impose, Paul would feel additional strain in his body. "It can only take so much. I am getting tired," he said. He felt flooded with intense sensations of heat and pins-and-needles-like tingling "through my organs," especially in the stomach, chest, face, and hands, which he related to big "growth jumps." My squeezing his hands intensely with my hands helped to release the heat energy blocked in his hands.

As Paul came out of the regression he said that he felt "a lot better," strongly "centered." We talked about the responsibility of his double identity and the energy associated with the information he was processing. Perhaps the shell we have constructed, even the destructive course, were not so formidable, he suggested. Pride, fear, "this ego stuff," were "dead ends" that start out with sensitivity and become "like a cancer" which "lock[s] everything" down. The alien-human "unification" might create a new balance, an evolutionary step, a kind of cosmic mutation in, in Paul's words, "the balance of creation and destruction."

As we reviewed the session, Paul spoke with wonder at the storage in his consciousness of "ancient memories," for example of reptilian intelligence, and the virtually unlimited perspective this could provide. But human beings have lost the "incredibly intelligent" power or utility of that memory bank. For example, "People just look at them, like, oh, the dinosaur has a small, little brain with flabby arms. It ate, slept, shaked. That's what it did. And they were killed by a meteor because it didn't have the know-how. And of course they're going to relate it to it didn't have hands like us and stuff, like, so it couldn't create a home like we have. It's, like, wow. That's so egotistical. You want it to be like you so it's intelligent.

You know nothing about their culture . . . you got some bones, you know. You don't know. You don't know anything about them . . . It occurs to me that this has been the way for a long time, but we know nothing about the animal kingdom. Nothing. Yet it's always around, and there's definitely communication . . . What about the intelligence of the energy making the form?"

I was out of the room for a few minutes, and the session concluded with Paul speaking with Pam of human domestication of animals into pets as an expression of our need "to control everything around us because of fear," the narrow perspective of human identity, and the "twisted," competitive, and intolerant culture we have evolved.

Our second hypnosis session took place six weeks later. Before the regression began Paul spoke of his desire to further overcome the inner impediments to his personal transformation and the fulfillment of his mission. More specifically he felt that throughout his life he has been immersed in dysfunctional systems, beginning within a family that often responded to his need for love and support with abuse and "manipulation to conform," and continuing in social and political systems that restricted his capacity to love. His dream has been to break down the barriers of fear between people and to create "a network of lines of communication" on the way to building new structures based on love and healing. But he fears the pressures that are directed against anyone who tries to push against familiar limits, "an incredible attack from the society in general upon you because you are pushing another boundary open." He expressed concern for what would happen to me. "Everyone's going to be scared shitless of what you're doing, and let alone the fact that it feels like it might be successful."

Paul spoke of his need to feel confident of the emotional climate in the house before "being opened" further, and commented on the uneasiness about the abduction work he had sensed in my wife, which was allayed somewhat in a brief conversation that he had with her before the session began. It became evident that questions of trust between Paul and me had developed, and, to a degree, I was being perceived through the lens of his disappointment with previous caregivers and also as an authoritative senior representative of the hierarchical social systems about which he felt so troubled. As we sorted out these concerns, trying together to separate reality-based concerns from distortions that Paul brought to the situation, he became better able to "trust you enough" for the session to proceed. I acknowledged that it was natural for him to need to test my trustworthiness, and he said that the "testing process" did not have to be "hostile" or "a violent proving. Then truth would be lost for sure . . . I know you're coming from a deep place," he said, "and I do feel safe in telling you what's going, what's happening."

Before we started the regression Paul spoke further about the intense struggle he had been undergoing as he confronted his own and other people's fears related to the abduction phenomenon. Pam had introduced him to several other abductees with similar difficulties, but, although this was helpful, his experience that "everybody keeps throwing things out" left him feeling as if he were "drowning." I spoke to Paul of the "hero's journey" on which he was embarked, and he talked further of "this outrageous doubt and fear that's here." In the regression he wanted to "go in" to the "incredible pain" he felt "right in my heart, right in my chest." Referring to his drive to my house that afternoon, Paul said, "I was crying my eyes out all the way here. I was just feeling everything. Just the pain of this world . . . When I pulled up in front tears were just streaming down my face . . . I have trouble crying in front of anyone." I asked if he had been able to cry with his father. "Probably," he said, but it would have been a cry of defeat. "I'm crying out of pain, out of power, coming here," he said. My last question concerned what he wanted to go back toward. He wanted to "access" that pain "for sure," he said. "It's going to blow out of my chest, and I'm in my own way right now. More than I have ever felt before."

Paul's first image in the hypnosis was from a recent abduction experience. A hooded figure on a ship took his hand and walked him through a door and down a hall to a dark room. In the room a light was on and he was strapped into a chair. The figure had a big pointer and showed Paul on a screen getting hit by someone in his family. Then "he's showing me the world" and "all these people are dying. He's telling me that I'm going to fix it." The figure said, "I already know how" and "it's in all of us and that's how it will just spread right across."

The scene changed and Paul went back to when he was a boy. He was about twelve years old in the cellar of his home when this episode began. "I'm fighting. I'm fighting alone. The thing that's fighting me knows that I'm right here, but I'm protected in some way, I guess. 'Cuz it can't just come out and just kill me. I think it wants to, though, definitely, he thought as it seemed to lunge toward him, only to be stopped by some sort of protective barrier. But it can't do that. It has to do it in other ways. It's going to try. It's going to take me apart, little piece by little piece . . . We've done this battle before. That's why I'm still here. He says he fights with everyone like this." The battle seemed mythic to Paul, as if he were "confronting destruction that wants creation stopped (some people call it Satan)," he said. "I'm yelling," he said, "but I don't think anyone's around."

In the darkness Paul perceived a glowing, nonhuman creature staring at him that he called "'symbolistic' . . . It controls humans to a great degree" and wanted to "destroy" him. But he was protected from "getting lost" by a "creational force" that held him. Once again he experienced numbness spreading through his whole body and helplessness. But the creature could not kill him because "I know too much about who I am and where all my strength comes from. It cannot cut me off." Death would come through "isolation," but Paul was "connected through my back" by "cords" that the creature was "trying to cut." Recognizing that he was speaking symbolically, Paul said that the cords, which were "connecting me to parts of myself," were cut by the creature. Experiencing great distress in his body as he related this, Paul said, "It's too painful. It's too painful to be here. It hurts."

No longer at home in his cellar, Paul now was lying on his back in his bed at night and experienced "'things' . . . moving around me." He could not move at all and saw "a thing in the closet" with a "horror face," like "that character from *The Exorcist*," that frightened him intensely as a child. The figure seemed to turn a light on, but receded further into the darkness of the closet. Paul wanted "to go after it alone." Surrounded by darkness now, he could still see the figure "huddled in a corner. I can feel it breathing over there. It's dangerous in a way, but I think it's just been beaten so badly." He chose to reach his hand toward the creature. A shift of consciousness occurred and Paul was once more back in his room. "Yeahhhhh! Yuck! Okay. It's so broken! It looks broken. Its arms don't move the right way. They don't move like my arms do. It's slimy. It doesn't feel like me. I can't, I can't, I can't understand it. It wants me to understand it."

The creature tried to communicate with Paul and to touch him, which "is freaking me out . . . It is telling me about myself. It's trying to tell me something about me." The figure told Paul that "it's me" and that he (Paul) has "the power to make this thing, and I can't see how I do. I can't see that I do." Once again Paul's body became numb and he was with the creature in the woods and they were talking. The figure seemed then to become smaller, "four feet or so," and looked "kind of like me. It's got eyes and nose, kinda," except "flatter . . . there's not much to it," and ears that are "just like holes in its head . . . aw, this is weird lookin'!" The head was large in proportion to the "thin, thin body." The figure reached out to Paul with a hand that had two or three fingers and a thumb, and "it just wants to talk to me. It doesn't understand why I'm freaking out!"

"Why are you freaking out?" I pursued.

"I'm, I'm afraid! It looks really weird! . . . It's just different than me!" The being "keeps touching me," and Paul could not understand why. "It wants me to understand what it has to say. It wants me to understand how to be me. It's trying to help me to be me."

Paul now believed he was about nine when this experience occurred. Still in the woods he saw a ship behind the creature. "It keeps tilting its head, but I can't, I can't, I can't talk to it. I don't know what to say to it." The figure held its hands out to Paul, and "it wants me to take its hands." But he was too afraid and could not "open to it like it wants me to . . . It's just so different." The figure pulled Paul by the hands into the ship. "Oh, my God!" Paul said, as he felt himself pass literally through the door of the ship—"it's liquidy in a way, but it's still there." Inside the ship it was dark at first. He was in a sitting position as several beings searched his body with their hands, as if "they're confused about something." Although he felt as if he gave permission to the beings to touch him, he resisted communication with them, which they did not understand.

The beings wanted Paul to lie down on a table, which he did. He had no clothes on, could not move, and felt cold. "I don't understand," Paul said, and felt terrified and confused. "They're cutting me open." Using what appeared to be some kind of light the beings made a seven- or eight-inch-long cut in his right leg above the knee. The "loose" flesh opened about a half inch, exposing muscle, ligaments, and bone but creating little bleeding—"It should be bleeding! Why isn't it bleeding?" The procedure did not hurt, but the sight of his leg opened up was terrifying to Paul. "They're just looking into it," he said. Then "they're taking a little piece of my bone." Using "just the light" the beings closed up the wound, and "now we're going to talk."

Left alone with his panic, Paul found he had difficulty breathing, which he experienced in our session. He felt that the beings tried, but could not understand why he was so intensely frightened. The beings explained to Paul that there is "some relationship between us" and that "I'm from them." At this point in the session Paul experienced a kind of split in his awareness. In his alien identity he understood that they were trying to help him, but as a human being "I'm having trouble understanding who I am" and "explaining to other people." The operation on his leg and "lots of" other things before and since have been done to him by the beings to "change things inside me" so that he can become "like a liaison" who can "introduce them" to me and other human beings. But he was afraid for his "new friends," afraid that "*they* are going to get hurt" because "everyone's too afraid of them."

The aliens, Paul said, have taught him many things, like "how I think" and "how energy works in me." I asked him to explain. "It's a very powerful thing . . . Your thought has great impact on where that energy's going to go, and they're teaching me how to be aware of where I want that energy to go. They show me how to use it. They show me how to feel it in my body. They show me how to feel it in other people and in other bodies. They're showing me their technology."

"Like what?" I asked.

"The way they heal themselves" when they get cut or hurt, Paul said.

In fact, Paul (like many abductees) has been psychically skilled all his life and seems to grow more so as he recovers memories of his alien encounters. What is unusual about him is his ability to communicate his skills to others in a simple manner. Pam Kasey has seen him use metaphors from common experience to enable people to move awareness from one place to another in their bodies or open to the solution of a problem. By asking a few simple questions, he can also help people become aware of a level on which they are already receiving information that they had until then been ignoring. His teaching ability, as Pam and others have observed, is extraordinary.

Paul explained then that "when they are exploring sometimes certain ones die," but they can be "collected again" and "brought back" (i.e., to life) by the energy of the other beings. "They make it [the dead one] absorb, like the energy that is the consciousness" of one or more of the other beings, because the one that died "wasn't supposed to die then." He then gave as an example when "the ship crashed" in "a desert" after "they were shot at by us" and "there's a couple of them dead."

At this point in the session Paul was puzzled to experience himself as actually *being* at the crash site. "I'm just here with them. I feel like I'm their friend. I know who shot them. Why? Why did we shoot them? Why did they shoot them? This isn't right. Men in uniforms. They're showing me who shot them. I don't belong with that group. It's the military. It's soldiers. They shot them. They [the beings] have been hurt. I can't help them."

"Then what happened?"

"All the jeeps are coming. We're taking off. We have to go. We're going to leave the rest." I asked if he was one of them or in human form, and Paul replied "I'm human." "We have to leave the crash site," he continued. The army's coming and they're going to take everything. They're taking the ship; they're going to take the ship." But he saw that his alien brothers were "hurt by the fear of those men . . . They [the aliens] have to show me this stuff," he said. "I don't like to see it." "Why?" I asked. "I don't want to be human. I don't want to be human. I'm sorry for being human. I didn't mean to hurt them." Paul explained that he was also nine when he was at the crash site, in another ship that went to rescue the dead ones.

Paul felt anguished that several of his dead "friends" could not be "collected" and had to be left behind in the desert. He had wanted to help them and felt sad that "they suffered all because of the fear of those other people . . . the ignorance of those human beings." His life, Paul said, is devoted to bringing "the awareness up," but in order to achieve that "I have to love me and let myself be here." In the session Paul felt his heart "opening up more . . . It feels warm," he said. "Things are melting," and he had the feeling of "peace and love . . . spreading . . . The planet [Earth] is going to grow," Paul said, "by caring for ourselves," but "it starts with me. I have to accept what I've learned," especially that healing begins in the heart "and then it flows out." The alien beings, he said, had shown him that "hatred of others" had caused sickness of the heart and that "the force to grow" was "all around me . . . They [the aliens] showed me to use that [knowledge]."

His own role, Paul said, was to function as a bridge between the aliens and the human world. "They want me to form a group that can meet with them. They need us not to be so afraid of them, to be open, to understand," to enter into an "exchange" of love. It is necessary for him and other humans to confront their fear if we are "to change this place we live in, that I live in . . . There is much to be done," Paul added, and "I need help to do this. And I need your help." As the session was coming near the end, Paul spoke further of his own "need to grow" and expressed his love for me and Pam. "I can trust you two to help me. Why do we find it so hard to love each other?" We spoke then of the connection between a personal history of hurt, as in his life, or even in the experience of the aliens themselves, and the unwillingness or inability to open the heart.

Before concluding we talked further of what appeared to be Paul's consciousness of an incident like what apparently occurred at Roswell, New Mexico, where a space vehicle seems to have crashed a few days after the first "flying saucer" sightings of our current era. The aliens,

Paul said, did not expect the hostile reception they received. "I felt they came with open arms and they got a spear right through them . . . they got blasted, it seems, and that really confused everything. They completely don't understand us. They start to now." But this initial reception made the relationship "really difficult."

Since in human biographical time Paul was not born until nineteen years after the Roswell crash, I asked him how he was able to be present, at least in his consciousness, at the site (if, indeed he is referring to that incident). "I don't know," he said. "I just, like, let go. I just kept letting go, I guess, which is a big thing I do in accepting information." I wondered about consciousness as a kind of "continuous fabric" that allows you to go "anywhere under certain conditions." Paul's response was complex. He agreed with what I said, but added, "At the same time, you get, like it flexes in. Your energy, you could, when you die, you'll like retract a little bit to that core consciousness, and the memories of who you are still very much here and very much incorporated into that energy. Very much so! And it goes back into the whole and the whole grows back again and then you come back . . . The memories are there, but it's kind of like you push back out again, and you take form again . . . you seem like you feel that independence because you're so focused into one direction, into one purpose, like coming back molecularly and drawing from everything, drawing from all because of what you're connected to . . . The lines that you have to where you're really from are ominous power. It's enormous! But, like, since you're pushing forward you forget what's behind you. That's what I think held me. I think those are those cords, in a way, that I'm feeling behind me, maybe. I'm not sure."

I asked Paul what "form" he was in during this episode. Was he embodied as a nine-year-old, or was it only a kind of consciousness. He said that he felt "very much like I was completely me and then everything changed around me to show me an example of what has happened, to help me understand it, right? And so I'm very much aware of who I am, but everything kept changing so radically that the information is so spontaneous." Paul had felt as if his body was literally on the ship during this event. "It felt like I was there. It felt very real."

Reviewing other aspects of the session, Paul observed "that slimy, broken thing" was his own externalized fear of the unknown. The image in our session was like a feared image from the film *The Exorcist* that had terrified him for weeks as a child. "When I met them [the aliens] and I touch them, and it's, like, Oh, you're slimy! And then cold and all this stuff, and that frightened me more." The aliens seemed to resist, even object, to the confusion of identity or false attribution that occurred here. For example, when they were taking Paul through the woods to the ship he experienced the communication from them that they had "reached out" and wondered "what's wrong? Like, come on. Like, I am who I am and you are who you are and, like, what is the problem with you? And they're trying to get their thoughts in. They don't want to take you over." He felt they were looking at him as if to say "Why aren't you communicating with me? How come you can't accept what you are?"

I spoke with Paul of the difficulty human beings have in accepting and acknowledging the source of power from which we come. He responded that "accepting another human being as a source of information is hard enough. But now [to] accept, like, nonhumans as a source of information for you, as a guru, as a teacher—I mean it's amazing what they have taught me now that I'm accepting it more and more. They may have shown me where the creational force is. They're the ones that assist in linking me up to it in a big way." Paul reflected on his experience of being "completely doubtful" about the reality of his alien encounters, which he related to the limited "definition of God" he had experienced in his upbringing as his parents went "from religion to religion." Yet Paul grew up feeling that he "understood the connection to a source. The terminology and stuff in between is immaterial." He reflected with awe upon the "unbelievable" technology he had learned from his alien encounters, especially the "flood of information" he has received about how they heal. "I've got notebooks full of this stuff," Paul said, "and it's very solid."

At the end of the session I had to leave the room for a few minutes to attend to matters in my house. Paul, feeling vulnerable, wondered to Pam if I felt "let down," and "John knows a lot more than he is saying," he went on. This emerged as a kind of projection, as Paul expressed his feeling that "we could have gone much further . . . We could just keep going. I don't have to stop." Pam reassured him that the work we had done was profound, that I was not disappointed, and that my temporary departure was based on other realities, all of which I affirmed when I returned to the room.

A few days later another abductee, Julia, a young woman with whom I have been working for nearly three years, called me and spoke about Paul, whom she had met for the first time in our support group. Although they had not known each other "on Earth," she felt that she knew him very well from meetings in the ships (it is very common for abductees to report that they have seen or been with other abductees who they know on the ships). She spoke of Paul's "overpowering" personality, which "exudes love." He is "rock solid," she said, and very "centered." In the environment of the ship Paul has a "presence" and a power similar in quality to "my doctor"—i.e., the principle alien figure in her abductions—or to that of any of the alien beings. In particular, Julia said, Paul has great gifts as a healer, and teaches people like herself to "pull despair and hurt from people." He does seem to her and others to have a great ability to carry the pain of people's suffering and to purge them of it, especially with the use of his hands. Julia had never talked individually with Paul, and, of course, knew none of the details of our sessions.

Discussion

Paul's case is illustrative of an increasing number of abductees who are not focused so much on the traumatic nature of their experiences but instead are seeking to communicate information that they feel they have received during their encounters with the alien beings. Our two regressions included traumatic incidents, especially the shock of helplessness and the lacerating procedure done to his leg, but these are of less interest to Paul than accessing the knowledge he has received during his abductions. This information is concerned with such matters as human fear and destructiveness, our resistance to change, the necessity of heart opening and the transformative power of love, space transporting, healing technologies he has learned from the aliens, the nature of consciousness (especially as the primal source of creation), and Paul's dual human/alien identity and role as a healer and a bridge between the two worlds. A central theme in Paul's material is the extreme and unrelenting quality of human destructiveness, which, although based on fear, remains bewildering to the aliens themselves. It appears to the aliens as if we have deliberately chosen death over life, and their experiments are in part an effort to understand our perverse, stubborn ways as well as a kind of intervention to move us along the path from destruction toward creation.

It is difficult to know how to evaluate the information that Paul has received. In the first place, as he himself says, it is hard to accept the power and knowledge of, let alone experience as "gurus," creatures such as these who are so odd in appearance and "shatter" our notions of reality. For Paul, like all of us who are exposed directly or indirectly to this phenomenon, the first task has been to accept the fact of his own experience. Furthermore, the accounts that he provides sometimes altogether defy space/time reality, as in his ability to be "present" as a nine-year-old boy at Roswell in 1947, nineteen years before he was born. Such space/time traveling can only make sense by conceiving of consciousness as a kind of hologram of universal sourcefulness which can create matter and form itself and to which Paul, and each of us potentially, has access if we can open and "let go" as called for into this primal universal information or energy fabric. In fact, much of the material of Paul's sessions was concerned with the form and identity-creating power of consciousness and the compelling necessity of opening ourselves to its infinite qualities.

What made Paul's communications so compelling and persuasive was the intensity of feeling and bodily movement and sensation that accompanied each new thought. In working with him it was as if the hypnotist's role was to facilitate Paul's access to knowledge that was stored within him and that powerfully affected his body as it moved into consciousness and could then be communicated. The idea of consciousness as an infinite source of energy and form to which each being has access makes it perhaps inappropriate to consider each communication of Paul's—as, for example, his presence at Roswell—in terms of whether it is literally factual or concretely "happened" in linear space/time terms. I appreciate that this will not satisfy anyone who would still hold to a view of reality that is limited to the physical four-dimensional universe. On the other hand, there may be some value in challenging our restricted epistemology and expanding our criteria for evaluating information to include the power or intensity with which something is felt and communicated and the potential utility the knowledge may have in relation to our contemporary dilemmas.

Applying these criteria, there is little doubt that Paul experienced the information that came to him in our sessions as having great power, as did those of us (Pam and myself) who received it. Furthermore, the relentless messages of the need for change, of the necessity of human mind and heart opening, and of the catastrophic consequences of our having mistaken the "shell" of our defensiveness for the whole of human identity—messages which comprise the essence of the information that Paul has received and imparted—are all communications of great practical value in the context of the present global crisis.

Finally, there is the question of Paul himself and my role in his development and personal opening. Those who know him outside of the therapeutic setting, such as Pam, Julia, and other abductees, all testify to his extraordinary intuitive and healing abilities. He has come to me to enable him to free his powers from the restrictions of the suppression (a word he used both in relation to his mother's uneasiness in the face of his intelligence and abilities and his former therapist's inability to deal with the abduction-related material) that was the result of several sets of forces. These include the repression of the ideas and memories locked within him (an adaptive response when one considers their intensity, Paul's need to function normally in everyday reality, and the absence of a supportive context in which it could be safe to open himself to such unusual information), the ontologically shattering nature of the information itself, and the sheer power of the energies involved, which required the creation of a strongly supportive and trusting context before Paul could allow himself to bring forth what was held within.

Paul experiences himself as a bridge between two worlds. He feels deeply that he has both a human and an alien identity. The task of integrating these two basic dimensions of himself—a challenge that many abductees who experience this double life must face—is formidable and is a central aspect of our work. For Paul, like other abductees who feel that they have accessed the source of creative energy in the cosmos, their human identity and participation is intensely painful, especially in the face of the destructive institutions or life systems we have created. "Home" for him, as it is for many abductees, is on the ship or with "them." Yet at the same time Paul feels strongly that he has been given, or chosen, a role on Earth to contribute, as an example in his own being of openness and love, to the evolution and transformation of human consciousness. My role, which Paul has shown me, is to facilitate his capacity to accept and live out the awesome responsibility that he and others like him have undertaken in the face of a culture that resists at every turn who they are and what they are trying to accomplish.

CHAPTER ELEVEN

Eva's Mission

Eva, at age thirty-three, was working as an assistant to a CPA when she read an article in the *Wall Street Journal* that described my work with abductees. She called my office and said that she would like to be interviewed because she "may be going through the same thing" as the experiencers described in the article and "it's important for a lot of people." In a follow-up telephone conversation Eva told Pam Kasey that she had been having "feelings night and day of entities . . . dreams" of beings in her room that are "still there" when she wakes up, and recalled incidents from early childhood and late adolescence when she could not move as her vagina was probed by "midgets" who had somehow gotten into her room.

Eva's unusual encounters made her wonder if she were "crazy." Although when I first met her she was already drawn to follow a spiritual path, Eva had always considered herself to be "a very logical person" and these experiences contrasted with accepted notions of reality. Nevertheless, she intended to discover the truths that underlay her experiences, a determination that fit with her sense of herself as a pioneer with a "global mission" to help others. Until she read the article in the *Wall Street Journal*, Eva felt very much alone in the struggle with her encounters. The day before her telephone interview with Pam Kasey she wrote in her journal, "I'm trying to cope with it in my own way. It's hard. No one to talk to. No one to cry with, to ask for reassurance, for understanding. It's a heavy burden to carry alone. How can I help Sarah [her daughter]? She's only six." Although her abduction experiences were disturbing to her, Eva sensed from the beginning of our work that they were connected with some sort of purpose, and that she was a "vehicle" through which information from another, higher, source might be transmitted. In January, February, and March 1993 we did three regressions.

Eva was the oldest of three children, born in Israel. Her father is a banker and real estate investor whose work has required him to travel a great deal. The family lived in England, Venezuela, Florida, and New

York as Eva was growing up. She married in 1980 and settled in the United States in 1985. As a child Eva felt that her creativity was suppressed because of the necessity of deferring to the needs of her father, whom she still respects but describes as cold ("He's not a touching person," she told me when I inquired about the possibility of childhood sexual abuse). Eva grew up as a conscientious person with a strong sense of wanting to please others, if necessary, at the expense of her own freedom and imagination.

Eva's husband, David, works as an electronic engineer for a large photographic corporation. Her marriage is somewhat traditional in that the care of the household and children falls mostly upon her, while David, as the principle breadwinner, works long hours at the company. Eva's "personal agreement with herself," related to her acute sense of responsibility, included the requirement that no one would be hurt in the process of her personal evolution. She did not, therefore, tell her husband of her unusual experiences or her work with me until after our second regression in February 1993, nine months after she first called my office. Her concern was not only that David would not understand her experiences, but also that he might be troubled by the information, which could create uncomfortable tension in their relationship.

Eva and David have two children, Aaron, age nine, and Sarah. After our first regression in January 1993, Eva spoke of her concern that Sarah was having her own abduction-related experiences. About three or four times a year she was waking up from "bad dreams." For example, one night a month or two before this session Sarah woke up in the middle of the night screaming for her mother. Eva went into her room and Sarah said first that she had had a bad dream, but then she said that she had seen a ghost "flying in the room" that was all white "and he wanted to take me" and "I didn't want to go." It was clear to Eva that Sarah was "fully awake" and "fully energized" by what had happened, though she did not talk about it the next morning. Eva does not believe that Aaron was having experiences, but "he's into computers and spaceships," which he builds in the computer, and "his imagination is so wild" that "if he would wake up in the middle of the night and tell me about a dream, I wouldn't know if it was real or not real."

The first abduction experience that Eva recalls occurred when she was four or five and living in Israel. She shared a room with her baby sister, who seems to have slept through Eva's experience. Further details of this encounter will be reported in the account of the first regression. Before speaking with Pam Kasey, Eva had written in her journal that she had started to read Whitley Strieber's *Communion*, but discontinued it so as not to be "influenced by anyone or anything . . . Then something triggered my memory," and she recalled waking in the night and seeing "three 'midgets,' about three feet tall." They had dark brown, wrinkled skin and triangular heads. They stood by her bed and touched her genitals with what felt "like fingers, probing, kind of experimenting," without force or a sense of sexual urgency or intensity. She felt helpless and could not move, and when she tried to scream, no sound came out, at least at first.

In her journal Eva wrote, "They walked through the space (about ¼") between the wall (external wall) and the door (like walking through a wall) and disappeared just as my mom entered the room. I told her there were midgets in the room, that they just walked through the door. She looked. Obviously she saw nothing. Told me it was a dream and that I should go back to sleep. I was scared. Didn't believe her. I was sure they were real. I saw them. Heard them. Felt them. The first time I remembered this was last night. I don't know what brought it on. I'm writing this now because I *feel* it. In my veins. As if it just happened. And I know it's true because I have goose bumps all over."

About ten days after our first meeting in October 1992 Eva wrote in her journal that she was driving to Boston listening to the tape of her session when she began to recall more details of this childhood abduction. As she listened to the part about trying to scream for her mother, "I jumped on my car seat" and "all of a sudden a picture flashed in front of me as a result of looking at the highway lights. I recalled a spaceship looking like this [drawing in journal]. It was huge, gray in color, metallic. It was hovering close to me. And then I saw a face of a female with big eyes (round and dark) and green (like eyebrows)—light green all around the eye area and coming almost to the edges [drawing in journal]. Then the picture disappeared." The recollection was brief, perhaps seconds, but very vivid with many details, including the lines, shape, and structure of the craft.

Eva recalls another childhood incident when she was about six that she now relates to her abduction experiences. She had developed pneumonia and was taken to an emergency room in a hospital. The bright lights frightened her and triggered an abduction memory. In her journal entry of May 22, 1992, she wrote, "It's not like a single dentist's light. It's a few of them. Above you. Like small projectors. And you're on the bed, helpless. And strangers around you. Touching, feeling, investigating, experimenting. STOP. No more." When the doctor in the hospital told her to lie down on a bed and take off her undershirt she refused. Her mother urged her too, but she still would not lie down or take off her shirt. Although she screamed in terror, "they forced me . . . I absolutely hated it."

Under hypnosis Eva recalled seeing a spaceship about a year later in the grassy area behind the apartment complex in England where the family was living. "It's very low, and it has like three, three things of fire coming out from the bottom. It's gray, and at the bottom there are, like, oh, what you would call windows, with, like, light coming out." Eva believes "they blocked my memory so I could not remember . . . You remember absolutely nothing," she said. "Otherwise it interferes with your daily life."

Eva believes that the aliens "have a tracking mechanism," and relates an experience when she was about nine and still in England to a possible implant. She was doing somersaults on horizontal bars, missed one, fell, and bumped her head "really hard." She says that she felt that "something moved" in her head, "something they could keep track of me . . ." I asked her if she could feel it move, how she knew. "I just know," she said. "They had their signal" from this accident, and they came back and "corrected it." Again I asked how she knew. "I know," she said.

Two other incidents occurred when Eva was nineteen while serving in the Israeli army. In one, which she recalled in greater detail in the first regression, she felt "as if somebody has given me a shot, a gas mask, I don't know what, to knock me out," and as she was waking she heard whispers in her room (she was alone in her parents' apartment at the time) and "there was one female and two or three males around. I was frozen again . . . totally frozen" and "felt something between my legs . . . I was very afraid, but at the same time I was trying to be logical and analyze the situation." She "felt the female and the male standing over me." At the time Eva did not think of this "as something extraterrestrial . . . I thought it was robbers and waited for them to leave." When she was able to move again and looked around there was nobody there.

The other incident occurred while she was on the night shift in the air control. At a slow time—perhaps three in the morning—she put her head down to doze and then "saw myself floating from the ceiling . . . My consciousness was up there. My physical body was down there." A "voice said, 'Come with me, it's good,'" and "I knew at that point I had a choice of living or dying." Although her heart was beating fast and "I was sweating like crazy," Eva was not aware of any life-threatening illness. She said, "I wasn't interested in dying, and I said, 'No, I'm not coming.'" Eva "knew" she could have died but does not understand why and found the episode confusing.

Eva had two experiences in the month before she read the *Wall Street Journal* article that probably sensitized her to its content. In the first incident, described in her journal entry of April 14/15, she woke during the night and saw a "violet rectangle, like a doorway into/out of somewhere not visible, maybe another dimension." She saw then "the upper parts of some people wearing white and they were standing at the front of my bed." She blinked her eyes, thinking "it's my imagination." But "when I opened them again they were still there. At some point I understood it's real . . . I felt as if they were 'bringing me back home'—whatever that means."

On May 6, eight days before she read the article, Eva wrote in her journal, "Last night when I went to bed I wanted so to meet them. I asked, begged, for an encounter. I volunteered myself (my body) for their examination so they can further their knowledge about us earthlings. I was just about to fall asleep when I felt a strange dizziness. A loss of gravity, as if I was swirling into a tornado, as if I was being sucked into somewhere. I knew I could stop it just by physically touching my husband beside me in bed. But I knew my wish was granted, and I didn't want it to stop. I suddenly felt (?saw) a light-blue light encompassing me. It was light blue, yet there was a white light inside the light-blue one. It was a soothing light, yet one that I know would lead me to greater knowledge. It was magnetic. It was the feeling I got from it that is beyond words. Words are too physically limiting. When I felt/saw the light the dizziness/twirling stopped. I went blank. I didn't sleep well. I know that much. I woke up 2–3 times that night, finding it difficult to fall asleep again. I was restless. When I woke up in the morning I was so tired! As if I went on a journey all night long. I hope I did. And I wish that someday I will remember those journeys and all about them so I can use the knowledge to help mankind."

The following morning Eva's husband, who seemed to have slept through the incident, told her that he heard a "big bang" during the night. She felt "full of energy" and filled with "love and hope." But at the same time she was frightened and wrote in the journal, "I felt I was really going crazy. I couldn't tell David. He'd think I was nuts. I was so scared. Couldn't fall asleep. Didn't know what to do. I have to get help?! Somebody I could talk to. Now it's me and the pen and paper. But I need somebody humanly to hear me out. With no judgments. No expectations. No accusations. Somebody with explanations maybe." She continued, asserting confidently that the beings mean us no harm, that they are here to help, and "I love them."

In the weeks before my first meeting with Eva in October 1992, she had several other powerful experiences involving the feeling of strange presences or the conscious perception of unusual entities, including "beings from a totally different dimension." During a hypnotic session with a dowser from another state who also did healing work she found herself going back in time to "160 or 180 years ago" and "moving from one dimension to another," experiencing a "difference in the energy vibrations" and located "in a different planet, star, galaxy—I don't know the name of it." In her journal entry of September 22, which she wrote after returning from an energetic healing class, Eva explained that she was "living two dimensions simultaneously . . . I have the gut feeling," the journal entry continued, "I was in a higher dimension where linear time is irrelevant."

After considerable difficulty in scheduling, Eva and I met for the first time on October 15, 1992. In this interview she provided much of the information recorded above and expressed the wish to explore her experiences further with hypnosis. "That's what I'm here for," she said. But again there were problems of scheduling and postponements, so that our first regression did not occur until January 18, 1993. But in the weeks before this session "things started to surface."

In her journal entry of December 6, Eva wrote of a powerful experience that had occurred the night before. "I was almost asleep. But not yet. I lay down on my stomach, my head to my left. My eyes were closed. At the tip of my eye I saw a gray spaceship [she drew the ship and symbols that she saw]. I panicked. I wanted to scream. I couldn't." She felt that she could break out of the experience but "I got courageous and convinced myself to go 'on the ride' but trying to acquire as much info. as possible for P.C. [should be K.] and J.M." After this she "blacked out" and the "next thing I remember was I was lying on a hard surface. Maybe two people in the room. . . . I kept my eyes shut very firmly because I was terrified to open them.

"I remember 'them' (?) or me, I'm not sure which, wearing a dark gray garment/robe with many buttons going down the back. I was in a fetal position, my back to them. They were doing something to my spine. My entire spine was stinging and cold. It was awful! It felt as if they were going inside my body with some very sharp instrument (syringe?) and inserting it between my flesh and skin. The stinging sensation persisted. At one point I started moving, resisting the situation, although I was at the same time scared of the consequences of doing so. I tried to get at some point more data with my eyes closed. The air was moist. The surface I was on was hard and a bit slippery. I had the feeling the room was not well lighted, but then again with eyes closed I'm not sure.

"I continued to resist, and at some point understood that they agreed to bring the situation/experiment to an end. Before blocking out again I remember I saw this symbol [draws it] in red. This is not the first time I saw this . . . Next thing I remember is finding myself in bed, hearing my husband. I was in a fetal position! I went to sleep on my stomach. I felt in total panic again, something that didn't happen before. I wanted to wake him up and tell him I was 'taken' somewhere. I also knew he'd never believe me. He'd think I was crazy. I thought maybe calling P.C. or J.M. Didn't want to bother them. It was Sunday morning 5:30 A.M. I was shaky the next day. Still am. I'm trying to relax, accept it, and somehow make sense out of it." She felt "edgy for a couple days" and tried to "put it in the back of my mind."

On December 22, Eva wrote in her journal of the reluctance she had had about writing down the details of the above incident as "writing would legitimize the entire occurrence, and I wasn't ready for that then." She developed cold or flu symptoms and wondered if these had been "brought about by 'the abduction' and being 'bare' on top, and the injection of whatever it was, etc. The clothesless me and the injection of a foreign substance might have caused a reaction in me. P.S. Nobody in my family had a cold/flu at the time." In our discussion of this incident six weeks later Eva said that the intensity of the physical sensations she had experienced convinced her the incident "was real . . . I felt the sensations. I felt it hurting. I felt it stinging. It was cold."

Eva came for her first hypnosis session on January 18, 1993. She felt anxious but curious. "I love the unknown," she said, and we spoke of her determination to proceed despite the scheduling difficulties over the holidays. Before inducing the regression we reviewed the childhood incident when the "midgets" had entered her room and the recent episode described above.

In the trance state Eva spoke immediately of lying on "something hard" with "something above me like hieroglyphics." She felt scared and heard herself screaming. A figure in black and green came out of what looked like a gray elevator. "It's cold in the room . . . He tells them to stop. They give me something, and I feel like I'm being sucked into a white light and I don't see him anymore. It's morning. I'm a child in Israel. I don't remember anything, and I just hear my mother telling me to get up and go to school, to get dressed."

I took Eva back to the nighttime experience itself. She recalled seeing "when they're bringing me back" a gray, "dome-shaped" craft with red lights over the balcony of her fourth-floor apartment. The bottom of the craft seemed to be a circular rim that turned and emitted "some sort of light or energy" and "that's their transportation method." She saw three beings that appeared like midgets, with brown skin and "all over wrinkled," wearing olive green and dark maroon suits with black belts. Their heads were hairless and pear-shaped with very dark eyes, like a "dead blue" and "mushed" noses. "One was shorter than the other two, and they were looking at me." Although her mother regularly checked to make sure that all doors were locked at night, when Eva screamed for her the beings disappeared as if through a crack in the door of her room. When Eva told her mother about the midgets and how they had gone through the door, her mother said "what are you talking about," the door was locked, and "it's a dream." Eva insisted she saw the beings going out, and her mother repeated "'It's a dream. Just go back to sleep.' And I did."

I wondered if Eva could recall the beginning of the experience. She said her father might have read or made up a bedtime story before she and her sister were left alone. The bed had a guardrail so Eva would not fall out. She was awakened by "a buzz" and the beings, who were shorter than humans, first appeared standing by the guardrail, "and then there were like searchlights in the room," which came from outside. The beings seemed to know "who to come for" because "they keep coming." She felt afraid, "even scared to call my mom," and shut her eyes and faced the wall in a fetal position ("the best defensive position") to "make it go away." She heard whispering, and the beings injected something that felt like a needle in her back to quiet her. This time—the first incident Eva could recall—she was flown from her bed in a horizontal position on what seemed like a canvas-and-wood stretcher. Then she was "sucked" into the darkness outside and into the bottom of "the ship" in the light that was "like a beam of special energy" emanating from inside the rim at the bottom of the craft. Eva felt a mixture of terror and confusion as she saw her balcony and the building next door as if from outside the house.

Once in the craft Eva found herself in an "examining room" on a table with "small people and lights" around her. There were "lots of green and red buttons . . . it would be comparable to computerized systems. But it was different." The little people in the ship reminded her of the dwarfs in Snow White and seemed lighter than the ones that had come into her room. One of the beings communicated to the others—not to Eva directly—in a voice that "resembled more ours" that their purpose was only to "experiment with me" and to cause no harm. She was "in shock" and defenseless as the beings poked her legs, spine, neck, and brow with "sharp things," as if "they were trying to understand." She could see a silver instrument with a round tip that was inserted in her forehead. A white or yellow fluid dripped onto her nose.

The beings seemed to Eva to be both excited and amused "because they kept communicating with each other" animatedly. Perhaps they were being too "intense . . . overdoing it," for the leader came in, communicated something and "slowly, slowly they stopped . . . He agreed with the original purpose, but it's like a teacher that walks out of the room and the kids in the classroom start to fool around and do their own thing." The return reversed the process of the abduction; Eva felt herself "like on a slide coming down back to my bed." Once she was in the bed "they were standing by the bed guard, making sure I was okay." Once Eva "regained my physical movement" she "screamed and they fled."

After recalling this experience, Eva had the sense that it was "not the first time." Although she could remember no details, she felt that something had happened when she was two or three years old. She feels certain that the beings are able to "keep track of me" and recalled the incidents from age nine described earlier in which they seemed to have "corrected" an implant that was dislodged during a fall. Eva has the sense that the memories of her childhood abductions were blocked by the aliens.

I asked Eva directly what the next encounter that came to mind was. She replied, "I was nineteen, in Israel, serving in the army, and I was sleeping. It was in the middle of the night in my parents' apartment, alone, and I woke up because I heard whispering noises, and I felt and heard like people walking into my bedroom and into the living room, and I thought maybe robbers or something, so I didn't move." Eva knew the windows were closed and the door locked, "and if it opens it makes a shrieking sound." She could not move as she saw "three beings standing there . . . They were whispering, and one of them went out of the room and the other came back, and they were touching me between my legs, and I did not understand it because I was not dreaming." In a panic now, Eva tried to scream but no sound would come. She experienced what felt like fingers, probing inside her vagina. "It was not pleasurable. It was incomprehensible," Eva said. She thought perhaps it was her own fingers, but when she checked her hands "I felt them on the outside of my thighs."

Eva does not know whether this experience occurred in the apartment or someplace else. Her eyes remained closed throughout, and although there was a lot of light that came through her closed lids, she thought perhaps it was early morning. The positions of her body she found confusing. "When I was aware of what was happening it was like I was on my side, but when it was happening I was on my back. I don't know." After this experience ended, Eva forced herself to forget it, and until the period when she contacted me did not connect it with her childhood experiences. She believes that she has had approximately ten encounters since age eighteen and that the beings are "more interested in the human after, at the adult stage, not so much as a child."

At this point in the session Eva shifted from the direct reporting of her experiences to speaking of the motives of the aliens and the meaning of the abduction phenomenon, based on information they have given to her. "Their purpose is to live in unison," she said, "not to take anything from us. They want to study us to see how they can communicate . . . There are different dimensions, worlds existing within worlds," she added, "and to go from one to the next is like a roller coaster. You need to speed up the energy, and then you go to another dimension where the reality is different. In the transition from one reality to another, you feel like you're contracting and expanding at the same time . . . It's like you become on the one hand, part of everything, and everything becomes part of you," but "at the same time you contract into an infinitesimal point." This, she said, is "absurd, because it's two conflicting ideas," but this absurdity contains the "secret of moving from one dimension to the next."

A shift in Eva's perspective occurred at this point, and for the remainder of the regression she spoke as if from the perspective of the alien community, using only the pronoun "we." "It's like I'm not speaking," she said. The intensity of her experience in this realm was physically difficult for her and caused pain in the hands due to blocked energies. The aliens were gone and she saw the frame of a white triangle. "It's too intense," she said, "they can cause damage" to the human body. The beings emanate "from different dimensions, beyond physical," Eva observed, "and they need somebody that's closer to the human being who is able to communicate somehow physically with them . . . The information they can relay," she said, "is of such high intensity that they need something to slow it down." The encounters with human beings do just that, i.e., slow down the transmission of information.

The information the alien beings bring, Eva said, comes from another intelligence, a realm beyond the physical world. But most people disregard it, dismissing it out of fear as "crazy" thoughts or just "imagination." In order to be receptive to this information it is necessary for human beings to be able to put aside their preoccupation with daily responsibilities like work, children, marriage—our customary "unconsciousness." Nevertheless, Eva insisted, it is important for us to overcome our need for power and control on this plane and to acknowledge that life exists elsewhere, though "not necessarily in physical form." One of the problems in reaching us, she said, is that human beings are anchored in the need for "physical proof" through the "five senses," which "we" are "trying to provide." This is difficult, for "we don't consist of physical data . . . We are not in space/time. We don't have any form . . . We are all. You could say I or one. It does not matter . . . We are an offshoot of I" or "what you would equal to God." Eva herself is "a vehicle through which we can convey that information." As the regression ended, Eva thanked me "for letting us be heard" and felt afraid to open her eyes.

After coming out of the hypnotic state Eva said that the experience seemed authentic to her. "It was me. I know it was me," she said, "But it was another me." She spoke of the difficulty integrating the world opened to her by the encounters with her domestic life. "What I'm going through now is alone," she said. Eva noted that she had not been open with her husband about the changes taking place. "He just gets bits and pieces," she said. I invited her to share what she was going through in my monthly support group, which she appreciated and agreed to do. Then she talked about her daughter's recent experience with the "ghost" that was flying in her room. As the session was drawing to a close, Eva spoke of continuing numbness in her hands but otherwise felt well. She described herself as a "pioneer" and a "warrior," who likes "challenges." But at the same time she feels "miserable with myself because I don't understand what's happening." She is reluctant to speak of the new information she is receiving out of fear that others will not understand. I shared some of my own struggles in learning to find ways of speaking about the information that I was receiving.

After the session Eva had a brief, intense headache, which soon subsided. She then felt well. She listened to the tape of the session, and found that she was better able to accept the reality of her experiences than before the session. She especially appreciated that she had someone to talk with about her experiences who "believes it's not make-believe." She felt full of energy, but wanted to be able to control it better. Writing in her journal—fifteen pages on one day—was particularly helpful. Another regression was scheduled for February 22, a month after the first one, aimed at integrating the abduction experiences with her everyday life as a wife and mother.

Before the regression began Eva spoke of the difficulty dealing with "all the garbage that's been accumulating," by which she meant the tension between her "conservative" daily life, and the abduction-related intuitive expansion of her self-knowledge. In having a son and a daughter Eva felt that she had completed "a circle," and was now ready to focus upon her "global mission," particularly her role as "a communicator between humanity and everything else—ETs, UFOs, call it what you will—higher intelligence." She spoke then of the suppression of her creativity and the need to please others growing up described earlier. Perhaps her mission is to be a healer, she speculated, enabling people to break away from the unhealthy effects of institutional systems. A few days before the meeting Eva had a vision of a downward-pointing white and yellow light triangle (commonly a symbol of the feminine principle or Great Mother archetype) with circles (usually representing universality, totality, or wholeness) within it.

Before beginning the regression, Eva spoke of her desire to remember more of her experiences, to "open up" in the service of her higher self and to "throw away the garbage." Again Eva defined the difficulty she has navigating between what we called "the world that defines reality for most people" and the new realms of her experience. "For me, both are real," she said.

In the first minutes of the regression Eva spoke rather abstractly about dimensions of reality, what it is possible to perceive and talk about, cosmic truth, and other such topics. But the bulk of the session concerned her profound struggle to integrate her daily personal and abduction-related lives, especially the problems of communicating openly with her husband. Her first image was of black circles surrounded by golden light, glowing "like sunspots" and "coming toward the earth." Some people could perceive this, she said, but for others it would not exist. She described the object as "energy" that "cannot be perceived by the five senses, but is real nevertheless." Such an "object can be perceived by those who can attune to that range of communication, and it will be invisible to all the rest." To perceive beyond the physical range people must want that communication, she said.

Shifting to her "we" mode, Eva spoke of the difficulty of conveying information about these matters that lie "outside of linear time and space." She seemed almost to be debating with me, as if I were a protagonist for a materialist philosophy. "You either perceive it and it exists, or you don't perceive it and, therefore, it does not exist. The same thing here [i.e., with the black and golden circles]. You are trying to perceive it in those certain limitations, but it's beyond it . . . It's like existing and it's not existing at the same time, and you're trying to make us say, does it exist, or does it not exist." The debate continued around Eva's "I" (personal) and "we" (alien/universal) identities.

The black/golden object, Eva said, was dense, and inside it were different energies and colors, green, yellow, and red. "The way I describe the object to you," she said, "is as though it has an inside and it has an outside, and it has boundaries, but it really doesn't. So it's hard for us to put her [Eva] inside or outside of it." These colors, Eva said, represent different frequencies or "levels of energy, but red doesn't really exist, nor yellow, nor green, or any other colors." We continued a bit further in this vein, as Eva talked of information relating to "cosmic truth," trust, communication through colors and vibrations, and more.

But she soon came down to Earth by observing that she suffers from an "old habit" of "not believing in herself" and the difficulty of accepting the new aspects of herself that have been "unfolding." She has "a very rigid, earthly doctrine," Eva noted. I encouraged her to tell me about that and she said bluntly, "I believe that, number one, my responsibilities are to my family and to my kids and all those mundane routine things. Number one. Having fulfilled those, I believe that I'm free to do anything I please or to manifest anything I believe in as long as NO ONE, and I underline and put it in bold, NO ONE gets hurt in the process, physically, emotionally, or mentally. And if those two things are fulfilled, then everything else is okay. Now, that's her doctrine."

With the shape of Eva's dilemma now sharply defined we were able to proceed with a deeper exploration of it. I said that her doctrine would not work at a cosmic level, that she would have difficulty navigating between her earthly responsibilities and higher self using such a rigid framework. She suggested that she might be "Superwoman," but I persisted that spiritual evolution cannot occur without pain, involving at times others that we love. The discussion continued in a somewhat argumentative vein for a while with considerable resistance ("dynamic tension," she called it) on Eva's part, denial of her struggle, and reassertion of her "personal agreement with herself" not to hurt anyone while at the same time restating the strength of her motivation to move ahead.

I encouraged Eva to notice what she was feeling in her body. She said that was hard when "you're a body and you're not a body" and that I was confusing her. I asked her to tell me again her husband's name, and she asked me why that was relevant and then noticed "thumps in my head" like "somebody with a hammer" and "my heart's beating fast." More denial of hurt, confusion, and an objection to being distracted from more abstract considerations in order to consider her body followed, but Eva did admit "challenges" to be "transcended into the next step." Then Eva acknowledged "she's [*sic*] aching on the right side . . . It's Eva's ache. I feel it now. It doesn't hurt, but it's an ache." But she said, "You created it!" We went back to her agreement to "not hurt anybody," and she admitted in rather convoluted language the problem of reconciling agreements reached outside of "incarnation" or "linear space and time" with those that operate "inside those limitations of space and time."

The breakthrough in the session occurred when I asked Eva what "cost" her not talking with her husband and children about her experiences had for her global mission, and she replied, "If it would be in dollars, you wouldn't be able to afford it." Then she added quickly, "We're joking." She spoke then of her husband's vulnerability, especially in pursuing a business career. He is a "great guy," but appears to have accepted the dominant "earthly perspective" and "belief systems" of this culture and is working in his job, and "here is this wife of his—or he thinks it's his—as most men in your society do—it's a possession—that it's his wife having these grand travels, cosmic travels . . . How is that going to affect him," she asked. I acknowledged the sensitivity of the question, and asked how "fragmenting" it was for her not to have told him of her experiences. "Sometimes it's tearing her [*sic*] apart," Eva admitted and spoke of how "miserable she" is in her job but does not leave because of the "financial situation."

I encouraged Eva to speak from an "I" perspective and wondered again at the sacrifices she was making in the service of her internal agreement. She objected to the word "sacrifice," but acknowledged that "something is not working." She spoke then of her plan to leave her job, knowing that this "scares" David, and reasserted her determination to help others. She will tell him, "This is me. This is part of me, and this is what I love. And if you cannot accept it, that means you cannot accept me for what I am." She wants "desperately" to bring forth her creativity, she said, and must find a "common ground—compromise" with David to be able to pursue her global mission. Eva spoke then of the tearful encounters she had had in the past when she had tried to convey to David her determination to follow her own path.

To tell David of her alien encounters will, she anticipates, be the most difficult challenge of all. "It's like those sunspots at the beginning of the session," she said, "that gold ring. There's not such a definite border anymore." She has tried to hide her encounters in "her own little bag" with "a lock on it where she keeps her journal and your tapings and all that stuff." But phone calls from our group were starting to come in, and she was gradually "letting go of that secrecy." David knew, for example, that she started seeing a psychiatrist. So the "borderline" between what "he is to know or what he is not to know" is "a matter of perception" and "that borderline will diminish and fade." She intended to tell David "briefly" of her encounters. "Not too much detail is necessary—just of their existence, and of higher communication and a general sense of the global mission, and that will be sufficient." I expressed doubt as to whether it would be so simple, and we discussed the tensions which lay ahead for Eva. Characteristically, she found this prospect interesting and challenging.

As the regression ended, Eva asked to be alone for five minutes. After this she spoke of a children's book, written in Hebrew, called *Soul Bird,* which concerns a bird inside every human being that contains many compartments, "a compartment for anger, a compartment for happiness, a compartment for jealousy, for love, for hate, and we're the only one that has a key to those compartments. And we decide which compartments we want to use at any time." Thinking a moment Eva said, "So maybe I should open up more compartments?" We spoke of timing, and then her thoughts returned to David. Perhaps in the service of her mission she might cause someone else pain. Perhaps it is what David wants. "Maybe it's something that he's creating for himself, so he can transcend, as a person, to the next level." She was thinking about telling him the truth of her experiences more fully, with the idea that perhaps he can only change his consciousness "with some sort of major challenge." As the session came to the end, we discussed the particular restrictions that the corporate environment places on the evolution of people's consciousness.

Following this session, Eva again experienced an intense headache and was extremely thirsty for a day or two, which she associated to the releasing of energy and her opening to "cosmic information." In a letter to me two weeks after the regression she wrote of her attraction to "the challenges of the (consciously) unknown." In journal notes which began with entries three days after the session, she wrote of plans to go to Israel for the summer and of past life experiences, including an abduction as a five- or six-year-old boy in the seventeenth century (described in the account of our third hypnosis session, March 15). She also described her sense of space/time collapse in relation to the encounters ("past and future are occurring now and forever"), her longing for global peace and understanding, and set down other philosophical and spiritual ruminations about the evolution of consciousness and "cosmic truths," inspired by listening to the tape of the February 22 session.

Writing in the voice of her cosmic we/she, Eva described the need of the beings to "adjust our communication from higher vibratory levels to those of earthly (verbal) vibration." To "slow down" in this way and "vibrate at more subtle levels . . . takes training . . . We are using Eva's body with her consent full time now. Earthly Eva has not left, but she has diffused with us so that her earthly powers have been greatly enhanced, so to speak." She warned of the limitation of using words to describe such profound experiences, especially the relation of the earth plane of existence to other realities.

In the journal she wrote of telling David about her experiences soon after our session. "He didn't show much interest. Support in any form was nonexistent. I wasn't surprised. Didn't expect otherwise. I'm not resentful. Totally accepting. My hypothesis: he's both in denial and hurt." Although David insinuated he is part of the game, Eva told me later that she is uncertain about whether his claims of involvement are genuine. He said they left a mark on his foot, that they are midgets, and he's seen them, but he didn't seem to show much interest. "Whether he's an experiencer for real or not, time will tell . . . However, he said they should contact him. He's got much info to give. He also suggested that to get rid of the fear I should teach them a game we could play together. And they can teach me one of their games. Interesting to note that from the following morning the fear subsided to almost zero. And the red spot on my nose appeared two days after that. It's like through a dream, except the suggestion was taken up by both sides and the fear worked itself out!"

Also soon after the February session, Eva had a dream of burning houses, which she interpreted to represent the "burning 'garbage' that needed to be burned up in relation to other people." She also sent two journal entries from the previous summer. In one she described being outside "planet Earth" and seeing a spaceship surrounded by a golden light. She was able to communicate telepathically with the entities on it and learned that she was much loved and that she would eventually be reunited with them. "I was ageless yet a young soul." In the other entry she described a past life as a teenager in the 1930s or during World War II. She saw a baby in her arms, and believed that this experience explains her love for all children. In a note on March 9 that accompanied a gift of fruit and other food in commemoration of the Jewish redemptive holiday of Purim, Eva wrote, "Thanks to you and others, I am learning to regulate the energies more productively."

We scheduled a third regression for March 15. At the beginning of the session Eva described how before a lecture at her energy and healing school she went to take a nap, could not fall asleep, felt restless, listened to music, "and then it started." She saw herself as a five- or six-year-old boy in the mountains somewhere in Europe. She lived with her father, a husky, blond-haired man, in a log cabin. She described their clothing, including the detailed designs on their white aprons. "We had something on our head like a yarmulke, like a cap or something." Turning to the left she saw a "saucer, spaceship." A few minutes seemed to pass, and "I was walking towards this spaceship and my father then and there was frozen. I mean, he couldn't move; he couldn't speak. He was like, you know, ice." As the boy, she saw "one of those midgets again coming out." Next she was in the ship, which took off as she looked down through the window at her father, who "unfroze" and looked up with tears in his eyes. "It's like he understood. It's like he knew all along," that "physically I was given to him through conception" but that in another sense she was not his child. "He just accepted what was going on.

"I remember those midgets again," Eva continued, the "same ones that I remember when I was four, five years old. The eyes again, were very dark, but I felt a lot of emotion in them, a lot of compassion, a lot of love. It's like we've come back for you, something like that, and then I remember a lavender color, and that's that." Somehow she knew the year was 1652. This experience further persuaded Eva that she was "not from here," not "earthly." The "ETs," she said, "have the ability to enter into our space and time dimension or leave it any time they want." She related this ability to the 1652 experience. "It's like I was brought to Earth—I don't know why I was brought for six or five years, or however old I was at the time, and then I was taken out to another dimension that has no space and time as we know it." Eva in this life is "an energy form given a body to carry out a certain mission," which has to do with some sort of full experience of "life in earthly terms."

There followed further discussion of the difficulties of perception and communication between the alien or spirit cultures and earthly forms and the choices that our souls make among the "infinite probabilities from which we can choose," one of which is to become embodied on Earth in a particular time and place. Eva then reviewed her communications with David about her experiences. With him and others she has been talking "about such things where I would never think of it before." David seemed to be "in a state of shock at the beginning," followed by "big denial." Now he speaks somewhat sarcastically about "your friends from the other side" who "are this, that, or the other." Eva described further abduction-related experiences, with "different procedures, surgeries, call it what you will" to remove energy blocks. She has discovered blue and red spots on her hand, breast, and other places on her body that have not gone away.

Eva hoped to learn more about the 1652 incarnation under hypnosis, and I cautioned about trying to target specific experiences. "Deep inside," she suggested, perhaps we "know where we're going to go already before we come here." She suggested that the ETs themselves "that we physically see" are "just a form they take when they enter this dimension . . . Wherever they come from," she said, they "don't live physically per se that way." Their souls can manifest in different forms. "That's why we get different pictures" of the beings she said. "Some people call it reds, grays, browns, you know, with wrinkles, without wrinkles, whatever—it's a combination of their biochemical energetic makeup and our perceptive devices . . . But there will be some common ground," she added.

Eva's first image in the regression itself was of being a four- or five-year-old girl swimming with dolphins that were her friends inside a cave. Some force pulled her out of the water, but "the dolphins will wait." The little girl is drawn from the "memory of the source" of her being into a domain of "growing up and responsibility." This loss of memory is necessary, she said, "because in the physical world if the memory of the source would be there, there would be no initiative to experience, so all would just lay back." I took her back to the little girl's experience in the cave. "She keeps coming back, incarnated for a mission," Eva said. Becoming quite abstract, she continued, "the energy makeup of the little girl will change with experience." Her energy will become more subtle and higher "until from a certain perspective you can say there is no energy vibration. It's all one and the process begins again."

Abductees, Eva said, "are souls that have, for their individual purposes and reasons, chosen the probability of physical form." But through their experiences they "are regaining their memory of source . . . The process of abduction is one form of such, of regaining of memory." The abduction "experience itself," Eva said, "is a mechanism to remove" the "structures that impede the reconnection with source," and "to purify the physical vehicle in such a way to serve to regain better memory and to bring knowledge to others." The "physical and emotional torture" of the abductions themselves, she said, is part of a balancing process. She herself "never really feared," and "if there was fear it was more due to not being able to understand what was going on, not fear as something horrible and dark and evil and unknown. In a way, the process always felt familiar."

I asked Eva what she meant by familiar. The abductions "felt familiar. It felt home. It's, it's never felt unknown," she said. I asked how far back her memory of abductions went. She remarked about past lives in World War I and II and "in Morocco long before that." Her "drive" in each instance was "to help mankind overcome blindness." I asked about Morocco. She had been a rich merchant named Omrishi in the early thirteenth century who was trying to "undermine" corrupt officials of the hereditary ruling family that dominated the local government of the village. Omrishi was well known because of his wealth and his reformist "ideas and ideals." He organized militia groups to obtain greater economic equality for the villagers and sought to infiltrate the local government with his supporters. Part of his plan was to create chaos within the village that would make it easier to overthrow the ruling family and cause its members to flee, but he was betrayed by a woman who overheard one of his plotting conversations and reported this to the officials.

Men on horseback dressed in black with white headdresses—"the bodyguards of the ruling family"—came to Omrishi's tent to arrest him. The women around him cried and the children hid, for they knew what was going to happen. He was taken to a white stone building, which smelled foul from people puking and urinating there. Omrishi was to be beheaded, and the people were told to gather in the center of the village to watch the execution, for they wanted "to deepen the fear in them." After his arrest Omrishi was taken from his cell to see where the beheading would take place. The following morning at ten o'clock "they took me, they got my head on the thing, and plop." The feeling was one of "release, freedom," of "going up, expanding, joy . . . There is no description," she said. "All I feel is white light, gold light." She saw a dove released from a cage "that's symbolically me . . . That's my soul."

I asked Eva to speak further of the "journey" of her soul and her thoughts returned to the little girl swimming with the dolphins. This represented "the path of the child's soul" that returns to the "physical dimension from time to time" for several reasons. One is to "experience physical life, physical body, physical sensations and perceptions, feelings, emotion, pain, and everything that the physical world offers." In addition, by "understanding through experience" souls can "return from time to time" to physical form "to help those who have not yet remembered." Omrishi, for example, had planted a "seed within people's hearts," and "through earthly time, the seed will sprout into a tree and the tree will bear its fruit."

I asked Eva the role of abductions in this process. "To clean the body, physical body, in order for more information to come through," she replied. "They [the aliens] have always been here. It's a matter of evolution if we are able to perceive them," she said. "There's times in the evolutionary process of mankind when they were around but we were not able, it was not right, the time was not right for us to perceive them." Omrishi perceived them not as coming in spaceships but through "higher communication" and meditation through which he received guidance.

"All people have guidance," Eva said, "but most don't listen to it." Abductees "are at the level where they are able to cleanse . . . to bring information through . . . Abductions are very real, physically speaking," Eva said, but people should not emphasize that aspect too much. "They should balance the data to comprehend it as a whole, and not try to prove that they exist or do they not exist." The focus should be on "the information given by abductees . . . That information should be gathered and developed on the physical plane to be of use. Time is wasted trying to prove or disprove the existence." (Eva spoke of my own extremes of intellect and unconditional love, a "cosmic tension," and advised me to "go to a retreat" in an isolated place without other people in order to balance these polarities and "connect your being to the cosmos." Picking up emotions of sadness and loneliness in me she said, "You need to know that you are never alone. Just ask for the connection and you'll feel us," i.e., "all of the nonphysical beings that have been guiding you all along.")

For Eva this session was the most powerful of all. Its impact, she told me two months later, was "totally beyond words to describe." She felt "very much at home" with what came up in it, like "the actor and movie seer at the same time." She saw her life as Omrishi as part of the evolution of her pioneering role in bringing about "peaceful change" toward "harmony between people." She doubts that Omrishi, from his thirteenth-century perspective, was consciously aware, "you know, of the bigger picture." He was simply seeking to do "something for his people in that place." He began a "communication," but the larger objective of bringing peace and equality to people remained for the future.

Discussion

Eva is a pioneer with a global mission of healing and peace. Her lifelong abduction experiences are a powerful vehicle for the evolution of her consciousness and they bring her in touch with the depths of her purpose. She experiences the abduction encounters as important sources of "information," emanating from dimensions beyond or outside of physical reality. She feels herself to be "an energy form given a body to carry out a certain mission." Like most abductees Eva has had disturbing, even terrifying, encounters with alien beings. But her determination to give herself to the process, to surrender the need to control and resist its intensity and meaning, has enabled Eva to move beyond fear and trauma to a place of greater inner balance and personal power. It is characteristic of her to write in her journal following an abduction experience that left her feeling very tired that she *hoped* she had been taken "on a journey."

In Eva's descriptions of her abduction experiences there emerges a consistent picture of the evolutionary purpose of the alien-human relationship, at least as it affects our consciousness. She repeatedly describes the access she gains during her abductions to another dimension (or other dimensions) of existence, an expanded reality in which human concepts of space and time do not apply. This realm abounds in paradox—with the sense, for example, of expanding to infinity and contracting to a single point at the same time. Although Eva is quite articulate, she finds her words fail to convey the ineffable beauty and power of this spiritual realm. Abductees, she says, are souls that have "chosen the probability of physical form," and the abduction experiences are a vehicle for regaining memory of the source of being, from which, at least in our culture, we have been largely cut off.

Eva herself, like many abductees, seems to exist in both embodied-human and alien form. During our sessions when she would move deeply into her abduction experiences, the alien or "other" identity would take over. Then she would speak from the perspective of a kind of cosmic "we" or "us" that was in touch with the possibilities of a higher consciousness which could translate into peace and harmony on the earth plane. Essential to this expansion of consciousness is the breakdown of boundaries, the overcoming of separation of human selves from one another and from the entities, including the aliens themselves, that populate the spirit realms.

From her "other" perspective, Eva is aware of the higher or more intense energy levels on which the alien beings exist and the complex problem of adjusting their intensity downward in order to manifest on the earth plane. Eva senses that even the physical forms that abduction phenomena can feature—humanoid beings, spaceships, travel of our bodies through walls—may represent adaptations of higher energy forms to the perceptual requirements of restricted human consciousness, a technology for reaching us in a language we can understand. The aliens, or the source from which they emanate, must create physical forms for us to know them.

Eva's access to past life experiences is part of the process of the expansion of her consciousness beyond a purely physicalist or materialist perception of reality. In this area, the alien beings function as spirit energies or guides, serving the evolution of consciousness and identity. The thread of her personal mission can be detected in the past life experiences that she has related to me. For example, as Omrishi, a well-to-do Moroccan merchant in the thirteenth century, she is already concerned with justice and equality, though on a more local scale, and as a small boy in the seventeenth century she discovers that her soul does not belong to her earthly parents but to a larger realm of being.

For Eva, the integration of her personal evolution, as experienced through the abduction encounters, with her domestic responsibilities has been a formidable task and an important aspect of our work. Her agreement with herself never to hurt anyone ran head on into the inevitable tension that arose between her spiritual life and her relationship with her husband, who was embedded in the conservative and practical demands of the business world. Eva had sought to keep these lives totally separate, but found the fragmentation of her sense of self increasingly intolerable. A storybook from her childhood called *Soul Bird*, that told of the compartments of feeling each of us carries within, seemed to capture the struggle to achieve inner peace that Eva was undergoing. She was eventually able to bear the distress of telling her husband, David, about her encounters, and though she was disappointed by his initial casual response, she was heartened by the fact that he offered a practical suggestion about how to communicate with the aliens and hinted that he too may have had some sort of contact with them. Quite isolated when we met, Eva is increasingly finding ways of speaking with other people about the truth and power of her experiences and knowledge.

Eva's last regression began with the memory of a little girl swimming with dolphins in a cave who must leave to assume other, more grown up, responsibilities. "It's dark, but it's not dark . . . We play together, and they are my friends," she said. This image, to which we returned later in the session, seemed to represent for Eva the soul's journey through the experience of time, the cycles of rebirth and death, of incarnation, and the return to spirit. The image is timeless and intimately connected with Eva's mission, an embodiment of the dream of peace, harmony, equality, and playfulness to which her life is committed.

Eva warned of considering the abduction phenomenon in too narrowly materialist terms, and discouraged us from wasting our energies trying to find proof for its reality by the methods of the physical sciences. Implants, for example, she wrote me after her first three regressions, are not likely to provide the definitive proof that abduction researchers are seeking. For to be sustained within our bodies they would have to be composed of substances that would not be rejected by our tissues, i.e., would need to contain elements with which we are familiar on Earth. And, I would add, it is hardly likely that a phenomenon of such intelligence, subtlety, and sophistication would yield its secrets to a method of investigation derived from a consciousness operating at a much lower level. "I personally continue to believe," Eva wrote, "that our focus should be on mutual communication of some form and on some level with our alien friends, learning, accepting, and integrating alien wisdom within our world and culture. Time, money, and energy spent solely on providing proof of alien existence is fruitless."

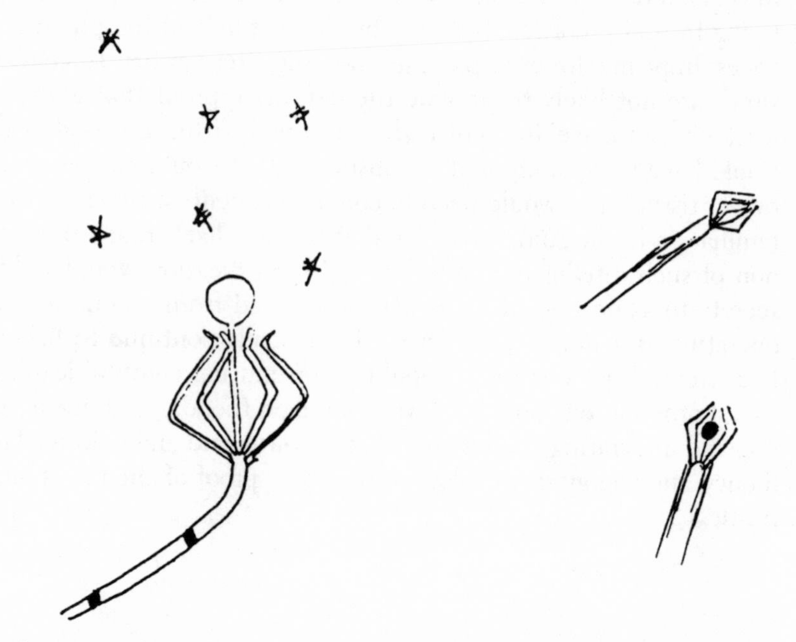

A probing instrument shown open, as drawn by Julia, and the same instrument shown closed, as drawn by Dave. The two drawings were made independently.

CHAPTER TWELVE

THE MAGIC MOUNTAIN

Dave was a boyish thirty-eight-year-old health care worker in an isolated community in south central Pennsylvania when he called me in June 1992 at the suggestion of his Korean karate and Tai Kwan Do (Chi) teacher who was familiar with my work with abductees and thought that his student's experiences might be related to the phenomena that I was studying. I was not available when Dave first called, and my assistant put him in touch with Julia, an abductee with whom I had been working for two years. Julia spoke with Dave several times and encouraged him to write to me of his experiences.

In his letter, written in July just before a powerful abduction experience in which he saw a being staring at him through a window, Dave told of possible abductions dating back to age three, an unexplained crescent-shaped scar that had appeared on his body, several missing time episodes, and a vivid UFO sighting at age nineteen. In addition, he wrote of his training in karate and his struggles to control the Chi experiences he was having in and outside of his classes through the work with his teacher, Master Joe. At the end of the letter he added without commentary, "I would like to be hypnotized."

On July 23, Dave and I spoke on the telephone and he reported more conscious memories of his abduction experience two weeks earlier, including the feeling of something being stuck in his anus, the compelling and controlling eyes of the creature in the window who seemed to be female and familiar, and how he found himself curled up by his wife on her side of the bed after the episode was over.

Dave's experiences include the traumatic elements that are common in UFO abductions. His case is of special interest, however, because of the intimate connection between the abduction experiences and his training in the opening and mastery of his Chi energy, which Dave defines as the "force which pervades the universe from which reality arises." As this opening has occurred Dave has been astounded by the number of synchronicities—events in his life which seem to be meaningfully connected—that seem to surround him. Possessing from childhood a lively, practical interest in the out-of-doors, Dave's abduction encounters and experiences with primal energies have filled him with awe before the powers of nature.

Dave has been drawn in particular to Pemsit Mountain, a place of Native American tradition and magic near his home where many of his experiences have occurred. The universe has become for Dave a place filled with mystery and strange intelligence. As he takes responsibility for the power and reality of his experiences, including two past lives we uncovered in our last regression, Dave is becoming a leader in his community in the exploration of anomalous experiences. Other abductees are attracted to him, and he is considering changing his career so he can use hypnosis with them and provide support by leading groups. He has a strong interest in nature and photography, and sent my wife and me several lovely pictures of wildflowers he had taken. Dave and I first met on August 13, 1992, when he came to Boston to explore his experiences.

Dave grew up in a small, close-knit, hillside community of about twenty houses by a creek in the Susquehanna Valley region of Pennsylvania—"it's the ridge and valley section of Pennsylvania. There's long, parallel ridges, and there's gaps in the mountains where the river runs through. I guess the river was already there, and then the mountains rose and made gaps." His town was at the base of the mountain. Dave's friends came from among the neighbors, and the mothers "were like all of our mothers." He could go out whenever he wanted and enter the other houses without knocking. "My grandmother lived down the street, my uncle lived next door." It was "my clan, right there."

Dave is the oldest of four boys. His brothers are three, six, and nine years younger. Their grandfather had a plumbing, heating, and fuel oil business and their father worked for him as a fuel oil salesman. Dave and his mother were always close. All three of his brothers appear to have had abduction experiences as does the son of one of them, who does not wish to be identified.

Dave's grandmother was an avid bird-watcher and taught him how to identify birds. As a child he began to develop a deep love of the natural world and after age nine or ten began exploring the woods and the mountain near his home. His father first took him fishing when he was five, and later he began hunting and trapping. When Dave was an adolescent he "spent a lot of time up in the mountain or down along the creek." He grew up feeling a strong kinship with the Native Americans of the region. "The Indians," Dave wrote to me, "attached special spiritual significance to white deer," and he also feels a strong connection with them. "I am, you know, spiritually attached to deer. Seems the deer is my totem animal," he has said.

When he was seven years old, Dave lost his right eye as the result of a "sword fight" with sticks he had with another boy in the neighborhood. According to Dave, the boys were jousting with pieces of a branch of a fallen tree. Dave's mother called to them to stop and Dave laid his stick down and sat on the ground by a tree. The other boy picked up Dave's "sword" and broke it against the tree. One of the pieces flew into Dave's face, slicing the eye and cutting his face below it. Blood poured from his eye, and he was taken to a doctor in the town and then to the local hospital.

Dave was operated on with ether used for general anesthesia. The surgeon was unable to save the eye, which was removed. Dave had not been told beforehand that there was a possibility he might lose the eye, and still did not know it was gone when he woke up with a patch over it. He does not recall asking questions about what had actually been done during the operation, but remembers that he had "terrible nightmares" during the nights afterwards, with loud screaming that led the nursing staff to walk him up and down the hall while the other children on the ward yelled at him to "shut up." Dave's father did not tell him until several days later that his eye had been taken out and he would have to "wear a fake eye." Later his father told Dave that he was so upset after he told Dave what had happened that "he went out on the front steps of the hospital and cried about it . . . He told me that I'd handled it better than he did," Dave said. It was characteristic of Dave's father not to show strong emotion until he had a stroke ten years ago. "Since then he's had trouble controlling his emotions and sometimes he cries." Dave himself never grieved the loss of his eye until many years later when he was able to cry with his wife about it.

Dave recalls resolving at the time not to let the loss of his eye affect him. "I can still see fine, and it doesn't hurt," he said to himself. He was in the second grade then and there were only six other pupils in his class and about seventy-five total at the small school, "so everybody knew what happened, and everybody knew me real well. So nobody teased me about it." But later, in junior high school, the classes were much larger, and he was called "cross-eyed," even by children who knew that he only had one eye.

Dave attended Penn State College for one semester when he was seventeen, and then completed his studies at a community college. While still in college he began on-the-job training as a health care worker, as there was no formal schooling available in this work at the time. When he was about twenty-five, Dave felt he was ready to marry and have children, "but I didn't meet the right person to marry until I was thirty-two." Dave's wife, Caroline, is the oldest of four children. They had thought to have children soon after marrying, but their first home did not have "enough room."

The couple borrowed money so they could build a larger home, but it ended up costing much more than expected. Dave also herniated a disc in his back while building it himself, which kept him off his feet and out of work for several months. By the time they moved into the house in June 1992 "we were both four years older, and at that point we didn't know if we wanted to have kids or not." Dave denies that his UFO/abduction experiences have interfered with his sexual life or are connected with his decision not to have children. His and his wife's basic relationship, he says, "is very good."

The first experience that Dave relates to the abduction phenomenon occurred when he was three years old. In his introductory letter he wrote that he remembered "three motorcycles coming down the street towards me in an unnaturally fast way. When they got past me there seems to be a gap, and then they were past me and raced up a footpath off the road towards a friend's house. I walked down there and they weren't there." In our first conversation in August 1992, Dave added that the motorcycles seemed to come at him "too quickly," that he had felt gripped by fear, and that the "riders" were "black." He also recalls being "astonished" that the motorcycles could have raced up a dirt footpath at such a speed, especially as the trail ended at a stone patio of the friend's house and went no farther. He also remembers having the same feelings, "a vibration of some kind, a tingling," that he had in association with later abductions. Dave and I explored this experience in detail in his second regression.

Dave's next recollections relate to when he was about twelve. He was exploring in the woods on the mountain near his home, as he often liked to do. He remembers finding himself on a path that led to an intersection with two other paths where there was mossy ground and a tree overhead. "It was a beautiful spot," Dave said. "I was looking at it in awe of it, saying, 'This is so beautiful!' I looked up at the branch of the tree, and then that's all I can remember." A gap in time occurred, and the next thing he recalled was "walking onto the patio below our house." Over the next week or two, Dave returned repeatedly to this area, but found no trails or ground of the sort he recalled in the experience. Looking back Dave recalls feeling that "the state of consciousness

I was in" before the time gap occurred was somehow "different," stronger, "more acute" than normal.

Another similar episode occurred in the same time period when Dave was at his uncle's summer cabin. He recalls walking along a path that followed an old railroad bed near a lake. Again he remembers looking up into the branch of a tree, and then recalled nothing else "until the next thing I knew I was walking back into the clearing where [the uncle's] cabin was." What to Dave was a walk of a few minutes had turned into forty-five minutes of missing time, and his aunt, uncle, and cousins were "real worried." His cousins said they went out to look for him, and after he returned demanded to know where he had been. But he could not remember anything that he could tell them.

Dave remembers that as a child he was interested in "flying saucers," which he thought might be related in some way to his experience at age three. Also, "My dad had said something about people seeing flying saucers, and I, ever since then, was, had a sense of wonder about them." In junior high school Dave was in the honors program. He was fourteen and wrote his only term paper, in preparation for entering high school, on UFOs, but he does not remember what he said. At the time he told his former Sunday school teacher about the paper and she said that she had seen UFOs land and take off from a point on the mountain near where she lived.

Other things happened during his teenage years that seemed odd to Dave. When he was fifteen, he and his next-door neighbor found a cave with an entrance about twelve feet across that was by "the end of the mountain where it drops off to the river." His friend wanted to go into it, but Dave felt afraid and said, "No, no, I'm not going in there."

"It's not that big of an area," Dave said, but he was never able to find the cave again. A year or so later, he was driving with friends at night in a hard rainstorm across New Brunswick, Canada, traveling at about sixty-five mph when a Greyhound bus "flew past us," going perhaps ninety mph. Dave was asleep in the backseat when he was awakened by his friends exclaiming "'Holy shit!' or something like that." They had just realized that "we just lost ninety miles . . . last time we knew we were ninety miles back down the road." None of the boys could account for the missing time, but they heard on the news of a terrible bus accident in which sixty-five people had been killed.

Soon after Dave started at Penn State, three weeks after graduating from high school, both he and his roommate lost more than a day's time that they could not account for. They went to bed on a Saturday night and woke up believing it was Sunday morning. But friends who lived on the same floor of the dormitory came by and said, "Oh, you guys skipped chemistry class, huh?" and shocked them with the news that it was midmorning on Monday.

When Dave was nineteen, he had a close-up sighting of a UFO, which affected him deeply. He was with his younger brother, Ralph, and a close friend, Jerry. His parents had moved a little ways out of the valley to an acre-plot of land on the side of a hill. It was a very clear night, and the boys were lying in the yard, "propped up on our elbows," across from a large farm field. Then "this light rose above the mountain" on the horizon and "immediately it made a right-angle turn and started to go out toward the river, and it stopped and started to go out back to the valley, and then it stopped and started to come towards us. It was coming towards us for a long time, and none of us had said anything to the other ones about it, and then I looked over at them and realized they were also both watching it." The boys considered that it might be a large jet, but they realized it was too close to the ground and silent to be an airplane.

Then the craft came to a stop right above them and a little to the right. It stopped again and seemed to pivot "so that its bottom was pointed straight at us. A blue-white light started to come out in a ring of dots (lights) on the bottom of the craft. They strobed and it shot straight away from us [he made a whooshing noise at this point] and then stopped. Then the blue-white light flickered out of the back of the craft and then began to glow very brightly . . . It took off like out of a slingshot, like out of a catapult. It just took off. It started going quickly right away. It just kept going quicker and faster and faster, and it went at an arc into the sky. It didn't go in a straight line like a plane. It went at an ever-increasing angle of up, and the last point we saw, it had to have been going up into the sky at a forty-five-degree angle, and nevertheless, it disappeared over our horizon within about ten seconds."

After the UFO disappeared Dave reported saying, "'Man, I wonder where that thing is *now!*' and Jerry said, 'Could be over China!' and I said 'Yeah, could be at the moon.'" Then the boys "ran pell-mell up to the house, and we got my parents and my brothers to come out." They yelled, "We just saw something!" and they all ran out, "but there was nothing there to show them." During the encounter with the UFO Dave had had the sense that "something maybe partly bounced back and forth between me and the craft . . . I felt some sort of gap there, some confusion at that point . . . I wonder if that blue light had some effect on us," he said. "It was a blue-white light. Very intense." A few weeks later Dave recalls reading about reports of UFO sightings in his area during the period when he saw the craft.

Jerry, who is a generally quiet person, could only say over and over, "Well, yep." Although Jerry, who acknowledges the UFO encounter, does not admit to abduction experiences, Dave believes that he is an abductee. Ralph was also powerfully impressed with the sighting. Dave remembers experiencing a strong sense of wonder at the time and had the thought "I didn't think that whatever was inside of it could be human." He also believes he saw a face in the craft "looking down at me," and when he later saw the picture of an alien on the cover of *Communion* he was shocked, for "that's what I imagined was looking down at me when I was nineteen." The large head and black, slanted eyes, in particular resonated with Dave's experience.

In his initial letter, Dave wrote that he "considered this craft to be the most impressive piece of technology" he had ever seen. He began to read everything he could find about the unknown, "hoping to find a clue" as to what was inside that craft. In particular he read the Castañeda books, and thought about "personally acquirable magical power." He also became interested in Tibetan Buddhism and discovered "that the Tibetan Buddhists purportedly knew all about UFOs."

When Dave was twenty-five he built a small cabin in an isolated area. It was still unfinished when he moved in. He had been living in the cabin for about two weeks when he moved into the bedroom for the first time. He had set a brown paper bag full of empty beer cans just outside the front door and lay down to go to sleep. He heard what sounded like an animal trotting up to the house and then heard a banging crashing noise as it seemed to run into the beer cans. Then he heard other noises that sounded like the animal kicking the cans as it ran off. Concerned about the mess that had been made he shined a flashlight out the window in the direction of the brown bag and was surprised to see that the bag was standing and nothing had been disturbed. We both noted the trickster quality of the incident. Dave wondered at the time if "a spirit of something" was involved, but did not relate the episode to his UFO experiences.

In 1988, soon after reading *Communion,* Dave had a dream which to him reflected the power of his Chi. In the dream a Hispanic man was holding a mastiff which was lunging ferociously at him. The man put the dog in a cage, so he decided it was safe to walk around him. Then the man put two fingers of his right hand on Dave's right shoulder and he was pinned to the ground as by a ton of weight. Then the man let the dog out of the cage, and "I resigned myself to die, in this dream, which I felt was reality. I couldn't tell I was dreaming." In his first letter, Dave wrote, "All of a sudden I was overtaken by a feeling like a rage that started in my chest and went down below my navel. It came out of my body at this location in the form of energy. It was like a rocket. It was incredible. I was hurled backwards at a high rate of speed. I felt the man and dog were flung off of me like they were nothing." Dave landed on the side of the bed opposite from where he had gone to sleep. He would have landed on his wife, but she had gone to work.

In the last few years Dave and his neighbors and friends have had a number of UFO-related experiences and "strange coincidences." For example, Rob, one of Dave's neighbors, was killed in an automobile accident in February 1990, nine days after helping firemen put out a chimney fire in Dave's house. Several years earlier Rob had seen a huge ball of light in the woods at the end of Dave's driveway and then had a missing time period of about forty-five minutes while walking to Dave's house to see it more closely. Rob had seen UFOs about six times over the field across the road from the end of Dave's driveway, and also told Dave that "he saw a UFO while we were watching the firemen and a strange mood overtook him and he ran to help them."

In October 1990, Master Joe, Dave's karate teacher, obtained a part-time job in the department where Dave works. Dave discovered that he was a seventh-degree black belt karate master and had learned "secret knowledge about this [Chi] energy from his master instructor, a Korean, who learned it in a Buddhist monastery in Korea." Master Joe is reported to have had someone drive a seven-thousand-pound truck over his stomach while lying on broken glass. Dave began to discuss Chi with Master Joe during the fall and winter of 1990–91, told him of precognitive experiences he was having, and started karate lessons that winter.

Dave and Master Joe have been working on opening Dave's Chi channels, which creates a tingling sensation. This power is said to emanate especially from an area the Koreans call the "Dungan," a region below the navel which is related to the will. According to Master Joe you "make power" in the fingers, and other body parts and control Chi with the eyes and mind. "My whole goal in karate," Dave said, is "to learn to control my Chi." In his first letter to me, Dave wrote that he had had about four Chi experiences while in the karate classes. In these experiences he witnessed himself performing unusual physical feats. One morning he awoke from a dream and was "ready to push my Chi out in front of me." To his dismay he ended up pushing his wife out of the bed without having any sensation of actually touching her. This experience helped Dave to resolve to learn to control his Chi energies.

In September 1991, Dave experienced a powerful synchronicity that Master Joe believed was connected with his Chi energy. Before going on a vacation with his wife to a national park in North Carolina, he dreamed of a girl who reminded him of one to whom he had once been engaged to marry. The relationship had fallen apart when the girl moved to Massachusetts.

Once in the park Dave met a young park ranger, Charlotte Hampton, with whom he felt strangely linked. She was working in the National Park bookstore and was first cold, then flirtatious with him. He found himself making some sort of strange, intense energy connection through eye contact with her, which led to a kind of blacking out on his part. Dave likened this to the way a sorcerer or shaman "hooks" a person with their will, guaranteeing that some sort of follow-up contact will occur. It turned out that this girl was the one he had dreamed about earlier, and she has had various psychic experiences that are connected with Dave's own abduction-related experiences.

Dave and I met for his first hypnosis session on August 14, 1992, the morning after our first meeting. Julia was also present. We decided to explore the abduction experience of the night of July 8, which had occurred just after he wrote his first letter to me. Before beginning the regression, we reviewed what he remembered consciously of the experience.

He had spent a frustrating evening at his parents' home twenty-five miles away, typing the letter on their word processor. He left their house at about a quarter to one in the morning and arrived home at about one-thirty. His wife was asleep in their queen-sized bed and did not wake up when he got into bed at about two or two-thirty. After he lay down, he heard a noise in the house, "a creak or something," and thought to himself, "Oh, they're going to come tonight." Shortly after going to sleep, "in the context of a dream" (an ambiguous phrase Dave used several times), he found himself "in our dining room. But it wasn't exactly like the dining room is in our house. The window was on a different wall, and so forth." There was a large woman in the room with him, and Dave felt "their presence." He was gripped by a feeling "that I now realize is familiar to me, like when the motorcycles came down the road and stuff." The vibration "then went from below my navel and out through my chest, and then it was real tingly. It was a real tingly sensation."

The woman pulled him down to the floor, as if to hide from the alien beings, and Dave felt he still had some control, was able to raise his head, and looked out through a picture window into the woods. Then he saw a female being looking back at him through the window. He remembered another abductee, his friend Randy, saying to him, "It's the eyes. They hypnotize you with the eyes . . . I didn't think about what Randy said until my gaze shifted to the eyes, and it didn't make me not want to look at the eyes. I was very curious, and I stared right at it." Dave could only see the "upper right-hand corner of her head," which seemed enlarged. Some kind of screen appeared to be covering the parts of the being's head that he could not see. "The skin looked very soft and light gray." I asked what feeling looking at the eyes gave him. "I immediately knew who it was," he replied. "It was this female being who's mine, and I felt that I knew her very well, and that I liked her very much, and that I was looking right at her."

At this point Dave said that he "blacked out, or lost consciousness, or something . . . The next thing I knew I was laying in bed on my side, all huddled up in a fetal position. I was huddled up right up against my wife, real, real close to her and all curled up into a little ball." Dave realized then "that wasn't a dream." Caroline rolled over and put her arm over him. He had a feeling "that there might be a being standing right behind me, and I was afraid to turn around and look. After about a minute the fear passed, and I propped myself up and looked at the alarm clock, which was behind my wife on the bedside table, and it was 4:00 A.M." Dave's thoughts at this point were confusing. He "figured that I'd been abducted," but had the sense also that the large woman in the dining room was "probably just my wife in bed with me."

In the regression, we began at Dave's parents' home, his frustration with typing the letter, and how "freaked out" he felt about "reliving through typing and writing this letter all this stuff that happened to me in the past few years." He felt rather "belligerent about the whole thing," and as he lay in bed and heard a noise in the house he thought, "Oh, they're going to come tonight." He then became anxious as he told of turning on his right side toward his wife and starting to fall asleep, whereupon he found himself in a room the size of his dining room but with a long table unlike the one in that room. Light from some source filled the room. He again described the large woman (in a dark dress now) and the big picture window ("just a sheet of glass, whereas the windows in our dining room open up") with the woods outside. Once more he feels "their presence," the "sense that they've arrived . . . It's like a total all-encompassing, very powerful feeling.

There's nothing you can do about it. When they're there, they're in control and there's no fighting it. I realize that I'm used to it. I'm ready for it again, for some reason."

Once more the woman pulls him down, and Dave props himself up and sees part of the face of the "visitor" looking at him through the window. The skin looks soft, leathery, and is light gray. "My gaze shifts to the eyes, and I know it's her and I know she's looking at me." Her stare is "just totally there. The eyes look like they're big, black, kind of liquidy." An inner struggle ensued in the session as Dave seemed to resist "what I'm supposed to do, what they want me to do." I spoke to his need to keep in control as a kind of strength. He likened the "trouble" or "danger" he now felt he was confronting with the Chi experiences and altered reality he sometimes discovered during his karate exercises. Even when, for example, during kicking and other moves he blanks out and discovers he is "no longer in this reality" he must "go on like nothing happened" and he cannot explain this to the other students for they would not understand what was going on. "It's the extreme strangeness of those experiences that I have to get used to. It's my mind not wanting to lose its grip on what was always pounded into what was reality, I guess."

This struggle, and what he called "stubbornness," began to occur between Dave and me. "This abnormal stuff" was "still too hard to accept, or some part of me doesn't want to give up control [of] the idea of reality that's been pounded into me." I took him back to the window, the being's eyes, and when he "blacked out" while looking into them, and asked him to describe the eyes themselves. "They're big, and they're black, and they're slanted, and they're real liquid looking" and they're "pointed at each end . . . I recognize the being. It's her. I know who she is. It's like she's mine, and I'm hers. I feel this is real. I really like her . . . I feel that she has to be special, even among them," and "I have the feeling from her that I'm special and I'm not living up to it." Dave and I talked then of his shame about wearing a "fake eye" and the pain of being teased about it.

The female being, whom he calls Velia, loves and accepts him unconditionally, he said, even with his one eye and his smoking marijuana sometimes, in contrast, for example, with a girlfriend he had when he was twenty who tended to be critical, formal, and possessive. I asked Dave repeatedly to return to his center by encouraging him to breathe deeply and focus on the breath. I assured him he need not feel ashamed that he could not maintain control. Dave continued to struggle with the block he felt looking into the being's eyes. "I look at her for a little bit, and then they or she then increased the control, or then let it be total . . . I wonder where they're going to take me," he said, and I encouraged him to let go. "I think there's more than one," he said. Weeping now, with a mixture of fear and relief, Dave said, "We're out of the house now. There's a mess of them, but I can't see them. I don't know where we are. I think we're gonna go to a ship, and it's out in the clearing I made five years ago." Dave explained that even before he built his house he made a large clearing about 150 feet from it, perhaps unconsciously inviting a UFO to land there.

Dave described the ship as "big and round. It's about sixty feet in diameter. I think it has a dome on top." Surrendering control now, Dave spoke of being taken into the craft through the bottom. His fear mounted in the session as he told of being forced onto a table on his back in a round, gray room in which there is an "earthylike" smell. Several beings gathered around "to do something to me." Dave was paralyzed now, able to move only his eyes, as he experienced some sort of "stern encouragement," communicated telepathically by the beings, regarding a mission that he was to fulfill.

The female being was there, "helping me out," but "I think this male guy's running the show." The female communicated telepathically that "it's going to be all right," which reduced Dave's anxiety greatly. "I guess they stick something up my ass, and I guess that's upsetting me," Dave said. "I feel they put my legs up in the air and spread them apart." At this point in the session I sensed Dave's shame and embarrassment as a man in being subjected to such a humiliating procedure and spoke to him at some length about forces in the universe over which we have no control, the potential empowerment in acknowledging powerlessness, and the inapplicability of conventional notions of masculinity in this context. He spoke then of a flexible instrument, perhaps four feet long, with "a little wire cage" on the end, inside of which was a small, spherical object. About "half" of this was inserted in his anus, as the female being continued to reassure him.

Dave expressed his feelings of violation and resignation as the procedure continued, as well as "tingling" sensations. "It swarms inside of me. It goes into more parts than just being up inside my anus," Dave said. I encouraged him to express his anguish and rage at what had been done to him, which he was able to do only to a limited extent. Some of his outrage seemed to be related to similar experiences going back to age twelve and perhaps to age three. After about two minutes, the instrument was removed. Dave believes that this procedure was some sort of "informational test" to "keep track of how you're doing physically, how you're holding up, what condition your body is in, if it's deteriorating, or if you're in good health or bad health."

Next Dave felt a sharp object was stuck by a different, tall being against the left side of his head near the center of his temple, which, surprisingly, hurt only "a little." The female being, who was "to my right" continued to tell him "I'm doing all right. It's okay. She looks at me and reacts to how I react." He felt a "total trust in this being." Next a suction-type device at the end of a tube was placed over Dave's penis, which he found difficult to speak about but not as humiliating "as the thing being stuck up my anus . . . They made me ejaculate," he said, which was "pleasurable as any ejaculation to any male being. It's just that the circumstances are distracting, distracting from any pleasure that you may derive from it." Finally, Dave believes that "they put something in my stomach," a "circular sensory device" about eight inches in diameter, for "checking something in there . . . It made a little bit of vibration. It wasn't unpleasant," he said.

This completed "the physical examination," Dave said, but "then she talks to me for a while or communicates with me," for example, "that I'm doing okay." This "exchange" occurred while Dave sat on the edge of the table with his arms at his side and "my legs dangling down." There was also communication about some sort of mission of Dave's to the effect that "we all don't have unlimited time on Earth. We're here for a limited time, and we have to make the most of it, and that would explain why since then I've felt this tremendous sense of mortality. I'm not going to be here forever." The female being supported the fact that Master Joe was Dave's guide, and "I know that I've just got to keep doing what Master Joe wants me to do."

After the above procedures and communication were completed "they then support me somehow off the table and down out of the ship." The beings "levitate me or something" and accompanied Dave down the path to his house and through the closed and locked door, and he was aware of "a glow back there" from the ship in the clearing. The beings floated him up the stairs, through the door of his bedroom, and "set me down" in the bed near his wife, who was still sleeping. He crawled closer to her and "curled up" as the beings left through the sloped roof wall at the side of the bedroom. Because of "what had just happened" to him "I just wanted to be comforted," Dave said. "I wanted to be real close to her when I woke up."

After the regression we reviewed what Dave had gone through. "It was hell," he admitted. When I noted that he wanted to skirt by difficult places rather quickly he asked astutely, "Do you think I could stand going through anything else?" He felt "a little washed out emotionally, drained" but amazed at how much he had been able to remember. "It wasn't like I was really there, but I was remembering being there," he said. The UFO, he now remembered, had hovered about ten or twelve feet above the clearing, and he was now convinced that "I made a place for them to come down." About two weeks after this abduction, Caroline told Dave that a week after his experience she had seen what might have been an alien face in the bedroom and thought to herself, "Oh, that's where they came in, or that's where they come in."

Julia, offering support, spoke of the brief depression that sometimes followed her hypnosis sessions, but said this "would clear and you move on . . . It's not all roses," she said, "and yet, for me, it felt good to remember stuff. It's freeing. It's wonderful." She was struck by the fact that she too had had periods of communicating with the beings sitting on the edge of the table with her legs dangling and her feet swinging. It seemed too frivolous somehow, "after they've done this," an "incongruent detail" that "didn't fit." Julia also recalled a flexible hoselike instrument with a "cage" at the end (see drawings, p. 264). Then Dave made plans to drive back to Pennsylvania, pointing out that he did not like to drive after dark. "Oncoming headlights bother my eye," he said. "I can't take bright lights."

Dave returned safely, and I talked with him on the telephone the next day. He said that he felt "empowered" by his work in Boston and talked of his experiences with his friend Jerry and with Caroline, who was somewhat troubled by what Dave had learned. He had surrendered some of his defensiveness and the experiences seemed "more real" to him. Two days later he talked with Julia, seemed cheerful, had developed "a fresh outlook on life," and was trying to deal with a "deluge" of questions from friends. She advised him not to reveal the details of his hypnosis sessions. He said he was planning to clean out the clearing behind his house, and she resisted the temptation to ask him if this was to make it easier for the UFOs to land. Ten days after this, Julia called Dave to find out how he was doing. He had been feeling slightly depressed after his Boston experience, but also because he wanted to do healing work, and keeping "dying people alive" through health care did not "seem right" to him. He said he did not fear the aliens and affirmed that clearing out the clearing behind the house was "to give them a nice place to land."

On September 9, Dave wrote Julia a letter which was the first of a series of communications that continued through the fall and winter documenting his abduction experiences, seemingly meaningful coincidences, and other significant experiences that were happening to his brothers and the people in his community. In a long handwritten letter to me he provided many details of the abduction experiences of coworkers and friends and wrote, "I know at least 15 abductees in this area, including my 3 brothers." Giving terminal care to people with whom he could not communicate had become more difficult and he had cut down his hours at work. He felt that his job in life concerned Chi, but it was difficult for him to accept "big responsibilities." He was a blue belt in karate now, "halfway to black belt." In October he told me in a telephone conversation that his wife had seen an alien being at the end of her bed.

In another long letter in January 1993 Dave documented in more detail synchronicities that we had discussed on the telephone. He enclosed pictures of a rare, partly albino deer with black coloration on the back of its ears and tail (called a "piebald") that he had shot in December. He said "if I would have seen how beautiful it was I don't think I would have killed it." When he took the hide to a taxidermist he said he had never seen anything like it. When Dave called Charlotte Hampton several days later, she said she had also seen a piebald deer in a herd of them that were "running around the place where she works." The piebald was the only one that did not run off when she approached them.

In this letter he also wrote of planning to meet Charlotte Hampton at Gettysburg and believes this desire may be related to a past life and "feelings I got several years ago that maybe I had been in the Civil War." He added that Julia had said "that the first time she ever saw me in person standing in your living room she saw me standing there in a Confederate uniform." Regarding his belief that he was abducted between two-thirty and three in the morning on the night of December 18/19 he wrote, "I 'came to' with a tremendous feeling of peace. I felt this was because you hypnotized me and I've resolved the issue within myself."

At the beginning of February Dave told Julia that "six beings appeared in the bedroom to Caroline" at about one in the morning while he was out of the house mailing the above letter to me. She woke sitting up with her arms crossed tightly on her chest and found the beings around her bed. According to Julia's notes four of them "departed through the wall behind Dave's dresser (in the direction of the clearing). Two others went behind the nightstand." According to Dave, Caroline was angry about this and it took her nearly twenty-four hours to tell him about it. Dave also said he was planning to begin hypnosis lessons to help other abductees.

In March, Dave came to Boston again for further hypnosis sessions. He wished particularly to explore childhood abduction experiences that seemed connected with Pemsit Mountain. The two sessions, which took place on March 11 and 12, were attended by Julia and also by Kishwar Shirali, an Indian clinical psychologist who has a deep interest in transpersonal phenomena and considerable knowledge of Hindu mythology.

At the beginning of the first session we reviewed the experiences, documented above, which Dave had undergone since his visit in August, and talked of the feelings of awe and wonder that we experience when such synchronisitic patterns or designs seem to unfold in our lives. Dave talked of what he was learning about Chi and its relationship to human spiritual evolution, superstrings, the primal sources of energy in the universe, the role of eagles in spiritual emanations, the capacity of babies to perceive other realities, and the openings that dreams give us to them. He said he was learning to open his Chi and control it through his hands, which he can make become hot. Dr. Shirali spoke of similar processes with which she is familiar in yoga and Buddhist meditation. Dave said he had talked with Master Joe about the possibility that the alien beings have mastered the capacity to communicate telepathically using something like Chi and may be especially interested in people who have changed their energy regulation or "assemblage point."

When I asked Dave what he wished to do in the regression he replied, "I want to find out about Pemsit Mountain" and "apparently it's all tied up with the Chi, and I feel the Chi is all tied up with the experience for some reason." "The one that haunts me most is when the three motorcycles came down the street at me when I was three years old." All of Dave's early experiences with "the visitors," he noted, occurred on the mountain, whose spiritual power and meaning to the Native Americans of the region he stressed once again. Many people who have lived where the mountain dominates their horizon also believe that it is a "UFO base."

At the beginning of the regression Dave's thoughts went to Sober's Hollow, a place he used to go fishing with Rob, whose death and UFO experiences we had spoken of in August, and also with another friend whose son was also killed in an automobile accident. This friend is an abductee and Dave had seen his son in an abduction experience two days after the boy had been killed, before Dave had heard about his death. Then Dave began to cry and remembered watching his little beagle, Spotty, being hit by a truck and killed as he stood on the porch of his childhood home. He expressed fear of "being with them . . . I walk out across the porch up onto the street. It's up about four steps. Right around the telephone pole. Here comes the three motorcycles whizzing over the top of the hill. They'd be about a hundred yards away, and they zoom towards me."

Dave again felt the struggle over control gripping him in the stomach. As he wondered what was so terrifying, the (now two) motorcycles with three black riders "turn into the beings." They "floated" him behind a bush on Shaeffer's (a neighbor up the hill) lawn and laid him down on the grass. The beings were tall and "skinny" with big black eyes that slanted upwards. He was paralyzed and terrified, as one of the beings, a female, was pressing something sharp against his head while she was "working something into my hair" with her hand as the other two, both male, stood above him watching. He was wearing only a shirt, and believes that they must have taken off the shorts he had started out with. Some sort of mind-to-mind communication seemed to dampen his fear. The female being also laid her hands on the lower part of Dave's abdomen, "checking me out."

Dave was crying in the session as he recalled the female being "saying there's something important about it." The tears were of relief, as "I guess I've always had it locked in there and it always bothered me as to what happened." A year or two later when he had asked his father about the motorcycles he had said, "You must have had a dream," but "I knew it wasn't a dream." The female being told him that "she missed me" since "I've been here," i.e., from the time of his birth. "I was sent here to do something," Dave said. Fear kept coming to Dave "in slight waves" as he spoke further about how he and the female being had missed each other, her "promise" to "take me up on the mountain," and his awareness that "there's more to us than what we know about here." She also told him that he would have a "hard time with my fellow man, but I'll come out for it better in the end" and explained that he was "living there on that mountain because they have a special place up there . . . I'll grow to love the mountain, she said," and assured him that "I'll find out all about them [the aliens] sometime."

Then they were "done" with him and became motorcycles again, disappearing up a steep path, upon which they could not possibly have traveled, that ended abruptly at Dave's friend's house. He looked for them, but there was nothing there. The female being had told Dave that "it would be hard for us to be apart, but they had some work to do with mankind and I could be an important part of that, help that out, so we had to be apart." He cried with grief as he said this, and I asked about the familiarity of the female being to him, even at age three. He said, "I guess I've always known," meaning "before this lifetime or something."

At this point in the session we were faced with a choice which I put to Dave, to pick up on "before this lifetime" or explore other experiences associated with the mountain. "I wanna go to the mountain," he said. The "end of the mountain" where "the rock layers that make up the mountain are exposed" is "a big power spot," he said. "That's why they go there . . . I think they go there for energy." Dave then felt a tingling energy which seemed to radiate from below his navel through the rest of his body and even outside of it, similar to what he had felt in association with his experiences at both age three and twelve.

Dave remembered how he would go to the end of the mountain when he was twelve. "I think the Indians used to go there. I'm a modern-day Indian." Then "all these memories want to come up." He was "back on the end of the mountain," above Spangler's Hollow, on a trail "above Matt's barn . . . All of a sudden I went to lying beneath the tree," Dave said, as he felt the tingling through his body. Several beings "come floating up around the bend" of a trail that was not ordinarily there, created now on a steep slope in the side of the mountain. His fear mounted in the session as he recalled feeling "shocked and surprised" as he was floated by the beings, feet first, along the trail. "It's only seventy-five yards where they start floating me until we go 'round the bend and I see the ship." It was in a hollow, suspended in the air. "It's amazing," he said, even though "I've been with them on ships" before.

Although it was daytime, the ship, spherical in shape, appeared to glow very brightly and Dave felt "fear and awe" at "just how powerful they are." (He remembered his Sunday school teacher telling him when he talked with her at fourteen about his UFO paper that she saw something that looked like the sun land and take off several times from the end of the mountain.) Then the beings floated him into the ship, through the bottom, and put him on a table which was held up by "a light-colored pedestal that just seems to rise out of the floor and it's all one piece." The room he was in and the ceiling were very bright, and Dave saw some sort of instrument panel off to the right. He was frightened, expecting some sort of "checkup." He saw about six beings, the familiar female one, several small ones, and a male who was in charge. The male being looked like the female, except that his eyes appeared more round.

Then Dave seemed to relive the fear, paralysis, and helplessness he had felt at the time. He believes he was thirteen, "in the midst of puberty." The worst part of "the exam," he said, is when they "stick something up my anus . . . They did it before," he added, "when I was twelve." Dave was determined to "focus" on remembering the experience. As the female was at his side reassuring him, one of the males spread his legs and "puts this thing, it's a couple of feet long, sort of looks like a thing that they root out sewer pipes with. It sort of has like a large end on it, sort of wiry or something like a wire structure–type thing on the end of it." They put it up "a lot farther than you could believe that it would go up. It's to check me out," for "seeing how you're doing," Dave added. He felt discomfort and humiliation but little pain and "like a zoo animal." Fear and awe "overshadowed" any anger he felt. "You can't do anything" for their power is "total." After moving the tube around "a little" for about two minutes it was removed, and the beings indicated they were pleased with his healthy condition.

To a certain degree Dave participated in their satisfaction with the results of the checkup, for he felt "somewhat akin to them" and "that I've known them . . . I admire their power," he said, and "feel like I'm as much or more part of them than human." I asked him about the "akin" and the "more," and he said that "they're more consistent to be around than humans." He recalled the change to junior high school when he was teased about his "fake eye" in addition to being singled out as "weird" for being in the honors program and smart. The female being told him that he would "go through a hard time" over his eye, but that "eventually, later in life, it wouldn't bother me." He was also told that he had lost his eye to learn in this lifetime "how to live with that, because I'd have to learn to live with a lot of other things." This involved a kind of warrior preparation, including his work with Master Joe in learning to manage his Chi. Master Joe, he reiterated, believes the aliens are attracted to Dave because of the strength of his Chi.

Dave wondered again if the aliens "have known us before, in other lifetimes" and if he had known them. Perhaps "it's an ages-long kinship," he suggested. "Maybe they help recycle souls or something, that they're not just concerned with our individual lifetimes but all our lifetimes as they relate to the development of our soul, the evolvement of the earth." Perhaps if he was born with "a higher energy" it is his responsibility to become a teacher, to "take advantage of it" and "to be an example to show man that there is more to us than what meets the eye, that the spiritual part of us is the most important part . . . if people would realize how important our spiritual part is we wouldn't have all the problems that we do . . . Chi is only limited by imagination," Dave concluded.

As the session was ending Dr. Shirali was impressed by the power of what Dave had gone through and the importance of such experiences for enabling us to look at things differently. She too felt tingling in her hands and feet during the session. Perhaps, she suggested, we need "otherworldly" experiences to wake us up to the reality of other dimensions. She was struck by the way the conscious use of the breath in the session created a line or thread between my "inner being" and Dave's. Julia was also affected deeply by the session. "You touch on a lot of things that I've been thinking about lately," she said to Dave. Again she spoke of how similar the anal probe Dave had described was to drawings she had made two years before of a similar instrument. "The instruments never have any resemblance to actual instruments that we use here," she observed. She said her eight-year-old daughter had also seen a spherical ship.

We agreed to meet the next morning to explore further the question that Dave raised at the end of the last session, whether the aliens have "known us before, in other lifetimes." Julia and Dr. Shirali were again present, but Pam Kasey could not be with us. Before inducing the trance state, I encouraged Dave to let his associations go to the familiarity of the female being when he was three and the feeling that they had missed each other.

The first thing that came to his mind after a long pause was that he was a Native American boy, given the name Panther-by-the-Creek, of the Susquehannock tribe living near Pemsit Mountain, and he was studying to be a medicine man. This was in a time "before the Indians knew about the White man," Dave said later. The boy lived by the river, caught shad, and dried meat for the winter. Eagles lived along the river in the cliffs. "The eagle is real special. The mountain's like a special place. Medicine men go up there to get visions, do journeys." When he was ready, he too went up the mountain to get visions, and it was there that he met Velia (the familiar alien female), who was a "friend and protector." He felt sad, for he missed being with her and also because something was to cut his life short. He cried a little as he remembered the territorial wars with the Iroquois and "a big fight around Hollowman's Island." He wore deerskin and carried a bow and arrow and war club and was just starting out as a warrior.

The battle was confusing, with a lot of screaming, and he was shot in the left side of the chest through the heart with an arrow. "It burned, and then it numbed. It was just numb." He coughed up blood, which filled his mouth and he choked on it. Then he blacked out and died. "The next thing I knew I was away from my body." He saw his body lying on its back below and also one of the Iroquois warriors bending over him and cutting off his scalp. Then he felt himself "floating up in the air" and "dissipating, spreading out all over like a fog of crystals . . . I just like went everywhere. I was spread out real thin. It was peaceful. I think somehow Velia was there after I left my body, when I was dying," he said. I asked at what point he saw her. He had not actually seen her but just "felt her presence." After he floated up in the air "I knew she was there," he replied. "I think I knew her before," he added. "I think for the Indians it's easier to die. It's more like a natural thing, just calming. Very calm. Death is just part of life, so it wasn't real hard to accept, not as hard as it is for us." Dave was somewhat surprised to discover this particular past life. "I never, never thought that that had been one of my life experiences." When he had first told of this incident during the regression he had been unable to remember the boy's name.

I asked Dave what his connection was with Velia after he died. He answered, "When I floated out of my body, before I dissipated, she told me that she'd be with me even when I came back." Again feeling sadness he said, "Then I was born again in Virginia." I asked him about the sadness, as he was crying softly. "I think I lived some place there and I was real happy," but "I had to leave." I asked him to go back and tell me first how he got from the "dissipated" state to being born again in Virginia. "I just kind of gathered together again and popped back out into the world." I pressed him to explain how he did that. Did he not have to become physical in some way? "I had to go into a woman's womb," he said. The woman's name was Mary Peg and she had long, dark hair. "My dad's name was John," whom he saw as a large man with sandy-colored hair. They lived in a little cabin and had a small farm from which they made their livelihood.

I asked why Mary was chosen and how he went into her womb. "I just went into her one night in the winter," he said, "everybody was asleep, and there was a little bit of a fire left in the fireplace. I went into her womb. I knew that's where I wanted to go. When I went in she became pregnant. She was laying there sleeping and I went into her." I stopped him at this point, observing that for a woman to get pregnant an egg must be fertilized. He explained that Mary and John had intercourse that night and "the egg was fertilized after John and her made love" and "when the egg was fertilized at that moment I went into her."

I let the possible discrepancy of her being already asleep when he "went in" go and asked, "Then what?"

It was "dark and warm" inside Mary's womb, her labor was short, and he remembers "great pressure" and coming out "headfirst." He was her first child, but later there was a little sister who died. Dave cried as he remembered Mary's sadness. "I don't think she lived very long." The sadness also concerned the Civil War and his own death. "We lived there on the farm, and I grew up. I was a big guy, and I think I had to go to war." He was smart, good on horseback, and served as a scout or spy on the Confederate side. He was captured by Union soldiers and hung at nineteen.

Dave suggested that these past-life deaths in war were related to the fact that he didn't have to go to Vietnam. "Some guys my age went to Vietnam. I didn't have to go this time. I was frustrated in those lifetimes" as he didn't get to become an adult. "I didn't get drafted," Dave wrote later to me in response to my question. "They had started the draft lottery for your birthday to determine who would be drafted first and in what order. My birthday was the 353rd date drawn and they never got that far. Besides, I would have gotten a deferment because of my eye."

I took him back to his relationship with Velia. "They" do not live forever either, he said, but Velia has been with him through these three lifetimes. He said that he loves her. "She's always there. She's real consistent." He believes his first encounter with her occurred when he was the Native American boy. He was perhaps fourteen, by a stream, when he saw her "floating up the stream by a path." He was surprised to see her, "frozen with fear," for "I didn't know who she was." I asked what she looked like. "Like she always does," he replied. "She had gray skin, big head, big black eyes, and said that I wouldn't really know about her . . . It was as though she was a spirit . . . She checked me out physically . . . I didn't have any choice. I just sort of laid down [later he said that she "floated me down"] on the ground." She "talked to me a little bit. Then she left." After he "woke up on the ground" he felt "happy that it happened—it was like being chosen."

He talked with a medicine man about the encounter with this being, and "he called them guardians." The medicine man also "described the characteristics that occur in the experience and that happened to the chosen ones" and explained that the guardians "try to make you more in contact with the spiritual part of yourself." I asked how they do that. He replied, "Just through being around, exposed to your inner being from their power. I asked if he remembered anything else about the encounter along the stream. "Velia said that if we were to lose the connection with the earth that that would be bad for us." It was like "the connection with the earth is part of our spiritual, part of the spiritual side of us . . . being part of the natural world, being part of the wholeness."

She told him (all of the communication was telepathic) that some day he would learn about eagles. "The Indians felt the eagles were important spiritually, that they symbolize something, part of the Great Spirit." She told him she "cared about me and she'd always watch over me." After the encounter with Velia, Dave saw a black panther and believes he received the name Panther-by-the-Creek "because," he wrote me later, "a panther was seen by the creek the day I was born. So when I saw the black panther by the creek after my experience, I took it as a good omen. It was a synchronicity in that lifetime similar to the ones I experience now."

I asked him if he knew more about Velia, who seemed to exist in some form whether or not he was embodied in one or another lifetime. This was difficult for Dave. "I don't think it's the same when you're dead—like she's not corporeal if you want to pursue her," meaning, I think, that even though she might exist in some disembodied form it would be difficult to find her, even after we (a human being) had died. Nevertheless, "she's alive all the time" but the embodied form "is not as important a part of their being. They're more, they're more spiritual . . . They want us to know," he added, "that it's not the material part of us that matters that much, that that's where man's problems arise." I asked him to say more about that. "When you're Indian and you're real close to the earth, you're real spiritual," he said, "and you get away from that you're more material, and that's to your detriment."

We were coming to the end of the regression, and Dave felt sad as he recalled his closeness to the mountain as a teenager and how he would try to imagine what it was like before the white men came. "That's what started me out in this spiritual quest in this lifetime, thinking about that all the time . . . I yearned for those times when it was all big trees and everything was pure . . . That's why," he explained, after seeing the UFO at nineteen he began to read the Don Juan books. He was seeking Native American knowledge and admired how close the Indians were to the earth.

After the regression, Dave connected the fact that he never wanted to leave home while his brothers were growing up, even though he felt "I should be out on my own by now," to his own early death in the two past lifetimes we had explored. "I didn't want to move away from my younger brothers, 'cause they were boys, you know. I wanted to watch them grow up, really wanted to watch them grow up." This "started being real important to me when I was nineteen." At that time he would think, "they're all going to be my age soon, and you change totally from when you're fourteen to nineteen. I wanted to see each of my brothers through that transition."

The day before the last regression, Dave disclosed that he had once had an image several years before of having once been a cavalry spy who was caught during the Civil War. He also said that Mary, his mother on the small Virginia farm, was Charlotte Hampton, but "I was afraid to utter that." This led to a discussion of the reality status of past life experiences. I suggested the possibility that consciousness is a "continuous fabric" and that potentially we could be identified with any object in the cosmos, depending on the evolutionary task at hand. Dr. Shirali spoke of the Hindu understanding "of the whole divine thing that's also within you. The Brahma, the whole, the part reflects the whole, and the whole is reflected in the part . . . There can't be a linear time." Dave resonated to this by saying that only "some part" of him had been with Velia before.

Julia, who had been remembering her own past life experiences, was deeply moved during the session. She began to cry as new details came to her from these experiences and she found herself "coming up with my own answers." She observed that more than one person "could potentially access the same life." She also had brought in her drawings of an anal probe, which compared accurately with a picture that Dave had made independently, except that hers was shown open and his was closed. "The only reason I saw this open," she said, "was because the doctor showed it to me."

Dave called this session "the capstone," which brought a lot together for him. I wondered about what potential connection there might be between his boyishness and the adolescent deaths in his past life experiences. "I'm thirty-nine, almost forty years old, and I don't act like it," he said. "Now I understand why," he remarked, he is "a mannish boy." Age nineteen, he said at the end of the meeting, was "always hardest." When first he, and then his brothers, reached that age he was afraid they "wouldn't get through it."

I spoke with Dave a few days after he returned to Pennsylvania. His experiences in the regressions had been profound and took time to assimilate. He was feeling somewhat isolated and was walking a lot, trying to "figure things out." His wife, he said, "can't get too immersed," and Master Joe had been quite busy, although he affirmed the validity of past life experiences in Buddhist teachings. In early April, Dave wrote me a letter which reflected his further integration of the material that came out in the regressions. "It makes everything in this lifetime make sense," he said. "I'll never look at Pemsit Mountain the same now," he wrote. "What it is about the end of the mountain is no longer such a mystery."

Dave connected his relationship with Charlotte Hampton, who was his mother during the Civil War, and his experience with her in this lifetime to "getting my Chi channels opened." On March 25, he had dinner with her in Philadelphia and "we got along so well it was uncanny." Julia told him that she herself had been the little sister who had died in his Virginia lifetime, which explained their connection now. "Your study brought us together in this lifetime," he wrote.

Early in June we had a long telephone conversation. Dave was feeling well and making progress in his work with Master Joe. He had "found his hands" in a dream, which, he has been told, is an important step in the mastery of his Chi, and enabled him to fly in his dreams. More people were coming to Dave for guidance, including a seventeen-year-old excellent karate student (referred to him by Master Joe) who has had abduction experiences since age five. A few days later Dave told Pam that a girl spontaneously told him that "the beings have a place up on that mountain," which "flipped him out." A book he was reading about a woman who was becoming a shaman and had abduction experiences was leading him to connect shamanism and abductions.

At the end of June, Dave wrote me another long letter, filled with "strange coincidences" and new connections between the people he knows in this life and in the earlier ones. Native American spirituality, shamanism, strange powers of nature, altered realities, Chi, karate, the mastery of dreams, UFO abductions, past life experiences, and a multiplicity of synchronicities are all part of a mysterious puzzle for Dave whose pieces, according to Master Joe, he is learning to put together.

Discussion

Dave's case illustrates particularly well that the abduction phenomenon cannot be considered in isolation. His abduction experiences are linked to a wide range of other natural forces and energies with which he has had a connection since childhood. These include a deep reverence for nature and its mysteries, an intimate association with Native American values and shamanism, a personal feeling for the spiritual significance of certain animals (especially eagles and deer), and a determination to master his "Chi," which he defines as the "force which pervades the universe from which reality arises." As Dave has opened himself to the actuality of these and other natural phenomena and spiritual teachings, the world has become a place of wonder and awe and the earth itself increasingly sacred. This opening has been associated with many seemingly meaningful "coincidences" (synchronicities) which taken altogether suggest a pattern of connections, a sense of design, in the cosmos that Dave inhabits.

Pemsit Mountain, close to where Dave grew up, is the focus of Native American tradition and lore, a place of special power, and also, for many local people, associated with UFOs themselves. A vivid, quite spectacular UFO sighting near the mountain when he was nineteen, witnessed independently by a friend and one of his brothers, and confirmed in the local media, was a turning point in Dave's life. He was determined after this experience, in which he believes he saw the dark eyes of an alien being looking down at him from within the ship, to read everything he could find about the unknown in order to learn about what was inside it. He even discovered Tibetan Buddhist writings that seemed to confirm their knowledge of UFOs.

Dave's abduction experiences contain the familiar traumatic intrusions, humiliating for a man, that many abductees have undergone. These include being removed from his home against his will, the taking of sperm samples, and frightening anal probes, reassuringly explained to him by the aliens as checkups, a kind of health maintenance program. But as these have been explored in the context of his accelerating personal evolution, Dave's terror and rage have been mitigated by his sense of wonder and awe and his willingness to surrender control before the power of a process he does not understand. In that context he has become increasingly aware of a strong, loving connection with a protective female being he calls Velia whom he has known since age three and, as we discovered in his third regression, possibly in previous lifetimes.

Velia appears to be a principle agent in the evolution of Dave's consciousness. Our investigations have uncovered information she has imparted to him telepathically about the danger to our survival and the fate of the earth, of our loss of emotional and spiritual connection with nature. Through Velia Dave's identification with the Native American reverence for nature has become linked with his contemporary role as a local leader in the reassertion of that connection. In that regard it is interesting that in our last regression he found out that the Native American medicine men knew of the alien beings and also looked upon them as "guardians" or protectors of nature.

Needless to say, none of this makes much sense in the framework of the Western ontological paradigm, which has no place for unseen forces of nature, intelligences in the cosmos that guard our destiny, beings that enter our world in physical form but are not of it, past lives, or even UFOs themselves. Yet there is little about Dave's personality that could be used to explain his abduction experiences. He is a practical, down to earth home builder, who is respected in his community and has held the same job for fifteen years. There is nothing about him to suggest a tendency toward psychopathy, delusions, or a proneness to fantasy. He, like many abductees, has intensely resisted accepting the reality of his abduction experiences. In his first hypnosis session he balked at "this abnormal stuff," even though he had had conscious memories of alien encounters before that meeting. One could argue that Dave's readings have influenced his experiences, but the process has been the other way around: he looks for information in books *after* he has had an experience that he does not understand.

In Dave's case we are finally left with the question that so many of the cases discussed in this book pose. How are we to regard consciousness itself as an instrument of knowing? There is a smattering of physical evidence that corroborates his experiences—UFO sightings with multiple witnesses, an unexplained crescent-shaped scar that appeared after one of his abductions, and a pattern of events that are too extensively and complexly linked to be attributable to chance alone. Yet the evidence for the existence of another world that is unseen and yet powerful in its influence depends in Dave's case largely on the reports of his experiences, the affective appropriateness and intensity with which he relates them, and the judgment of the investigator as to his sincerity and the genuineness of his communications. Of this last I am quite convinced. Dave leaves us finally with the choice of rejecting the entire body of his experience as the product of some sort of mental aberration or collective influence, or of considering the possibility that consciousness is a valid instrument of knowing and that the view of reality provided by the empirical methods of Western science has been too limited.

Dave's past life experiences deserve special comment, for they provide an alternative explanation to certain aspects of his life and personality that would otherwise be based entirely on his biography in this lifetime. I discovered nothing in his family life or personal history that could account for a certain adolescent quality that seemed unusual in a man of thirty-nine. He did not seem, for example, to possess dependent qualities that would help us understand his reluctance to leave home before each of his brothers reached the age of nineteen. Yet his violent adolescent deaths in two previous lifetimes, relived with strong emotional conviction in the third regression, provide a possible explanation for this anxiety and his own fear of becoming fully adult. One could argue from a traditional point of view that his past life experiences were related to an overactive imagination in relation to Native American life and the Civil War. But the opposite is also possible, i.e., that Dave's past life experiences, which continue to permeate his consciousness, are an important source of his imagination.

Dave's case richly illustrates one of the more interesting mysteries of the abduction phenomenon, the creation or staging, presumably by the alien beings or whatever intelligence guides them, of alternative physical realities (see also Catherine's and Carlos's cases, chapters 7 and 14). His story abounds with rooms, like the dining room in his first regression, that are not quite the actual rooms of his house, and landscapes with caves and trails that are not there upon later searching. There is something rather frightening about this. For it confronts us with just how arbitrary the physical reality within which we happen now to find ourselves is. All that is required for this to be abruptly changed is the choice on the part of some other intelligence with a power greater than our own to do so. We have, as the abduction phenomenon seems repeatedly to tell us, little control over the reality that surrounds us.

Finally, Dave's case is about power, the immense power or energy, both spiritual and physical, that resides in nature. Through his study of karate and Chi, Dave is seeking to gain control of the expressions of that energy within himself and to provide an example in his life of its constructive mastery. Perhaps this is the central teaching of his case. In his April 8 letter to me he wrote of Pemsit Mountain as a "big power place. That's why the visitors go there . . . Power," the letter continued, "is a mystery we will never understand. We can only learn to handle it. Maybe that's what mankind needs, to learn how to handle his power."

CHAPTER THIRTEEN

PETER'S JOURNEY

Peter, a former hotel manager and a recent acupuncture school graduate, was thirty-four when a fellow student who had heard me lecture on abductions at The Cambridge Hospital told him about my talk. "I might have had that too," he thought to himself and called me. This was January 1992, and we first met on the twenty-third of the month.

Peter's case provides one of the most dramatic examples of the way the nature of abduction experiences can be transformed in conjunction with the evolution of the experiencer's consciousness. His abductions were initially intensely traumatic as recalled in our first hypnosis sessions. Gradually these experiences, together with our exploration of them, became a central element in a spiritual journey that has led Peter into perceiving other dimensions or realities beyond the manifest world. He is one of the abductees I have worked with who has discovered a dual human/alien identity. In his alien self, he has become aware of having participated willingly in the alien-human hybrid breeding program, and his case raises questions about the ontological status of that process. Peter is also a leader among abductees, having decided to "go public" with his experiences, speaking at conferences and on television and radio programs in order to play a role in educating the community. Between February 1992 and April 1993 we did seven hypnotic regressions.

Peter has also been afflicted by vivid, disturbing apocalyptic images of the destruction of the earth, and we have explored whether these are to be considered as literal prophecies or as metaphors or warnings of possible futures. Peter's abduction experiences have represented a kind of "other" life of compelling power and meaning for him. His wife, Jamy, has been a steadfast partner throughout Peter's personal journey; yet he has felt that it is inevitable that he assign priority to his abduction-related life. This has created the kinds of strains in their marriage that occur under the best of circumstances when one member of a couple is deeply involved with abduction experiences. Peter is also one of the few abductees in my sample who has undergone an extensive battery of psychological tests. The reason for selecting Peter for these tests will be explained in the context of the referral.

Peter grew up in a Roman Catholic family in Allentown, a steel town in eastern Pennsylvania. His father developed a left-sided paralysis, muscle weakness, and a limp due to poliomyelitis he contracted at age two, but was able to work most of his life as an office manager in an auto body repair shop. He was college educated and attended medical school for one year before dropping out to support his family. His father was eighty and fully retired when Peter contacted me. They write regularly to each other. Peter's mother, who was born and raised in England, worked as a knitter in a knitting mill in addition to taking care of her family.

Peter told the psychologist Dr. Steven Shapse that he became the class clown, was rowdy, and began drinking and smoking marijuana at an early age. He has two sisters, Linda and Corinne, six years and three years older than he. Peter feels close to Linda "without knowing why." Linda entered a convent to become a nun when she was in the ninth grade and remained there through high school. She has seen a UFO and believes what Peter says about his abduction experiences, though she does not say she has had them herself. Corinne has no such recollections.

Peter attended a combination of public and parochial schools and graduated from Allentown High School in 1975. He earned a Bachelor of Arts degree from Penn State in Vocation Industrial Education, completing a six-year program in 1981 that enabled him to become licensed as a professional cook and teacher of culinary arts. From 1982 to 1984 he worked in a new hotel on the Big Island in Hawaii, where he met his wife, Jamy, who is three years older than Peter. Jamy is a practitioner of Shiatsu, a Japanese deep-tissue massage technique, and, like Peter, works as a healer and therapist. She has a master's degree in counseling psychology. They have a close, warm, and confiding relationship in which they can talk through difficult matters fully.

The couple has decided not to have children, at least for the present time. Peter attributes this in part to the fact that Jamy was the oldest of seven children in an alcoholic family. "She had to raise a lot of the children herself, and I don't think she really wants that again." Peter would enjoy having children, he says, but adds that possibly "there is some destiny for me, or there is some predetermination in my life that is tied in with this alien thing, something for me to do. It may preclude children." From 1986 to 1990 Peter and Jamy were managers of a twelve-room hotel and a restaurant on a private island near St. Thomas in the

U.S. Virgin Islands. In the spring of 1990 Peter came to Boston with Jamy to attend the New England School of Acupuncture from which he graduated in May 1993.

In our first interview Peter said that he has always known "there's been guardian angels. I've always known there's been beings . . . I'm very spiritual, and I've always known that I could commune with God." He has also always known "that there were UFOs" and "extraterrestrials . . . It's just been something in my mind from the time I was a little kid." In the context of his regressions, Peter has had hints of encounters with the alien beings in infancy, and in his sixth regression he had images of himself as a child as young as four playing with hybrid children, which continued until he was perhaps eight or nine. In our first meeting Peter told of remembering consciously going to a storage space at the end of a long hallway in his home and feeling afraid of what was on the other side of a window, which he used to sit by. He also remembered the beings watching him play with other children or "just observing me be a little kid." At the beginning of the third regression he had memories of being happy to see the alien beings as a little boy, of feeling chosen by them, and of floating "right through the window."

In the third regression, Peter also relived an intensely traumatic abduction from age nineteen or twenty in which a sperm sample was taken against his will. In the course of our work he has also had recollections of visitations during the period that he lived in Hawaii, where he and Jamy lived in an open, isolated part of the island. Peter remembers seeing "something outside the window" of their home. At the time "I always used to think it was an owl" which "used to call me and tell me it was 'time,'" but now he believes "it wasn't an owl" but an alien being. "I always felt a kind of communion with this owl."

The most powerful experience that Peter recalled consciously before we met, occurred in the Caribbean during the 1987–88 period. During this time he remembers he would sometimes go to sleep afraid and then be awoken by a touch or something "hitting me right at the base of my spine." In our first conversation he recalled experiencing terror, rage, and a loss of control as light filled the room and he felt a "presence" around his bed. "I remember my whole body vibrated and shook maybe for a second, two seconds, three seconds."

On at least one occasion Peter saw small, hooded beings in the room and would shout angrily at them. He also remembered consciously that under their influence he would walk outside on the patio and, "bathed in a light," he would be "lifted up" to a "round ship" with "a dome on top" and flashing white, red, and bluish lights "spinning around it" that was visible "above the treetops outside the house." At the time, according to Jamy, Peter described that "they had this laser beam, and it went right through here [points to center of forehead], and it was so bright that his eyes, you know, like it was hurting." While Peter was undergoing these experiences Jamy would be "out cold." After one of these experiences he remembers finding two small, red lesions like healing pimples behind his ear that were distinct from insect bites in the rapidity with which they healed and the symmetry of their arrangement. Peter's terror in the conversation grew to the point where he felt "stuck," unable to go further, and we decided to explore his experiences in greater depth under hypnosis.

Two days after our first meeting, Peter told Pam that the only reason he believed what he had told me "was real was because there were emotions." He found that he was "distancing" from what he had told us and "wants to believe that it's his imagination." The first hypnosis session was scheduled for February 13. Peter was apprehensive before the meeting and had had a difficult time sleeping the previous few nights, especially because Jamy had been away. In one of his dreams he communicated "with these beings" about how "we have inner knowing and a knowledge and power that goes beyond intellect." He had "the feeling" that the beings are afraid of "the power that we may possess."

Peter chose to explore an incident in the Caribbean, probably in February or March 1988, which began with two beings appearing by his bedside. He became anxious, "like I want to strap on the seat belt," as his thoughts drifted back to that time. I assured him that I would not leave him alone with his experience in a way that he would find unmanageable.

Under hypnosis Peter recapitulated the setting of his house and the hotel restaurant where they'd eaten that night, what he had for dinner, and going upstairs to bed feeling afraid "something was going to happen." He remembers waking up as Jamy slept, feeling that he wanted to cover himself (he slept naked at the time). Despite feeling anxious and vulnerable, for some reason Peter got out of bed and walked over to a couch on the other side of the room and then saw "this little creature . . . It's happening again. It's happening again and again. I walk over and I'm humiliated. That's what I feel most. You're [the being] looking at me and I'm naked." He also felt "out of control," "inferior," powerless, and enraged. "I'm paralyzed. I want to kill it, and I can't do anything." He had no will, for "they shut me down."

Peter saw two beings, the one "that controls my feelings" being slightly taller than the other. The top of his head came up to the level of Peter's chest. The beings were thin and wore close-fitting bodysuits with hoods that seemed to be made of something like Lycra or latex. His anger appeared to intensify Peter's lack of control. Compelled by a force that seemed to lift him, Peter lay down on the couch. His fear was mounting now and I encouraged him to concentrate on his breath and to relax. "I know they're going to hit me with this light, and after that that's when it gets bad . . . The light hits me in the forehead, and that's when I jump and after that things become peaceful . . . It's the moment where I lose consciousness, where I lose remembrance, where I lose control, where I lose that I'm part of it."

The smaller being was holding an instrument that looked like "the flashlights policemen hold with a head on it and it's pulsing." Yelling now, Peter cried out, "Now it's going to hit me. It's going to hit me." The larger being "knows my consciousness. He knows what I'm feeling. But he's detached" and "doesn't want me to know what he's about." This being controlled the smaller "drone . . . the little shit that does all his dirty work for him." The smaller being lifted up the light, "holds it there and hits me in the head with it." After that Peter felt cold, shaking and shivering on the couch in terror as control of "my functions" was "shut down." A shift occurred then—both at the time of the incident and in the session—and he felt more peaceful. "My body feels like it's cut off from my neck, from my head." Despite his nakedness, the fear and sense of humiliation were also gone.

Then "the little guy walks beside the couch. He does something with that thing again, like waves it over me, under me. How can he get it under me?" The light lifted Peter off the couch and he "felt really light." With his arms held up as if by some force Peter found himself moving toward the door. He looked over at Jamy, who was still sleeping, "and I know she's going to be safe," which the beings reassure him about as well. He got a closer look at one of the beings ("It's like now he's my buddy. Now I'm not afraid") whose face seemed ugly and distorted. "I can only see half his face. It's wrinkled. It's like one of those animated Disney characters . . . He's got this big eyebrow. It's like a really big eyebrow." The skin of the upper part of the face seemed thick, with "lumps" and "like three ridges, three furrows. It's almost comical looking." The eyes were not big, but very dark, "like blue and black," and deeply set in the face, "like an animal's eye, like a raccoon's eyes." The nose was pushed in "and then it widens."

Next Peter was floated through the dining and kitchen area and out the door to the porch or deck, which was illuminated by a soft, white light that revealed the trees in the background. As he left the house Peter could see that the source of the light that he was bathed in was a small spaceship. As he and the beings were "floating up" Peter could only see white light as he looked up, but was "aware of the island in the distance," of the ocean and "all of the trees down there." When he looked down he saw the tin roof of his house and wondered why he did not feel afraid as he is generally scared of heights. Now he is feeling that "this is an adventure" and "really happy Jamy's okay." He passed first into a smaller ship which then drifted up through "a hole in the bottom" of a larger ship. Then he was aware of darkness, "like the inside of a house," and the light was now below. All was quiet except for a kind of humming sound. Peter saw benches and uniforms or "jumpsuits, like the things speed skaters" wear "just laying there and I think, 'Why aren't they hung up?' . . . I almost feel like I've been invited to their home."

Above a wraparound bench that seemed to be made of a hard, molded plastic were little lockers or compartments. Peter walked about, as if on a kind of tour, and felt "honored . . . like I'm someone special" as "they let me look around . . . I want to say that there are other humans there, but I don't know." He looked out the window from "like the control room." He could look across to another island, and "it's beautiful, 'cause the moonlight is shining on the bay and then there are the lights of other hotels. I can see them clearly." Peter saw "another being that's got a control table or something," and he observed more of the suits hanging on a bar. They flew higher and as he looked out a window, the earth became "just a pinpoint now." Peter became worried, and felt confused and lost. "Where am I? Where's Earth? How am I ever going to get home?"

Peter was incredulous about what he was recalling and began to question his experience. I encouraged him to "tell it straight" and we would figure it out later. Curiously he no longer felt naked, though he was wearing no clothes. He walked through a kind of "French door," which slid part way into the wall. Peter walked down two or three steps and found himself in a sunken room where there were perhaps a hundred men and women. "I don't see them as naked . . . I see them as in a flesh-tone outfit, this flesh-tone, one-piece suit." He has a feeling like "walking into a cocktail party, and you're not sure who you're going to go up to talk to." The taller being he had seen originally came over to Peter and telepathically communicated, "these are all people from Earth. You can get to know them. They are all here for you," as they have all "had the same experiences that I'm having." Peter felt he wanted to walk over to speak with one of the men who was talking to a woman. The man acknowledged Peter but said, "Not yet. You have to go. Not yet."

"This is the part I don't like," Peter said. "I don't feel good right now." He asked to go to the bathroom, and when he returned said, "I feel really scared, John." He turned to the right and went into another room where one of the beings "was working some dials or something . . . I feel like I'm going to cry," Peter said. "I feel like I'm really afraid. I feel like a little kid. I feel like I'm going to get abused or something. This is not nice. This is not fun." I encouraged Peter to breathe deeply again. "This room has a real sterile feel to it, a medical, ominous feel to it." The floor was jet-black, like obsidian, and there was a glass wall with human beings suspended, rather like in the movie *Coma*, with helmets on their heads. "A couple of beings" monitored what was going on in a "control panel" or along a wall. They wanted Peter to go up some steps, which he did, and then a "silver dome" or "helmet" was brought down over his head as he stood by a table.

He did not want to do so, but Peter got onto the table, which was "the most comfortable table I've ever been on in my life. It's like it molds to my body . . . It's the greatest examination table ever made." The table was tilted to a forty-five-degree angle as one of the beings began to issue commands to another. "I hate this guy. I hate when he gives these commands, and I know it's going to hurt me," Peter said, practically shouting. "I don't want to tell you what happens now." His legs were held together and strapped down while a female being communicated to him "that it's going to be okay" and he was aware that "I'm just one of hundreds of thousands. I'm not alone." Then he felt "like my life force is being sucked out of the top of my head" and "I don't know what's going on anymore . . . I want to stop," Peter said. "I just want to relax."

Meanwhile several beings were observing Peter, "watching to see how I react." Then he was taken to another table, a "cold, icy metal" one contoured to his body, "a great medical invention," he said, with mixed awe and irony. There was more measuring and checking, including "endorphins" or "something in your brain." Then, using tools that reminded Peter somewhat of a dentist's fiber optical instrument, the beings probed his groin, "right here where the bone comes in," going "right through the skin." Although the instrument seemed to be passed deeply into his abdominal cavity, Peter curiously did not feel pain at this point. His genital organs were "just off to the side," and evidently not involved on this occasion. "I'm amazed at some of the simple technological things that they've got that we haven't thought of yet," Peter said rather sincerely.

The most humiliating and frightening part followed as "they're groping at my legs" and put a tube in his rectum to take a stool sample. "These guys don't know how to touch people . . . like, get some bedside manner." The tube was passed deeper into his rectum and Peter felt that they left "an implant" or "an information chip" inside of him. "Why do you have to do this to me?!!" Peter shouted, "I'm tracked now [almost crying], I can never get away. That's how I feel. I'm stuck for life. I feel like a tagged animal. I feel like they put something big up my anus, and spread it and then stuck something else up through it and then they left it. It's way up inside of me."

Feeling defeated and humiliated now, Peter said that "what pisses me off the most is that they told me they were going to do this. They held it up for me and showed me." Now he felt "like I can't get away. I feel monitored. I feel like one of those polar bears with a collar on." He suggested that this happens "every time," and that the beings repeatedly remove or put back in this and other implants. This abduction, Peter thought, "was the one that indoctrinated me. This was the one that made me one of theirs, one of their beings or one of their animals . . . I feel really alone now, and isolated and afraid," he added. "I feel defeated." I argued as best I could that these were altogether understandable feelings, but distinguished between them and actual defeat.

As the session came to a close, we talked further of Peter's feelings of humiliation and violation. Despite it, he feels that he is somehow "a willing participant" in this process, "willing to help these people make the bridge between their world and our world." Furthermore, he believes "the other people, those other hundred humans that were in that room, are willing participants in this too." Although he felt violated and traumatized on "some psychological level," on "a complete other level I understand that it hasn't harmed me." He likened the process to a woman who goes through the pain of labor. "She accepts it, and she does it. She's not mad at the labor; she's not mad at the baby. There are fits of anger. There are fits of hatred, fits of all that stuff. But it's accepted as part of the process of birth."

Peter was not able to say clearly at this time just what or who was being served. Perhaps it was to "accept beings of another planet" in order to raise our own consciousness. "It's about evolving to a higher consciousness," he suggested. Perhaps beyond the feelings of violation, abductees may come "to a place of feeling good about it." In a card to me, written four days after the session, Peter expressed gratitude for "the wonderful opening experience" and wrote that he felt "less scared" and understood "a little better what these 'experiences' are for me."

A second regression was scheduled for March 19. Jamy accompanied Peter to the session. Four days before this, Peter had another abduction experience while staying at a friend Richard's house in Connecticut. He had just completed an energy healing workshop and was scheduled to return to Boston that day. Though he was inside the house at the time of the experience, three women colleagues who had also spent the night there had a vivid sighting of a small UFO over the house while taking a walk. One of the women, who I have interviewed and who is herself probably an abductee, wrote down the same day what she saw.

> As we walked, not five minutes from the house, there came an amazing sound behind us to the right from Richard's house above the river—my thought was the dam [water coming through a nearby dam]—this was not an airplane or helicopter, and no engine I've ever heard. But it was very powerful and accelerated down the river, and by us, at 'warp speed' and disappeared. I saw a shape, rather like a crescent, some darkness underneath, and maybe orangeness to it, zap by above me to the right.

In other notes this woman wrote that she thought of the wings of "a huge bird," rather than an engine, and of the excitement she felt at the sighting. She felt certain that "Peter was in the ship," but when Pam asked her if she thought of looking in the house she answered, "Peter wasn't there. To me Peter wasn't there. No one was there," and she felt no desire or need to check! Peter had recently "woken up" from his abduction experience and was intensely shocked when the women came into the house and told him of their experience. Only upon driving back to Boston did Peter tell the women of his abduction history. The other two women were also impressed by the sighting. One of them, according to Peter, was "moved to tears by the whole incident."

Before beginning the regression, we reviewed the details of the sighting and Peter's decision not to go on the walk with the women. The house was in a beautiful area, and Peter had planned to go on the walk. He had slept poorly, was somewhat exhausted from the workshop, "and then I just decided not to." The house was cool, there was a fire going in the fireplace, and Peter sat by it, covering himself with a blanket. He "prayed a little" and quickly fell asleep at what he estimates was about 10:15, as the women left for the walk at exactly 10:10 A.M. It was 11:05 when he "awoke," as Richard, the owner of the house, was walking through the door.

In the four days before the regression Peter had recovered several conscious memories of the abduction. He had seen the alien beings beside him, remembered feeling "that we had to hurry," and recalled details of the inside of the ship. He felt waves of terror but wondered "why doesn't this hurt" when one of the beings "was doing something to my eye and I can still feel it . . . My eye really ached the next day," he said later. "It's amazing." The day before, he had told his therapist about the experience, and he had had Peter lie on the floor and express his feelings in a bioenergetics exercise. This brought back details of playing in the storage area in his home as a child and "a very clear recollection of the ship over our backyard, over the roof of our house, and that I used to go out there."

For the next twenty or thirty minutes we talked of the strains that Peter's abduction experiences were placing on his and Jamy's marriage. Jamy felt excluded from much of what Peter was going through and expressed fears that he would leave her. Peter spoke of how his abduction life "rocks the very foundation of my whole belief system," and had so radically changed his life. Jamy spoke of feeling ineffectual, "so nothing I can do can help." I spoke about how difficult it is for spouses who are with someone who has an all-consuming mission and tried to be supportive to Jamy in what she was going through. Peter said that "there's a part of me that wants to live as normal a life as I can and keep all this in context," but appealed to Jamy to be aware of how much support he needed in the face of what he was going through. There was no clear resolution, except a commitment on both their parts to stay connected and try to be sensitive to what each was experiencing.

The session had been scheduled before the March 15 experience, but we decided to explore this incident further under hypnosis, as it was so much on Peter's mind. At the beginning of the regression we reviewed in greater detail the events of the night and morning leading up to the abduction experience, especially his decision not to go on the walk. Peter had felt a compelling need to be alone, but "I'm afraid to think, John, that I was communicated with," that "somehow they communicated with me to stay." As he sat by the fire, he "just nodded out." It was not clear to Peter whether he fell asleep or not, but he remembered being awake but "in another state of consciousness—it's like a small part of my brain becomes aware and that's what's alive," he observed. "What happens is your body goes totally asleep and then something clicks on in your brain that connects you with the beings."

Then he heard sounds. "I can hear the vibration," he said. "I know there's a ship behind me." Now he was paralyzed, conscious of "what's going on" but "not having any control of it." With his eyes closed Peter felt "a presence . . . Then I look up. There's a being standing on the other side of the couch, and I know I have to go. I know I have to hurry." He got up, the door was open, and "there's the little being in front of me. He's the blue people. He's the blue man. He's blue. He's very dark." The ship "looks like a sunset," and Peter floated up with one being "on my left and the big one's on my right," as his arms hung down by his side. After he was inside the ship it "moved really fast. We're just going off. We're flying, zoom."

Then Peter walked down a kind of ramp and sat in what looked like a wheelchair. He was afraid at first, but then curiously relaxed as a "retractable" arm, "like a dentist's chair arm," was "coming at me," with the instrument entering above the inner corner of his left eye. "It's going up inside. It's going up inside. They got it! It's twisting around. It's twisting. And now it's coming out. That's it, they're excited. They're happy. They got what they wanted . . . I'm just a piece of meat to them," he said. "I hate them. I just hate this. How can they fucking just do this? No regard for me. I just hate this." He felt he wanted to get angry with the beings, "but I can't."

What Peter felt they "got" was information. "There's a thing inside of me that they put in to see what's going on and to track your memory . . . that somehow records everything in the mind." His notion was that there was "a little black chip" in his brain, perhaps from a previous abduction, and that the flexible instrument, which looked metallic, long and thin, "retrieves the object" (the chip). Then they have "a moveable panel" which reads or deciphers the information. Peter has the sense that the beings are interested in what he learned at the workshop and "want to see if I'm a worthy leader. It's my reactions to situations they want to know that are recorded in my brain, that are recorded on this thing of theirs."

At this point in the session the timbre of Peter's voice changed to a kind of monotonous droning and he shifted to speaking from the alien perspective. "We," he said, want to study the chemical reactions of the brain, and how people will react in order to "know when it is time to be present . . . For as we measure the impulses," the voice continued, "we will know at what level the shock will come in, so we will be better able to control it so we will be in tune with the [human] beings as they go through this shock process, as they go through the unfolding of seeing us for the first time." Returning to his normal voice, Peter said that the beings are working with us, measuring what is in our brains so they will be able to tell how we will react to their manifesting before us.

I returned Peter to his feelings about the eye probe. He expressed further resentment, and then said that he felt he was "being primed for something." "This has been going on since I was a little kid." I asked him about his leadership role, and he spoke of a process occurring "throughout the world" whereby "common knowledge" of the abduction process was developing and that he was one of the people who would "stand up" and be "comfortable with the possibility of alien beings coming to this planet." Human beings "all over the globe" have been abducted, Peter said, and these individuals, especially certain leaders, will help reduce the shock that would occur when the alien beings manifest themselves on Earth. The monitoring that he described was "measuring to what degree consciousness can become available." Memory of abduction experiences would open up, Peter said, at a rate consistent with the shock that he and others could integrate of knowing of the beings' existence. "They were excited" on this occasion, he said, because "I had reached a certain level."

Peter did not recall the specifics of his return from the ship. He was "put down" on the deck outside the house and then "the tall one" walked him through the door. Peter sat down again in the chair, put the blanket over him. "I looked at the being, and he looked at me. I sat in the chair and just went to sleep, and the next thing I know I wake up and look at my watch." After the regression we spoke of the channeling of the voice that had come through during the session. He recalled that this had begun four years ago in the Caribbean, and that it feels like an expansion of his own energy which allows the "alien energy" to come through him. This can occur, he said, when he can "surrender my mind, surrender my ego." He (like other abductees who have this capacity—see, for example, Eva and Jerry) does not altogether trust the information he receives in this way, but does feel that "it's coming from a higher consciousness, from the spiritual plane." It was occurring in present time, he said. "It's live broadcasting."

As the session was drawing to a close, Peter spoke of the possibility that the recognition "that there's beings on other worlds, and there are other life-forms" could help us look in a different way at how we deal with one another on the earth. "A subtle shift" on this planet was occurring, Peter said, which is bringing about changes in how we think of who we are and the way we live. "By them showing up, it's going to make us appreciate what we have." His own task, he said, was "to surrender to that divine plan . . . It's almost like I feel like the aliens are watching out for me. They've got a plan for me," he said at the end. He did not fear the changes that were occurring. Though in many ways this process was taking over his life, Peter felt "it's also going to support me, somehow." His consciousness was "shaking," and "some days it's terrifying, and other days it's sure and secure."

A third regression was scheduled for April 2, two weeks later. Peter wished to confront his fear of the beings more directly and hoped "to be able to consciously communicate with them or be in their presence without being in such terror." This session turned out to be one of the most dramatic I have witnessed. For the investigator it was the kind of experience that cannot help but convince him that something actual and important has happened to the experiencer, whatever its source may prove to be. Before beginning the regression we reviewed Peter's life situation. He felt supported by Jamy and felt some confidence in his emerging role as a professional healer, but continued to feel quite isolated and alone "because of the experiences that I'm having going through all these regressions." Although Peter had in mind that we would explore childhood abduction experiences or those that occurred in Hawaii, we decided to let his "inner radar" choose where his unconscious would travel.

In the trance state Peter did in fact begin with a memory of a visitation in Hawaii where, surrounded by light, two beings floated him upwards. But quickly "another vision" replaced this one. He was a child now, four perhaps, and "happy to see them . . . I love it," he said, as he was floated down the hallway, then puzzled and excited "'cause I can go right through the window." Again the scene changed ("so many memories, John"), and he was an older child, "about eight maybe, and now I realize that we're not going to keep playing anymore. Now there's other things to do. There's more work they want to do, more experimenting. They want to do something with me. Now they want to put something in me. And they watch me. I've been chosen now. I have to lead the other little kids playing, and I'm not sure what's going to happen . . . It's like I'm old now and can't play anymore. I have to do something."

Next Peter remembered being "on the ship. There's a glass wall where all the kids play, and on the other side of it is where they watch us from. It's like a big playroom. And now I have to leave." The beings bring him over to them. He notices they have long fingers. "The tall one, his fingers are touching my eyes, opening my eyes and looking inside my eyes. They tell me not to be afraid. They communicate with my mind."

Relaxed now, he is in "like a dentist's chair," where the same machine, "the one I remembered last time," was used to probe inside his nose. "They drilled. They moved something. They put something in my nose, [gesturing] it's way up in here. It feels like there's something there." This reminded Peter of how he had refused to go to a hospital when he broke his nose in a car accident at the time he was a freshman in college. "It just dawned on me that this is why I didn't go." It was not only the association to the fearful procedure, but also "I think I knew there was something in my nose that I shouldn't go to the hospital."

Again the memory shifts. Peter is older, fear is mounting, and he feels he is "avoiding something; it's really on the surface." He spoke of major changes that the beings have told him are going to take place "on this planet." They have the ability to see into the future, he said, and "wish to help us to avoid what is about to transpire . . . John, this is too weird for me," he said, as more images of future possibilities flooded him. I encouraged Peter to breathe deeply and to permit whatever information came to the surface to do so without judgment or interpretation. His fear became more intense again, and I urged him to go back inside, let his body sink deeper, relax, and name whatever was "wanting to express itself," including the judgments themselves.

What followed was one of the most disturbing episodes that I have encountered during a regression. Peter screamed in terror and rage as he recalled an episode from about age nineteen in which he was on a table, "just laying there," as a cup was placed over his penis, ejaculation was forced, and sperm taken from him. "They have control over my genitals, over me," he said. "They're draining me. I'm relaxing. They're just happy. They got what they wanted," he exclaimed bitterly. I asked if there were other times. "They do it all the time," he said. "Every time they come they take it. They take my sperm. They take my seeds. Then they just let me go." Were these not the same beings who played with him as a boy, I asked. "It's different," he said. "They don't care as much . . . They don't want me. They just want my seed. They want my essence." Crying and shaking now, Peter recalled how cold the room was and the utter helplessness he had felt. "I want to fight them," he said, but "I can't do anything. I can't fight it." In order to reduce his feeling of humiliation and shame, I assured him that indeed there was no way he could have fought this.

"All the cells in my body are vibrating," Peter said. "Everything is moving. Everything is vibrating." I asked where he was now. "I'm lying flat. It's like I'm in suspended animation. I'm suspended. I'm traveling." His mind had evidently taken him back to the start of the abduction or an earlier one from the same period. He spoke of the terror of being lifted up, paralyzed, with the light around him. "I'm suspended!!! They've got me!!!" he screamed, breathing heavily, recalling now "more clearly" being taken into the ship against his will and being able to move only his head while he felt the intense vibration and shaking "inside my skin." If he had been free, Peter said, he would have lashed out at the beings. He focused on "this little being off to my left . . . I'd rip his head off is what I'd do to it. I'd kill it. I'd struggle with every ounce of my being," he burst out angrily.

Peter continued to scream, venting his fury, as he spoke of being taken into a room with an "operating table with all the lights on it" from which he could see another room "off to my right with glass and there's people in there suspended. They're on tables, and they've got those things on their heads again, those metal dishes on their head . . . and now I'm just suspended." Calmer now and breathing more deeply, Peter told of being on a table and subjected to a procedure similar to the one he described in the first regression. An incision was made under his testicles on the left, "they're searching around in the cavity," and semen was withdrawn from inside by an instrument like a huge needle or tube. One of the beings stood over Peter and told him that "they won't have to do it this way again if I relax." But whether they took his semen by this surgical-like method again or not, they certainly "took my semen the other way" (i.e., by the suction cup method) "a lot of other times" during Peter's youth.

After the procedure was over "they thank me. So now I'm just laying on the table, resting . . . They know it was painful," but they tell Peter "it's okay" and "they want to work with me to not be so upset." He was becoming tired now and moaned softly as he recalled a procedure in which there was "rooting around" under his arm and "I'm tagged or something . . . There's a lot more probing that they did," he said, including "something going down my trachea," but "I think I've had enough, John. I'd like to cool it for today."

The remainder of the session was devoted to a discussion of the impact of these experiences on Peter's consciousness. Much of the terror, he explained, he related to "expanding my consciousness" to "the point where I could accept the beings and accept what was going on." The terror, he said, was to "stretch my reality." I found this puzzling, as it had seemed to me that in his experiences he certainly had accepted the reality of the beings. Peter responded that "the path we [abductees and perhaps others] are all on is to bring this to consciousness where we can remember our experiences consciously without this deep hypnosis."

Peter tried again to tell me what he was getting at. "The terror," he explained, "has more to do with having an experience that goes beyond my accepted perception of reality than it does with the physical experimentations or diagnostic, whatever they do to me." Abductees tend to become stuck, he suggested, "at the point of terror that is beyond our perception of reality." The physical aspect of the experience, Peter continued, is essential to the shifts of consciousness involved. "You have to experience it in the physical before you can accept it in the psychological," he said. "A person holds the experience in their body. They have the physical sensation of the experiences actually happening, and then it becomes part of the reality. Just accepting reality does not do it . . . They need us to expand our accepted reality for them to be able to communicate with us, for them to be able to come here," Peter added. "It's an evolution of consciousness."

Playing the devil's advocate I suggested that perhaps they just wanted his sperm so they could go off and propagate on their planet and it was only a "spiritual conceit" to think they were concerned about our evolving consciousness. He objected strongly to this. "There's no question about it. I know that they are helping me, expanding my reality, and they want me to do that in order for them to communicate with me . . . They need my consciousness to expand. They need more of my brain to be awake," Peter continued.

"For what purpose?" I asked.

"So they can interact with our planet. So they can interact with our life force," he replied.

"Toward what end?" I insisted.

"So they can prevent our beings from becoming extinct," he said.

Our interests and theirs seemed to come together as Peter spoke of the beings' "vested interest" in our planet, and their unwillingness, "like an older person seeing a child who's about to make grave mistakes," to "let the child hurt" himself. "We're heading towards some cataclysm," he said, "and they want to help us." But "until abductees come to the surface and can accept it and live with it, they can't do it." My last question involved the source of the beings' intention—was it their own or was there "some sort of spiritual force that's behind everything." Peter's feeling was that "they are acting through a greater power," but he did not know what that might be. He has "the sense" that "these are real, physical beings that have mastered an awareness of time or space or whatever. They can see what's going on. They can see what the possibilities are, and they're here to help us, it's almost like—steer us, to prevent us from hurting ourselves!"

As the session drew toward its conclusion, Peter reiterated the centrality of the physical aspect of the experience for expanding our perception of reality. "The body memory makes it real," he said. "There are two separate things going on." The semen taking and tracking are distinct from the stretching of consciousness that the terrifying physical experiences bring about, although both purposes "are intertwined." For human beings, Pam suggested, having become physical, the body is "our avenue for learning." Peter agreed with this "absolutely." The beings are like "God's angels in a very roundabout way," like "messengers, or He's working through them, just like He's working through you and me and Pam . . . Pure form consciousness on the spiritual plane does not have the experience of the physical," Peter observed. "So it had to embody physicalness in order to have all the experiences of the physical. It's like consciousness said, 'Well, what do I want to learn today?' and it chose the earth as a place to learn physical things."

A fourth regression was scheduled for six weeks later, May 14. The last session had helped Peter come to terms with his experiences, but at the same time left him feeling more alone and isolated. Yet "going through the depths of that real agony" had brought him "to a place where I'm happy to be alive." He had accepted now that "this is real," not "my imagination anymore," and at the same time "accepting that I'm not a freak." He felt more "whole" and the experiences were a kind of "gift"—as he put it, "it's given me a piece of myself." Acknowledging the uniqueness of what he had been through allowed Peter to tolerate his isolation and aloneness. Before, he said, he had "felt isolated, and I felt disconnected from people. This work really brought that to focus." Now, he said, "I feel singular in my existence, but I don't feel lonely. I feel alone . . . I feel connected to the whole but yet still alone."

In that context the abductions themselves were no longer "so bad anymore," so "horrible and traumatic and mean and cruel and all these things." I asked Peter to say more about the nature of the connection he had found. "It feels like I'm part of a continuum," he said. "I'm part of a thread. I'm like a bead on a thread of pearls or something. I'm one of the pearls, but yet I'm connected to the whole process. I can be happy with my aloneness . . . It used to be really important for me to be connected," he said. "Now I know I'm connected. Somehow I just know it. It's just a deeper spot . . . This work has accelerated my whole process somehow." Once again we agreed to let Peter's "inner radar" take him where he needed to go in the regression.

Once in the trance Peter was afraid again. He was in Hawaii, before living with Jamy, and four beings were around his bed. His roommate was in another bedroom. He began to cry, overcome once again with fear and rage as "They're going to do it *again*. They're going to do it *again!* I'm so fucking sick of doing it again." He remembered being given a choice to remember the experience or not, but "they're going to get me either way." It seemed like they had returned after a "great void of time" and wanted to see "how I've grown, and who I am . . . It's like they want to see how the experiment worked out." Once again he was in the ship where experiments were being performed on many other human beings.

Panting now Peter remarked that there was "this leap I've got to take." The beings "want me for who I am," he said. "They want me for what I represent, what's inside of me." Putting his hand on his heart he said, "It's here, it's here, and they want me for here . . . It feels like they're grabbing something that's in there. It's inside of me. It's like they want my love . . . They want me to like them. They want me to care about them." I observed his fear and hate. Still panting, he said, "I've got to get beyond that." The process had gone beyond the experiments, Peter said. "They want me to understand them. They want me to understand what they're about. I'm starting to vibrate." I encouraged him to let the vibrations come. "They want me to understand," he said, "that there's understanding here, truth. There's a meaning to this."

Despite these positive assertions, the experience became once again distressing for Peter. "I want this to stop," he yelled, as the fear and intensity of vibration in his body increased. Letting out bloodcurdling screams he said, "I don't want to lose it. I want to just go off. I don't want to go into the depths." I tried to comfort and reassure him, and had him recenter through focusing on his breath. Then he spoke of feeling peaceful as he was floated by the short beings out of his house in Hawaii to a "big ship." As they lifted off one of the male beings, who was "always with them—he's the one who cares about me—" assured Peter that he would be okay. "I trust him," Peter said.

It was important for Peter to convey to me that despite his own distress, the beings were "not here to hurt anybody." They have been here a long time, he explained, and "know us really deeply . . . They have this wisdom to tell us," he said. "They can see what's about to happen to us, and they're just watching us." As he was saying these things, Peter's mind was "flashing" on the pyramids of Egypt and the "faces of Mars." But once again Peter was overcome with fear and rage as he was taken to the black-floored room and forced to lie down on a table. A being with four "clawlike," long, bony fingers put his hand on Peter's chest as he stared into his eyes. "Those eyes. Those eyes," Peter said. Once again a tube was inserted near his testicles, with the same outcome, though Peter was calmer now and had "learned that I don't have to be afraid." Then he walked out of the experiment room into a room with about twenty people, many of whom were women and children.

What he relived now seems to be his return from the abduction in the Caribbean reported in his first regression. He went through a door of a larger ship into a smaller one which descended. Then the bottom of the small ship opened up and Peter floated down to the deck of his home. He watched himself enter the house and lie down on the couch inside. All this, he said, "has to do with my maleness for some reason. It has to do with me. They want me. They want my heart. They want what makes me tick . . . They work with you throughout your whole life," Peter observed. "They like open people," whatever their age, he said, including old people. "They want to figure out how to get us to be open to them." They are trying "to figure out what makes one person more open and less afraid than another, and they genuinely care about us." The beings fear us, Peter said, because "they don't have the, I want to call it the 'killer instinct' that we have, and the ability to fight and kill each other. That's what they're afraid of . . . that is the most terrifying thing for them, our ability to hate . . . It's like my rage that I have at them for doing these experiments is so foreign to them. They can't understand it. They're afraid of us."

Peter noted that "it's our perception of what's happening. None of us have been really hurt." Some abductees say "they have scars and psychological damage, and I want to say, 'Can't you see you're confused' . . . The beings that I've worked with," Peter said, "are here to find out how to communicate with us, and how to find a common ground and openness." The ones that he feels connected with, at least, begin with children and "nurture them, and they work with them, and they study them to figure out how it is they operate and what it is about them that makes them open and loving and caring, and those are the qualities that the beings want."

A lot of older people, he continued, have had abductions as young children "and just don't remember." Then they come back and study the same people when they are older to figure out the qualities of openness they have in common and what can foster them. "They're trying to study a way into us. They're aware of our anger and hatred and our ability to hurt them." They have "evolved out of" knowing "how to fight . . . It's almost as if they're so open that they know we can hurt them more easily . . . They want our love and how it is we love and care and have such compassion. They also are terrified of our anger and our ability to hate and kill and all that stuff, and they're trying to get the two apart." They are trying "genetically" to take "the higher human qualities and separate them from the lower human qualities and somehow, I want to say, reincorporate them into our race . . . The whole thing with the sperm and the reproduction . . . is really trying to help us evolve to the higher qualities of humanness." They could not manifest more openly now on Earth, for we might "choose to attack them." If we were to "react out of a military place, or an anger place, then everything would be lost. So they're very cautious. They can't come here."

I asked Peter how our loving or open qualities might serve the beings. "They are human too," he suggested, and in their own evolution have "followed the path of almost rational intellectualizing" and "lost much of their emotions, and they want to get that back. And it's through our planet and through our race . . . They are humanoid," Peter said, and "We've evolved from the same place," but "we've stayed in our emotions and our emotions have ruled the planet . . . How we react as a race comes from our emotions." In their evolution they "chose to react from intellect or from mentality . . . They're willing to share their intellectual growth with us if we can share our emotions with them." But at the same time they are afraid of the destructive part of our emotions. "It could be a beautiful marriage almost," Peter observed.

My last question to Peter in this session had to do with the quality of looking into the beings' eyes. The intense eye looking, Peter said, relates to "a longing to connect." As the being "looks at me . . . there's a longing that it has to connect with me, with us, and it's just like, I wish that you could understand us. I wish you could feel what we feel." I asked what the beings long for from us. "It's like a brotherhood that's been lost. It's like a friendship that's been lost, and we just don't understand." They long for the love, compassion, and joy, "all the healing they see in the planet, all the goodness that they see here . . . that ability to connect human to human that they just have so little of." I asked if he could see that in their eyes. "In the eyes there's a coldness," Peter replied. "There's a coldness and a pure blankness, but behind the eyes there's—I almost want to say there's another eye behind the eye. And the other eye behind the eye is, I want to say it's sad and longing, and it's trying to comfort me, but at the same time it wishes I could help it. It wants to connect with me. It really wants to connect with me. It's almost like it's looking at an infant [sobbing now]. It's just, if you were only a little older and a little wiser and we could have a relationship or something."

I asked Peter about the sadness he seemed to feel and his sobbing. He said that "the sadness is such that I'm weak and crippled and I can't do it. I'm so afraid that I can't give them what they really long for, and so my sobbing is like someone who couldn't protect my loved ones . . .

It's almost like a mother and son," he added. "It's like a mother who can't love, who can't really connect." We were coming close to the end of the session. "It's gone beyond the experiences now," Peter said.

I noted that for many abductees "the resistance gathers around acknowledging the eyes." He responded that "there's a real longing there in their eyes. It's almost like a love that doesn't know how to come out. They have this tremendous compassion in their eyes for us . . . When you first look at the eye," Peter recalled, "it's like mechanical and cold and nonhuman. It's cold and noncaring, but as you stare at it, and as you look at it, and as I allow myself to look at the eye, I can see deeper in the eye. It's just like looking at a human. You see the eye, but after you keep eye contact, then you see the person and that's what it is . . . On the other side of the eyeball is the being, the creature," he concluded. In that sense, I noted, the beings have "not yet been found by us." He reiterated their kinship, our brotherhood, their humanness. "I just have the sense that they're just a little bit removed." They "chose a different way to create . . . They have emotions," Peter emphasized, but "not as highly developed as ours, or else they had it and lost it and now that's what we have to offer, our emotions . . . They're not really different from us," he concluded.

I had made the decision in May, even before this last regression, to ask a Ph.D. clinical psychologist colleague to administer a battery of psychometric tests to Peter. Psychological testing is time-consuming and expensive, and it is not practical to have all, or even many, of the experiencers with whom I work, tested. I chose to ask Dr. Steven Shapse of Harvard's McLean Hospital to test Peter because I thought the results could have scientific value. By this time I knew Peter well enough to be impressed that there was no manifest or underlying psychopathology that could explain, or even shed much light, on the experiences that he was having. Some of his abduction-related experiences, and the various adjustments that he had to make in relation to them, were obviously stressful. But Peter was, nevertheless, a high functioning, stable, and likable person who had excellent relationships with other people.

My thought was that if this impression were affirmed by a qualified clinical psychologist, it could help to dispel the notion that the reports of abductees are the product of some sort of psychiatric disturbance. For even if some abductees were to demonstrate significant psychopathology, the fact that some, or even one, did not, would go a ways toward ruling out a psychiatric explanation of this phenomenon. By the time I asked Dr. Shapse to test Peter I was doubtful that a psychiatric explanation would be forthcoming, as I had found none in any of the cases that I had seen up to that time.

In May and early June, Dr. Shapse met with Peter twice and, in addition to his clinical interview, administered the Wechsler Adult Intelligence Scale-Revised (WAIS-R), a standard intelligence test; the Bender Visual Motor Gestalt Test (BVMG), which tests for organic brain dysfunction; and the Thematic Apperception Test (TAT), Minnesota Multiphasic Personality Inventory-2 (MMPI-2), and the Rorschach Inkblot Test (RIBT), projective tests which reveal the nature of psychological functioning and structure.

Dr. Shapse found that Peter was "highly functioning, alert, focused, intelligent, well-spoken, and without visible anxiety." There was no organic neurological dysfunctioning. "Situational stress was noted to be high, and his experience of this stress to be severe." Peter expressed sadness on the TAT and seemed to be "battling evil forces." Dr. Shapse concluded, "most significant is the absence of psychopathology. No psychosis or major affective disorder is diagnosable . . . Significant was a moderate level of sexual preoccupation. His particular profile suggests that he may have been sexually abused . . . It is felt that underlying psychological themes do not impact in a dysfunctional manner at present." No psychiatric diagnosis was made. "Psychosocial Stressors" were said to be moderate: "Recollection of unusual and disturbing experiences." When I asked Dr. Shapse privately what in these findings might be relevant to the abduction history he replied, "I don't see anything that would account for it." The suggestion of sexual abuse is interesting in the light of the traumatic procedures imposed by the aliens. There is nothing in Peter's history to suggest sexual abuse by human beings.

Beginning in June 1992, Peter began to take a leadership role in the abduction community, speaking publicly of his experiences and appearing on television. On June 15, he participated in an abductee panel at the Abduction Study Conference at MIT, organized by MIT physicist David Pritchard and me. In his five- to ten-minute presentation Peter spoke of the details of his traumatic experiences as he had learned about them in the regressions and of how he had "gone from tremendously deep anger and resentment to an understanding." He talked as we had earlier of the beings' intense interest in our "ability to feel, to have emotion, to have compassion and caring, and our deep spirituality," the "qualities that make us human and permeate through all races on the planet." The virtually exclusive focus on the traumatic dimension in abduction investigation and treatment, he said, was due to our inability "to integrate these experiences into our conscious understanding of perceived reality because we have not thought it possible." He told the audience of eighty or so investigators that he thought the aliens did not intend us harm and their purpose was to "get us to the point where we can interact with them consciously and not have it be so frightening for us."

On August 15, Peter was filmed in a conversation with me at my home by producer David Cherniack for a Canadian Public Broadcasting one-hour program on the abduction phenomenon called "Sky Magic" in Cherniack's *Man Alive* series. In our discussion Peter spoke of the "animal instinctual" way he had reacted to his experiences, being taken to a place where you are so "stretched that your mind would explode," the transformation of the quality of the abductions, the memories of them in the regressions, the shattering of what he had thought was possible, how "hair-raising" he had found it was to "run afoul of consensus reality," and other ground that we had traveled over in our sessions. Most dramatic was a ten-minute segment where an audio tape of the most traumatic part of the April 2 regression was played, and the camera focused on the fear, tears, and quivering of Peter's face as he relived the horror of those moments. The program was broadcast in Canada in October and has received favorable reviews. It has not been shown in the United States.

In the next few months, with Peter's permission, I used that video segment several times, including in a presentation to the Department of Psychiatry at Harvard's Brigham and Women's Hospital, to enable audiences to experience the emotional power of the abduction phenomenon. I stopped doing so when it became apparent that many people, including abductees themselves, found the video *too* disturbing and would sometimes erect new defenses rather than becoming more open to the experience. When I showed it at a dinner meeting of the Skeptical Inquirers of New England in November, several of the attendees became so distressed that they had to turn their backs to the screen and then proceeded to concoct outlandish explanations for the abduction phenomenon.

On December 2, Peter and I spoke at the Harvard Divinity School on "The Alien Abduction Phenomenon" to an audience of about 250 people. I gave an overview of the phenomenon and showed the video segment. Then Peter talked of his experiences for about ten or fifteen minutes, followed by an hour's discussion with the audience. In preparation for the presentation we had talked on the telephone a few days before. Peter spoke of how he had felt "naked in front of God" in the utter terror and helplessness of his experiences, controlled and beyond choice, yet connected with something greater.

In his presentation itself he said that his "journey" had enabled him to discover his "place in the universe." He had felt abandoned by God and reduced to "nothing more than a sperm sample." Yet after his regressions he had had "feelings of tremendous expansiveness." In his total aloneness he had discovered his oneness with God and spirit. "The abduction experience," he said, "allowed me to feel this aloneness yet to feel totally connected with that one individual source." He spoke of the "likeness of God" among peoples all over the earth who share a common humanity and "all struggle to understand the mysteries of the spirit world and our connection with it."

Further, he went on, "I began to know on the most cellular level that we are not alone in the universe, that God created many creatures in His likeness." The alien beings "are not so unlike ourselves. They too struggle. They too question existence. They too are inquisitive. They nourish themselves. They procreate and they pass on. They wish to be accepted for who they are . . . God created a lot more in the universe than we can ever imagine," he continued. "Like thousands of others like myself who have had the experience, have seen God's creation in other forms, I know in the deepest part of myself that the beings are like us in many ways." Through his abductions, Peter concluded, he has discovered "that I am connected to a creation process that is far greater than anything I have ever been asked to imagine in any of my other previous exposures to any spirituality."

Between May and November 1992, Peter told Pam and me later, he had felt at peace, relatively comfortable with his abduction experiences, settled within himself and had "worked out all the pieces" with family and friends. But ten days after the Divinity School presentation Peter wrote me that since late November he had wished to "remember more" and "go deeper" and asked to do further regressions. He wrote that he had also passed his national acupuncture examinations. Our fifth hypnosis session was scheduled for January 14, 1993. At the beginning of the meeting he said he had "wrestled" with it for a time, but felt ready to "remember something bigger, something broader." As we planned the focus of our work for the day Peter remembered a time in Hawaii when "three people" whom he identified as "just Holy rollers, you know, selling Jehovah's Witnesses, or something" came to his home. "I can still see this woman's face," he said, "just looking at me like she knew something about me."

Peter connected this memory vaguely with an abduction experience that had occurred in late August on Nantucket Island. He had been visiting a friend, Craig, who leads fire walks. During the night Peter heard a noise that frightened him, saw lights, and had visions in his mind of a being floating across the room to Craig's bed and coming back to Peter and saying, "'Okay, it's time to go.' And we left and we came back." In the morning Craig said that he too had heard a high-pitched noise that he chose to dismiss as "crickets" but had been afraid to get up to go to the bathroom. Before beginning the regression we reviewed what had been happening in Peter's life, especially Jamy's progress as a therapist and their thinking about having children. Peter himself was feeling more trusting, that "I'm being taken care of on some level" and "the road is being cleared for me so I can do my work."

Again we agreed to use an open-ended approach, not focusing on any one episode. Peter's first memory was of looking out at night from a condominium where he was living in Hawaii at "this huge ship hovering over the golf course." At the same time three people, "like three Jesus freaks or something," walked into the door of his living room. "I think they're talking to me about God and about religion, and wanting to convert me or something . . . It's them!" Peter exclaimed. "There's a connection between them and the ship. It's like their human form. They've come again. They've come again. I can't fight 'em. I can't fight 'em."

His attention shifted again to Nantucket. He sees a being floating toward Craig's bed and becomes anxious for him, but "then it's time to go . . . I'm always cold when this happens," Peter said, but also fog came in the room, and though "it chills me," it also "calms me down." Two beings have come for him, but it is "subtler now." A lot of the "formality is gone now," and it is "not necessary to be afraid." This is "like a training," an "initiation into accepting what's really true." Nevertheless, Peter sat on the bed shaking with fear. As if his "spirit" were up in the corner of the room, he could look down on his body on the bed. The beings waited for him and "they are not forcing me." He is reassured that Craig will be safe.

Peter's fear deepened now as he recalled being given a choice of whether or not to accompany the beings, "to walk through this wall, to walk through this second-floor ceiling." This choice seemed to take on metaphoric power in the session, involving, in Peter's words, the "freedom of going to the next level. It's the next level of going with them." He spoke of a "great web" of connection, a "consciousness of the whole . . . I am them, and they are me and there's the ship!" His thoughts returned to the woman at his door in Hawaii who "kept looking at me . . . I wasn't ready to accept God or Jesus or something . . .

What frightens me," he said, "is that they took human form when they came to me. They walked right up into my door . . . It frightens me that anybody I could walk into on the street could be a fucking alien or something."

Peter's attention returned to Nantucket as he sat rocking on the bed, crouched on his haunches with his knees up to his chest. The choice of whether or not "to get up and dive" through the angled wall or pitched roof of the house now represented passing through to "'another dimension' . . . It's not the same as like physically walking through a wall. It's like stepping into an energy field . . . They didn't carry me through that wall this time. I walked through it . . . I'm looking down, and I can see that I got up, I took one step, two steps, third step I went into the field. One, two, three," he whispered.

I asked, "What about the fourth step? Was there a fourth step?"

It is difficult to convey what happened next in the session. Taking this "fourth step" took on great symbolic power. Meaning and action, past and present, physical event and metaphor, seemed to be condensed in this choice, which came to represent willing participation. As he sat on the edge of the bed, Peter was facing "the terror of accepting, of accepting the responsibility." At stake was "a step into evolution, leaving behind the animal." He felt that by walking through the wall he could "transcend time and space." He wanted "to let out a battle cry" and "charge," to "go with them." But the struggle continued for many more minutes. "Nobody can help me," he said. "It's not even terror anymore. It's a choice," but he was afraid that "if I make this step there won't be anything there." Unlike in previous abductions the beings "aren't going to do anything . . . There's nothing but just this moment in time of me." I kept taking Peter to the edge of what loomed as a lonely choice, but could do little more than be with him.

The fourth step took on the meaning of a leap of faith. What if he were to take it and "there's nothing there" he asked over and over again. "What if there's nothing on the other side of that wall." This signified "failure" to Peter, that "all this is for nothing, all that went before me is for nothing . . . What if I can't make it! What if I get stuck! What if I only go halfway! What if I can't go! How can I trust you?! How can I trust that I can go through there?! What if there's nothing there?" Peter kept repeating these and similar questions and I could only encourage him to go more deeply into his doubt and fear. He screamed and moaned and his body began to shake as he approached what increasingly became an acute moment of existential choice. The moment lightened a bit when he quipped, "I'm scared of heights. What if I fall?"

Finally, he said, "I drifted right through this wall." But the moment lacked conviction and his consciousness returned to the bed. "Okay, I'm gonna get up!" he said. "I'm walking, I'm walking! I'm not even running." He let out a loud hissing sound as the process reached its climax. Peter's doubt continued. "I'm bouncing back and forth now," he said, as he kept returning to the bedroom to check whether he really went through the wall. "Did I go through that wall?" he asked. I burst out laughing at this. "It just happened in a split second," he said. Eventually he felt "totally convinced and calm that I went through," like "Peter Pan," and saw the pine trees below.

Outside now Peter looked down. "I realize that I'm floating, and I see the ground. I see the side of the house . . . I see the ship above us," he said, "and I make my way towards it," and "next I'm in the ship." Inside "all the babies are there, all the children are there." He was standing in a room that was black, "like black marble or something." Along a curving wall of a corridor that seemed like the outer perimeter of the ship, there were little lights at about hip level. He walked into a room where he saw three chairs and a table and "another chair for me at the end . . . Images and images" were coming, but Peter's consciousness kept drifting as he seemed to want to avoid what occurred next.

He saw three beings. One, directly in front of him, had a "really big forehead" and seemed older than the other two. The one on the left was an alien female. The third being was an alien as well. These were the three who had visited him in Hawaii. Peter was "there to learn about the future," he said. His large forehead was knotted with a kind of split in the center. Peter said that "he's the same one that oversaw all the operations . . . I feel like he's smarter, I mean he's definitely the boss." The female being was to be his teacher. "She's going to be my guardian, or something. She's gonna watch for me. She's, there's something that's going to transpire. Oh, my God."

"What?" I asked.

"That we're gonna . . . we're gonna fuck." Fear, not desire, filled Peter at this point. "I'm going to breed with her," he said. "I'm getting that that's what this was all about," which was communicated to him mainly by the old one, whom Peter also called "Mr. Know-it-all" and "Bubblehead."

"I didn't want to know this," he said. "It just kind of blew me away."

Peter was shocked to be told that the babies he saw upon entering the ship were his alien or hybrid babies and that he was "breeding with aliens . . . That's what my sperm has been doing," he said. Peter's fear grew again, as he considered the implications of what he was discovering. He realized that he had repeatedly made love with an alien female. "It feels like she's my real wife—I want to say on a soul level. She's the person that I'm really connected to, and she's the one I'm going to be there with, or be with, or something." This information was communicated to Peter by "Bubblehead" and the alien woman herself, and he felt its truth with great conviction. But as he spoke of this Peter experienced his "consciousness being separate from my physical body, like up here, looking down at myself sitting at this table."

This was all becoming "too much" for Peter. "It's not even about making love or sex," but "breeding with her," he said. "Oh, God, John. My hunch is that this woman is a human also, and that they've been taking her eggs and my semen and putting them together and breeding them up there." As Peter thought about what he was saying he said, "I don't think I can put into words the depth of how shocking, about how much I've been taken aback by what I just saw . . . The most frightening thought for me," he said, "is that she may be an alien and I'm seeing her as a human, and that I'm making love with an alien." I asked Peter what was "so horrible" about the idea of breeding with an alien. "They want me to like her as a human would like or love something," he replied, "like I would love making love with my wife . . . It's about my feeling of attachment to her and my love for her as an alien," he continued. "What about my life on Earth, John? What about my wife?"

Peter spoke, in addition, of a certain physical revulsion. "I can't get beyond the fact that they are these horrible ugly creatures that I've been terrified of with this slimy skin, like cold skin, that I'm making love to . . . I don't fucking believe I'm making love to an alien." Peter talked further of his worry about becoming separated from Earth and his "earthling family" and losing his connection "with all that I love here." He became confused as he contemplated "how many children" he must have, considering how many sperm samples had been taken from him.

Despite his distress at all this, Peter looked upon this process as a project to which he had agreed. "The first step," he said, "is simply the creation of children, the zygotes, the infants. The second step is the pairing of the parents," an Earth parent and an alien parent. A bond "is created between these two so that they can raise the children in a fashion of both Earth and alien." We talked more of his shock at the idea of being an "alien father," although he recognized that the hybrid children needed "a mother and a father . . . With the advent of the destruction of the earth as we know it," he said, these will be the children that will "repopulate" the planet. They are like "poppy seeds . . . pods" that will be "scattered around the earth."

In the light of the strangeness of this information I asked Peter about the level of conviction he was experiencing. "One hundred percent truth," he said. "It's about learning. It's about the future, those were the feelings. It wasn't just knowledge, but it was real, it was feeling." We reviewed the staged conference, with the three aliens on the three chairs—the old wise leader, the human/alien female, and the "original alien" on the right. It was as if they were reasserting his commitment to what he now experienced as "an arranged marriage" with the female being. "It feels like two people that are bonding for a purpose that's bigger than themselves," he said. "If I don't my race will die out . . . It's gone beyond taking my sperm. Now it is about do you want to father? Do you want to be a conscious participant in this breeding?" This decision might have been made "back ten lifetimes," he suggested.

Peter was convinced that his alien/human mate was the woman that came to his home in Hawaii. She was rather plain, with auburn hair, "not homely, just kind of neutral in her appearance. Not good-looking. Not bad-looking. I don't really remember her figure that much." This led to a discussion of his struggle to live "in two parallel worlds." In some respects his existence in the other world has a stronger power for Peter. "I'm going more in depth into my reality there," he said. "My biggest fear right now," he continued, "is going home and telling Jamy: 'Well, how was your regression today?'; 'Well, I saw my alien wife and our kids.'"

As the session was drawing to a close Peter talked about the potentiality of his life's work and the complex responsibilities he had taken on. Taking the fourth step, "walking through," represented a deepening of his commitment to the breeding project, especially his fathering role, which included a loving relationship with the female being "because the parenting process is something that comes from two people." It has to do with "what comes energetically from the parents, from the feelings, from the connection between parent, between husband and wife, or between mates, and then was transferred to the child." Peter stressed the importance of the parenting process for the hybrid children after the beings have "raised embryos" on the ships, or in whatever domain they are growing.

This had been a powerful session, and at the end I reviewed what had occurred and raised questions about the literal and metaphoric aspects of all that Peter had gone through. For Peter it felt that his struggle was concerned, above all, with an evolutionary step, and that "accepting and choosing, consciously choosing, to walk through the wall," represented his commitment to that step.

A few days after the regression Peter told Pam that he had gone over with Jamy what we had uncovered in the session and she was "perfectly fine" with it and supportive. He felt relaxed, trusting, and "strong enough" to "go deeper." The regressions, he said, had had the effect of cutting his "ego down to size." Then, in the place of terror and anger, he could find "space" for the connection with the alien beings and the "breeding program."

A sixth regression was scheduled for February 11, four weeks later. Honey Black Kay, a therapist who was learning about the abduction phenomenon, was present during the session. At the beginning Peter reviewed for Ms. Kay what had occurred in the last session and Jamy's acceptance of his process. It had been difficult for him to accept "the depth of it," while at the same time he noted a certain "absurdity" in what he had discovered. "I've gone from a victim role to a participant role," he said, and talked about the importance for that shift of overcoming his fear of stepping through the wall in Nantucket during the January 14 regression. He approached this session with a mixture of eagerness and anxiety about what he might discover.

In the trance state Peter returned first to being on a table "back when they used to take my semen," then to when the three alien/humans had visited him in Hawaii, and back to the experience in Nantucket in August that we had explored in the last regression. Once again he experienced the emotions related to his decision to "go through the wall" and what that had meant for him. Once more he felt the vibrations in his body as he sat on the edge of the bed and the surrender of control that going through the wall had represented. In addition to the breeding program he also had experienced a certain "loss of my identity" in the discovery that he was, in a sense, "part alien." In the session now, he felt that "I'm starting to vibrate on that cellular level."

Peter let out a long, loud scream followed by rapid shallow breathing as he discovered himself once again on a table, beside which was his female alien/human partner who said to him something like, "You're going to know later" and "this part isn't so important." He felt his mind opening up now and experienced the "freedom of choice to go to any memory I want to go to." Once again Peter experienced walking down the corridor to the room with the three beings. They told him that "since I was a young child I had agreed to do this, and I chose to come here [to Earth], and then they first came to me, and I chose to play with the babies, the other beings, the aliens." The other aliens "watched me interact with them, and because I didn't show any fear, any problem when I played with them, they asked me if I wanted to continue." He said that he had agreed as a four-year-old child to continue to play with the hybrid children, which he spoke of as "more of an interaction with me just sitting there with them and there's a communication."

I asked Peter to describe the hybrid children. "They had big heads, and, like, wisps of hair, and they have bigger heads than bodies," he said. "They have skin that's kind of like our skin. It's a little rougher—more fleshy. It's not like baby fat. It's like old-age fat, and the arms are really fragile, but they have big bellies. They're funny. They're cute. They look like little babies."

When the three beings visited Peter in Hawaii they asked him again "if I wanted to continue." Although he was not "ready to remember" at that time, "there's always been a choice I've made to go forward." A new choice seemed to be developing now, "something else," Peter said, as he recalled being led down a hallway where "they're gonna show me things." As he walked down the hallway, Peter was shown pictures on either side. He saw nuclear explosions, sections of Europe and the United States destroyed, "a lot of people burned, a lot of people upset . . . the human race changing" its "form" and "texture." Peter and the woman and the thousands of babies they had produced together were "part of that change," of the "repopulation of the earth." I asked Peter why this repopulation was necessary. "Because of the destruction of the earth, because of what's happening, what's going to happen."

I asked Peter to say more about that. He said that there was "a battle going on" over who was going to "get control of the earth" between "beings from all over the universe," not only "the beings I'm associated with." This "has been going on" for perhaps two thousand years, he said, but was now "all coming to a head." It was not quite the same as "good and bad," Peter explained, but rather "different possibilities of going forward." The "whole reproduction process" in which he was participating was playing "an integral part" in this unfolding. Peter spoke of prophecy, "revelations," and people being "taken up" as in the Biblical raptures. But to him this had "nothing to do with religion," for there were "ships in place for that to happen." He has looked into "this black void" of the future and seen the earth below "with people going up . . . It's not so much seeing it as knowing it," he said. At the same time he felt the "loss of everybody I've ever known, and every connection I've ever had to Earth."

Peter objected to the word rapture, preferring to speak "of the beings coming to help us to the next place of evolution . . . The slate is going to be wiped clean," he said. "It's a new millennium of the earth" that is "supporting a whole other world." He had experienced a "bargaining going on" with "greater forces actually negotiating the future of the planet," its "next two thousand years." As he lay on a bed, Peter saw before him naked human beings, standing or "suspended in animation." He began to comment on his perceptions with words like "weird" and "bizarre" and I asked him to put aside judgments and commentary for the time being and just report his "raw" experience. Then the year, he thought, might be 2010, and once again Peter saw "people coming up" during a period of time when "a lot of shit happens on the earth" and the ships make trips "back down."

Peter's role in all this is to "make the babies" with the alien/human female with whom he has been connected, to bring the whole process "into consciousness," and to work with the female in "taking care of the kids." He would become a leader of "a new, original tribe," a "new race of humans." Although Peter said he had accepted the repopulation plan, he found all this quite disturbing, especially "the destruction of the populations of the earth," which will happen "in a blink of the eye." Though nothing could be done to prevent all this, Peter said, on the positive side this evolutionary process could create "a second chance" for humankind. The hybrid tribe, he said, would come down like "sections of people" to be placed in various areas of the earth, basically an advanced "transplanted population" with knowledge "from another world," prepared to start "new life . . . A whole system" would be transplanted.

I asked him about the fate of us "original" humans. A lot of humans would be left, he said, but plagues, pestilence, and "all this stuff" would destroy the "infrastructure of the civilization of man as it is today." The entire society would crumble. I wondered what I or we were supposed to do with this distressing information. Peter replied that it was "no coincidence" that I had been active in the antinuclear movement, which implied the possibility of some sort of preventive efforts. But he said we were seeing the future "on the horizon," the situation was hopeless and he felt "like watching the ship sink and you're on the lifeboat."

Peter felt heavyhearted about all this but resigned at the same time. Despite the fixed nature of his vision, which to him felt totally real, he said that "on some level there is negotiation going on about the future" which was "about all these different possibilities for the earth . . . If there is consciousness in the universe we have a stake in wherever we are," he observed vaguely, and "if the world comes to an end as we know it and another, another human consciousness, inhabits the earth then we've progressed." After all, he noted, the alien-human interbreeding was "interbreeding within our own species on some level."

The regression ended here. Ms. Kay noted a certain powerful energy she was feeling in her spine through all this. Peter, seeing our glum moods, said, "I feel like I ruined everybody's party." He reiterated, nevertheless, how powerful his feelings had been and real his "story" had seemed. The remainder of our discussion concerned the reality status of Peter's prophecy. He resisted some the notion of "probable realities" and of our power to choose a different future. For in this regression he had gotten the strong sense that "we're not the only consciousness that has a say in our future." By surrendering control "to the greater forces" Peter said that he felt "much freer" and "more comfortable."

In a telephone conversation with Pam three weeks after the regression Peter said that he was upset afterwards for a while, especially because he felt so isolated with his information about the "earth ending." Nobody wants to know about that, he said, but he continued to experience the evolutionary process he had seen in the regression, with aliens and hybrids coming down to Earth, as "very real." Peter continued to experience intense new healing energies which he related to his connection with the alien beings, and found he was able to transmit this energy helpfully to other people. He asked her if we could schedule another regression.

We met for the seventh regression on April 22. Peter began by telling of a session with his therapist in which he had lain on the floor in a fetal position, hugged a pillow, cried and rocked, and experienced himself as an alien being or consciousness that had entered a human body. He wondered, "Who am I?" and felt he was "losing myself to some degree . . . I am an individual soul that just came into this lifetime" and entered "Peter's body." In other lifetimes, past and future, he would be "something else," but "I was always Peter. I'm always this soul . . . My soul is actually alien . . . I'm not sure what that actually means," he said, whether he was "created by them or them coming into this body." If that were the case, "then I feel lost, like, who am I? Am I just a vehicle for their use? Am I just a machine now? How much of this is my free will? It puts a whole other twist on it."

Peter then described in great detail the change he had touched upon with Pam concerning the development of his healing energies. This came about in a class at the Barbara Brennan school when the teacher, seeing something in his auric field, touched the side of his head and he felt "something just explode inside my head." He experienced intense vibratory changes in his body that the teacher and others in the class perceived as powerful pulsations in his energy field. After returning to Boston, Peter found that for two nights his whole body was "vibrating and shaking." One night he awoke hearing a guidance that "you're supposed to transmit energy of a certain vibration for people. That's what I'm supposed to be doing." Peter had mixed feelings about this. Although to a certain degree he accepted and had some effectiveness in using his enhanced healing powers, he did not "have any real desire to stand up and say I'm transmitting energy from aliens."

Peter's intention for this session was to "open up to a deeper knowledge." He had "come to grips on the physical level" with the abduction phenomenon. He wished now to work "on more of a soul level." Although he felt he was "on the edge of a deeper understanding," Peter was also "afraid to find out what I'm going to learn."

In the regression his first images were of alien beings in his room, of children, and of feeling like a baby or fetus, which he found disturbing. He felt next that his own body was "exactly as if I'm in an alien body. My head feels big. My neck is real thin. My body is thin. My fingers are long. My body is low. I have a real thin waist. Everything is long and skinny, and I feel like I'm standing there talking to this other alien." He was "one of them," and the conversation occurred before he came to Earth. But "future past" was "all the same," and he was "looking at the past and the future at the same time." Peter knew he could transcend his body, "receive information and work with them. I am them . . . It feels like I came here [to the earth] for a reason," he said.

The conversation was with his alien mate whom he had known for "eons." They were saying good-bye, for this was "the last time we're going to see each other in this form . . . I don't want to leave her. I don't want to leave," he said. "I'm starting to be afraid now. It's all new." But he had to leave, for he had "chosen to do this to help her, to help all of us." The helping project was to impregnate his alien female partner "with my sperm as a human, as a male, as an Earth person." Then he was "laying on the table and they're taking my sperm. But now she's standing there. She's looking in my eyes and she's comforting me." There is also "something that they want to alter—my molecular structure or something," Peter said.

At this point in the session Peter felt mounting fear and screamed and moaned loudly as he struggled with his "willingness to look at" whatever was to come next. A cold feeling came over him and he experienced a "fine vibration" in his body. Images began to come of vast

Earth changes that were to occur before the year 2002 as he lay on a table with "two aliens off to the side who were talking to me" while other beings were "working with other people in Europe and Africa and in the East. "But," Peter added ambiguously, "we are the aliens who have come here. We're the ones that have come here." Then he felt he was "down in the ocean" as great shifts occurred in the continental shelf and a great tidal wave engulfed much of the East Coast and the Gulf of Mexico swept across the South.

I asked Peter questions about the source of this information, in what capacity he was gaining it, and what his role in the process might be. He said that the information had come through the alien beings, that his job was to prepare for the future by changing the "energetic structure" or vibrations of people with whom he would work, and as an alien in human form to mate with his female alien/human partner to produce a hybrid race. He explained that "when I'm taken back to the ships as they're called, and the semen or the eggs are removed from us, us being the human physical body, then they can be used and mixed with our reproductive system."

I observed that from what we had learned the hybrid beings seemed rather listless or lacking in life and wondered how they would function in a repopulating process. He said that this was, in effect, a false perception based on a human perspective and the limits of what we could accept. "They are simply lifeless in your terms . . . They're really us, the aliens, and we're not lifeless. We're only lifeless in your eyes . . . We are formless. We are not like you," Peter continued. "I'm seeing this as a human and as an alien at the same time. It is important for you to see us as we truly are, which you would interpret as lifeless because it is not human." I was not able to see their "spirit" or "being," he said. Somehow I was not satisfied with this and wondered how it was that we could not. Peter's answer was confusing to me, referring to my biases in what I interpreted as life, "the embodiment of soul, of God. It is just you are seeing a different species of man, that is all."

I commented upon how "needing of nurture" the hybrid beings seemed to be. What "better way" could there be, Peter said, to get human beings to "accept us [he is speaking now in the first person as an alien] than to show them us in our most needy form . . . The goodness in their heart," is the "common theme of all humans," he observed. "The heartbreak is the memory that they too are us, and that they're connecting with their species so to say." Human abductees, Peter explained, have an alien identity as well, so that when they connect with the alien children they have created "there is a conflict between remembering their connection with their species of origin [presumably alien in this context] plus their human connection with that baby, that child, that infant." I noted how devastating this is for the abductees, and he agreed that it was "cruel, in your eyes." He too felt "split in two," and his "human part is just resisting the alien part, the part that knows, the part that has so much information to bring here."

Lines were being drawn, spokes or tunnels through which people would pass into the future. There will be a "shift in time," and it will be like "a veil that people are going to walk through . . . The species of man will continue," Peter said, but in a different form. "The earth is going to open up to a place of interaction with other beings," including the hybrid children "now being created." Once the humans who are "part of this plan" move through time and "through the shift of consciousness, there will be no anger or hatred or resentment" among abductees who are "shown their children . . . That which you feel is cruel," he continued, is just the human mind trying to understand." I was not altogether convinced by this, but it seemed time to move on.

Peter returned to the theme of human evolution and became "scared" again as he developed a kind of scenario of salvation. "Not all of us can go forward in the new being," he said. There would be a sort of "straining of those individuals that won't fit into the next place." The change would be quite beautiful, as consciousness opened "to a much finer and higher vibration." His fear related to leaving behind "what I know," but "the higher part of me knows that all of those beings, all of those humans that die actually transcend." The "next step," Peter said, was "the process of getting these people ready to move through the veil when it's time." He saw three forms that would interact without veils between any of the groups: humans embodied as we are now, a mixture of humans and aliens, and the aliens themselves.

Peter's vision was of a kind of "Golden Age of learning and openness and opportunity." Peter himself remembered that deep inside "almost my soul is the alien," and in his embodied form his task has been to use his reproductive systems to create an intermediary species and to "somehow change the vibrations of other humans that are here when the changes happen." People his age were the "first generation" in this process. "Our [hybrid] children are the second generation . . . The scary part," Peter said, was "the radical shift in my waking consciousness" involved in "actually embracing the work," embracing the "conscious awareness of who I am, at least in this incarnation." He sensed the "vastness" of all this and felt that this information probably was changing me also in some fundamental way, making me think "beyond just this clinical paradigm of 'what is the experience.'" He thought we might be on the "same edge . . . You're one of us," he said. "You're one of them."

Peter found this session particularly difficult and was quite withdrawn for several days afterwards. He found particularly troubling the vivid sense of himself as an alien, which seemed "crazy" to him, and wondered if other abductees perceived themselves in this way. Pam shared some information with him about others we had been working with who had similar dual-identity experiences. He asked to meet with me to try to integrate the information that he had opened to in the last regression.

We met on May 19, and Peter shared the "oceans of doubt, despair, and questioning" he had gone through after the last session. We wondered together on what plane of reality to view what he had reported. He said that he viewed me as a kind of "midwife," and that after each regression he had come to a place of resolution, of "transformation." He was more convinced than ever that "there's a power out there greater than me, or greater than us, and they have some control over my destiny and over the planet's destiny . . . The Western mind, the social-economic structure that I grew up in, doesn't hold for me," he said. "I'm a different man than I was prior to my regressions," he said. He is now "trusting in the universe . . . waiting for a greater call, waiting for something to shift."

"I'm connecting to the beings," he said, "and I have a sense that they're connected with God, whatever that God is . . . As intermediaries, he said, "they are doing the same thing that we would do if we came across a species of anything that was on the brink of extinction." We would try "to help them without direct intervention." The beings were "acting with God," he suggested, to breed for the "highest qualities of mankind." At the same time, he said, they seemed to have "been able to transcend time and space and see what the possible future evolution of the planet is," using "the mythology of UFOs" to have us "realize we're part of something greater." Peter felt forced to turn inward, and was planning, he said, a ten-day vision quest, a solo camping trip in Montana at the end of July, to explore more deeply the questions that had come up for him this spring.

The remainder of the session was devoted to the practical implications of the changes that had occurred. He continued to feel isolated and alone with his experiences. Though his transformational process had had a certain inexorable quality about it, he feared losing his connection with Jamy, and the "nice secure role in society" that he would have as an "educated young man" and a "professional" acupuncturist. He was still seeking the support of his father and sister, but had received no response from them after he sent videotapes of his talks and after they had seen him speak emotionally of his experiences on television. He and I, Peter said, "and other people in this group [of experiencers and investigators] are being pulled away from the safety of our normal world into something else, and that's what's troubling." Nevertheless, he said, he was committed to continue his work in bringing about "subtle vibrational change in our energy fields."

Discussion

Peter's case takes us to the edge of the mystery of consciousness, the evolution of human identity, and the apparent purpose of the abduction phenomenon itself. His regressions seem to contain a progression, changing from more "standard" traumatic intrusions to complex spiritual experiences involving the opening or expansion of his consciousness and visions of possible human futures. Memories of more purely physical events, especially of one-sided invasive actions by the alien beings, are largely replaced by communications between Peter and the beings that seem to be part of a complex, reciprocal learning enterprise.

Interestingly, seeds of the next regression seem to have been planted in each preceding one, although the selective processes at work on the part of Peter and the investigator may be operating here. As is probably always the case in this kind of work, the directions of the case's progression depend on the mutual opening and learning of both the abductee and the investigator. This is not a linear phenomenon, like "leading" a witness or client. It is a rich, partly unconscious, interactive or intersubjective emerging, creative in its unfolding and flawed in the limitations of its objectivity.

There is a curious problem inherent in the notion that the regressions reflect some sort of progression. For most of the abductions—the examination of the Nantucket experience of August 1992 in the fifth regression is a clear exception—occurred before Peter and I met. Therefore if there is an evolving of consciousness evident in the sessions, it means (1) Peter is perceiving differently experiences that have already occurred; (2) his psyche is bringing into consciousness those abduction elements that will serve the current moment of his evolution; or (3) a more outlandish thought—his changed consciousness is actually altering the nature of the past experiences themselves. Since we do not know in what reality these events have occurred in the first place, it is difficult to choose among these possibilities.

There is a problem of tense in the reporting of the sessions that should be noted. Peter and I are talking presumably about events that have occurred in the past. He, however, like most abductees, is reliving these experiences so vividly in the current moment that he speaks in the present tense most of the time. If I adhere strictly to the past tense in my contributions to the narrative, I may be introducing a distorting voice. For Peter's reenactment, his "live" communication in the present, may be a more accurate depiction of the place of the abduction as, simultaneously, both a real and a psychic event (again, this division may not hold strictly) than my rigorous sticking to a past tense narrative form.

The first regression focused upon an experience in 1988 in the Caribbean in which Peter, and apparently many other human beings he saw on the ship, were subjected to humiliating anal probing and the implanting of some sort of implant which he believed was a tracking device. Even in this first session, Peter had the sense that he was in some way a willing participant in this process, however traumatic these experiences had been. The second regression and the events surrounding the sighting in Connecticut were particularly shocking to Peter's belief system, for there were independent witnesses of his UFO abduction experience. The apparent probing of his brain, though emotionally traumatic, also was connected with Peter's leadership role, a testing of his capacity to be an intermediary between Earth and the alien realm, should the beings choose to manifest themselves more directly.

The third regression, involving a dramatic and highly disturbing memory of the forced taking of a sperm sample, explored the relationship between bodily experience and consciousness evolution. Peter needed to experience fully in his body, with all of the accompanying terror, the sheer fact of what had happened to him. Only then was he able to acknowledge the existence of the aliens themselves and thus become able to accept what he would subsequently learn through further communications with them. In the fourth regression we began to explore the mutual longing for connection between the aliens and us, a relationship of unrequited love, felt most acutely in the contacts through the eyes. The hybrid breeding process appeared to be selecting out human loving and caring qualities from our destructive tendencies. From us, the aliens would rediscover a lost emotionality, a capacity to feel love in a more embodied form. From them we would learn of a larger reality and open up our earthbound consciousness. After his fourth regression, beginning in June 1992, Peter became a leader among abductees in communicating openly and publicly about the facts of his experiences and their meaning for his personal and spiritual growth.

The fifth regression was the most complex. Peter's passage through the sloping wall of the house on Nantucket became a powerful metaphoric expression of a transition of consciousness, a kind of leap of faith into another dimension of reality. Once he took the fateful "fourth step" across the barrier, he opened more fully to his chosen participation in the alien-human breeding program. He discovered, again with a certain shock and even horror ("What about my life on Earth, John? What about my wife?"), that he had an alien female mate with whom he had been participating willingly in producing hybrid offspring for the ultimate purpose of repopulating the earth after "the destruction of the earth as we know it."

In the sixth regression Peter discovered that he was "part alien" and began the difficult task of reconciling his human and alien identities. The session was principally concerned with his opening to—and imparting information to us about—the coming apocalyptic destruction of the earth, the cosmic "negotiation" that was going on about the future of the planet, and the role he had agreed to play in creating a race of hybrid beings that represent the next step in human or human/alien evolution.

In the seventh regression, in April 1993, Peter experienced the collapse of past, present, and future, and felt that his consciousness could transcend his body. We explored further the mixture of human, alien, and hybrid beings that would manifest on the earth after the changes that were to come. Beyond the destruction, he had a vision of a kind of new millennium, a future "Golden Age" of openness and learning. In addition to his role in breeding for this step in evolution, Peter felt he was receiving a gift of special energy from the alien source that, as a healer, he could transmit to others toward changing their biological vibrations and enabling them to move "through the veil" into another future.

Peter's case seems to answer certain questions about what the alien abduction phenomenon is really about while, of course, it raises others. It seems to tell us that alien-human interaction is for the purpose of evolution, both biological and spiritual. A new breed or "tribe," a hybrid form, is being created between the alien race or races and human beings. Peter, and other men and women like him with a dual alien/human identity, appears to be playing a vital role in the creation of this tribe or tribes, breeding with an alien or hybrid mate to produce offspring that would be able to survive in some sort of postapocalyptic future. At the same time Peter and others like him are experiencing a process of consciousness expansion or transformation that will enable them to move out of a purely earthbound existence and become "children of the cosmos." Peter also has a role as a teacher in changing the consciousness of other human beings who are participating or will take part in this evolutionary process. Whereas all of this seems quite purposive, there is another, more purely relational aspect of the process, as human beings and aliens struggle with the task, both traumatic and joyous, of merging their qualities and identities.

All of this, of course, poses profound ontological questions. In what domain of reality is this breeding program and Peter's dual human/alien identity, for example, taking place? Although to Peter and many others like him, the process is all too real, we do not know "where" or "when" any of this is occurring or will happen. We do not even know whether such words apply. It is as if beings that are semi-embodied and derive from another dimension, have taken on the task (or have been recruited to do so by some other, "higher," intelligence) of becoming embodied to the degree that they can combine biologically with human beings. But we have no knowledge of what this complex process, operating as if on the margin of biology and spirituality, really involves. If it is "genetic" at all in the sense that we know this word, we have no information whatsoever about what the genetic alterations that the aliens create might be which could permit our species to merge.

From a purely Western scientific/philosophical perspective all this would have to be dismissed as nonsense. Yet to Peter, whose psychological health seems to have been established by my time with him and the formal tests he underwent, these experiences are so vividly real, so richly textured and consistent, and accompanied by so much corroborative physical information, that to dismiss them all out of hand would, it seems to be me, shift the burden of epistemological responsibility to the side of the skeptic.

The same problem of ontological location and definition applies to Peter's apocalyptic visions. Are these true pictures of what is to come in a finite time period? Or are these prophetic pictures powerful embodiments projected into human minds (for many abductees have been experiencing them) of what in a biospiritual sense is already occurring on our planet in the form of an emerging eco-disaster whose impact is, evidently, having its effects on a cosmic scale? Stated in different words, do these images, which affect experiencers like Peter so vividly as to seem literally predictive and even depressing in terms of the losses of connection on Earth that they forebode, represent possible futures based on a kind of extrapolation of our current course? Are these, in effect, the offering of choices by a higher intelligence whose methods of intervention are subtle, inviting us to change by offering a relationship with quasi-embodied intermediaries who reveal us to ourselves at the same time as they change who we are?

On August 26, Peter and I did an eighth regression which provides a kind of postscript to his case. In the session we explored an abduction experience that had occurred a few weeks earlier, at the end of his vision quest in Montana. He remembered a deeply emotional and sexually exciting connection with his alien partner, who had assumed a kind of appealing hybrid form. This was no longer simply a pragmatic exchange, a bringing together of sperm and eggs to create a new species. It was an enjoyable union, awkward in some ways, between a fully embodied human being and another being obviously unversed in the densities of sexual love. Yet Peter felt disgust and revulsion after the experience ended, at the time it occurred, and once again when he was coming out of the trance state in which the experience was remembered.

What this suggests is that the alien-human relationship is something far more complex and complete than a program of hybrid procreation. It appears to be a halting, difficult attempt on the part of an intelligence of which we know very little, to create a merger of two species who seem to need and long for something each has to offer the other. It is an experiment, which, as far as we can determine, is as formidable, frustrating, and demanding as anything its creator has ever undertaken. For it involves nothing less than bringing races of beings together whose principle homes until now have been in separate ontological dimensions. Yet, to make the matter still more complex, there is evidence in Peter's reports and those of other abductees that we and the alien beings have derived, or split away, from the same primal source, and that our deep longing for one another grows out of the desire to rediscover a lost brotherhood (see the fourth regression, p. 312), to return to one another, and thus come closer once again to the cosmic "Home" from which, in the experience of Peter and other abductees, we both once emerged.

CHAPTER FOURTEEN

A Being of Light

Carlos* is a fifty-five-year-old man, a husband, and the father of three grown children—two sons and a daughter. Carlos exercises his creativity almost daily as he is involved in drawing, painting, the writing of poetry, drama, academic essays, and a novel; he is involved as well with theatrical production and direction of plays. A fine arts professor, he teaches extremely popular classes in a small southern college and frequently offers extra courses to meet the demands of interested students. He has contributed significantly to the cultural environment in his county and state by volunteering in the state prison system and working with handicapped children, the mentally ill, and the elderly; he also has worked to address regional environmental issues.

Carlos wrote to me in July 1992, upon the recommendation of two of the men involved with the Allagash (Fowler 1993) account, about a period of lost time on Easter Sunday in 1990. "Six (or more) afternoon hours disappeared during a hike over a mountain slope" on the Inner Hebrides island of Iona located in the straits between Ireland and Scotland.

Carlos previously explored his encounter experiences in several hypnosis sessions, involving seventeen hours, with a psychiatrist near his community whom we shall refer to as Dr. James Ward. He spent several days with me in August, during which we talked for many hours and I conducted two hypnosis investigations totaling six hours. Although I use the term "abduction" in our discussion of his experience, Carlos is emphatic that the experience is an "encounter," and refers to himself as an "encontrant," and not as a victim or an abductee. He infers that in some respects he is a participant, a co-ceptor (rather than a receptor) of the image process and the experience partaken. Carlos believes this variance in language speaks to a difference in the manner by which a

*This chapter is the result of an unusual literary collaboration between Edward Carlos and me. Indeed, he is the co-author of the chapter.

researcher-hypnotherapist relates to others in the ufology field and the way in which an artist/encontrant perceives the phenomenon from within.

Carlos's case touches the edge of several of the mysteries that surround the abduction phenomenon. While shedding light on some dimensions of the phenomenon, his experiences open profound new questions. As with so many abduction cases, the objective correlates are suggestive and tantalizing—UFO sightings, burned earth where a UFO landed, unexplained cuts and scars appearing after abductions, and, above all, dramatic photographs of a light beam reaching from the clouds to water in a bay at Iona. But the weight of the evidence depends upon the reported experiences while under hypnosis of a highly intelligent, sensitive, and sincere man, suffering from no apparent mental illness or distortion of thinking and perception, who to the best of his ability is seeking to understand the events that have overtaken him.

The information that Carlos's case provides can be thought of in two ways, which at first may seem contradictory: Either he allows us to gain insight into technologies that we can only imagine but which a more advanced intelligence has mastered, or we are being opened to alternate realities, domains of being which are not part of our accepted universe. But on closer scrutiny this distinction will appear to have little power. For advances in technology and the expansion of our notions of reality are inseparable. What is unique to the investigation of the abduction phenomenon, and well illustrated by this case, is the necessity for human consciousness to expand in order to allow us the capacity to conceive beyond our present technological abilities and perceptions of reality.

Through hypnosis Carlos—he prefers to be called by only his last name—has recalled many encounter experiences, beginning from when he was three and a half years old. It is somewhat difficult to sort out the memories of distinct times and places of encounters that Carlos has been conscious of throughout his life as opposed to those that were recovered with the hypnosis sessions that began in February 1992. As an artist, Carlos has a powerful visual sense. This has enabled him to be extraordinarily sensitive to the light and energy transmuting forces that are central to his case and perhaps to the UFO abduction phenomenon in general. Carlos has had a history of what he calls "visionary" experiences, that he has become more aware of through his almost thirty hours of hypnosis, and that he is beginning now to associate with his history of encounters.

In November 1970, Carlos was invited by an Anglican priest visiting his hometown, who had seen an art exhibit of his on St. Michael and the Fallen Angels, to do a work of art at the Church of St. Michael's and All Saints in Tollcross, Edinburgh, that would also be based on the Archangel. After a week in Scotland, the priest arranged for Carlos to accompany him to the island of Iona. Although he did not quite know where Iona was or how he had ever heard of the place, Carlos felt that he had always wanted to go there. He had images about the monastic history of the island, images with which he felt a sympathetic association. They traveled by train, and on the way there, while painting watercolor landscape sketches from the train window, Carlos felt an almost empathic connection with the land rolling before him and found himself crying, feeling as if "for the first time in my life I'm going home." Although he had been raised a Roman Catholic, no one in his family knew of Iona and Carlos himself had no previous historical interest in it. Nevertheless, he was deeply moved and excited both en route to the island and as he arrived there by ferry from the neighboring island of Mull.

Before he had left Edinburgh an elderly woman, a well-established British author who had written several books about Iona and various historical legends about Scotland in general, invited Carlos to her apartment to discuss his impending journey. The author told Carlos that when he went to the island he should go to the beach by the Bay of Seals and sing hymns to communicate with the seals, who, according to local legend, contained the dead souls of monks who had been killed in the Viking raids. Modern day monks believe that the Scottish royalty, like the monks themselves, who were buried on the island, returned as mermaids (as the seals looked so much like women with large eyes and long hair coming out of the water). His first morning on Iona, Carlos traversed the one-mile width of the island, from the town of Iona to that particular beach, and, "just for the hell of it," sang Gregorian chants in both Greek and Latin. To his surprise a seal came close to the shore and followed him for what seemed about half a mile as he walked along the beach, and it continued to follow him when he walked back. Charmed by the seal's friendliness and the legend's apparent authenticity, Carlos conceded, "It was a very beautiful new experience."

The next night he attended a weekly, traditional village dance held in the small schoolhouse gym in the little town of Iona; the music began around midnight. After about two hours of continuous dancing with women visiting the island who were staying at the abbey a short distance from the community, he was feeling very free and open, "very wonderful actually." With exuberance and an energetic vitality, he ran with chest forward, arms out to his sides "like a cross," down the dark, narrow lane in the heavy, dense, and cold rain to the end of the dock at the edge of the small town, the site of his initial arrival on the island. Although the night was pitch-black and "pouring wet," Carlos saw, over the sea before him, a pink haze like a bubble, perhaps twenty-five or thirty feet across and "luminous from within." The luminous mist seemed to form externally before him as a spot growing from within his vision. There was a subtle flash to its appearance and then it developed large in front of him over the water and in the night sky and was eventually enveloping. Carlos says that he ingested no intoxicants at the dance and that the bubble was so substantial he believed anyone else who would have been there could have seen it, not questioning its reality. In "a second" he was in the bubble form and the scene shifted.

He was now no longer a thirty-four-year-old man but was a twelve-year-old orphan boy on a beach at the end of the island past the abbey, somewhat far from the dock. With him was a close friend who had modeled for him as St. Michael, as Adam in the Eden story, and as a fallen angel. The friend is twelve years younger in "real life," but now was also an orphan and the same age as Carlos in this other reality, which Carlos calls a "vision." The vision was of the sixth century, the time of St. Columba, who is historically associated with Iona. In the vision were two of the monks of the community and the two adolescent orphans whom the monks were raising. Carlos recognized one of the monks as another young man who had modeled for drawings of both Dionysus and Christ for him at home. The second monk he did not then know.

In the vision, which to Carlos was altogether real, a small boat came toward shore—a small, squarish beach defined by two large stone structures or "gates"—from a larger wooden Viking ship (the Vikings had, in fact, sacked Iona several times during the community's existence). The monk who was the man that Carlos knew suddenly fell dead on the beach, followed almost immediately by the other child who fell over into his lap "and I realized he was dead." The events were very sudden and it is not clear to Carlos how the two were killed, whether or not, for example, they were shot with arrows. The surviving monk then ran toward Carlos, who jumped off the dune on which he was sitting, and grabbing Carlos by the hand, yelled for him to follow him along the beach. They ran until they fell beside a nearby large stone jutting out of the beach on the dune side of one of two huge rock formations that acted as a natural gateway to the beach. As they fell, the monk with his cape "fell over" Carlos to protect him (Carlos did not know if he was alive or not). The Vikings, "thinking we were all dead," ran by the survivors and up from the beach into the fields. The vision ended with a blackout and Carlos found himself standing on the dock soaking wet, at which he returned to the small hotel where he and the priest were staying.

The next morning, before they left the island to return to Edinburgh, Carlos felt compelled to return to the Bay of Seals where he sang chants again "to see if the seal would come back." The morning was overcast and stormy, the rain having continued all night and into the day. Carlos "climbed out over a treacherously wet peninsula of large boulders and rocks to stand overlooking the dark waters which seemed to boil in a great upheaval beside me. With the storm all around me, I sang the 'Dies Irae' and the 'Asperges Me' from the Gregorian chants. A huge, white sea lion with large, white tusks rose up out of those black waters and scared the hell out of me. I didn't understand this. Everything was too much!"

Carlos returned to Edinburgh and completed his project, *Michael's Triumph,* painting a series of translucent hangings on chiffon and satin that captured the qualities of light on the island and depicted a passage from the Apocalypse. Included in the hangings were several life-size male nudes representing the fallen angels displayed behind the central figure of the Archangel Michael. For almost twenty years, the church authorities displayed the large hanging (measuring twenty-five feet wide and about ten feet in depth, with each panel hanging about seven feet in height) for two months each fall, from St. Michael's Day to All Saints' Day, during the time of an annual pilgrimage to the church from other parishes throughout Scotland.

Two months after Carlos returned to his hometown the young monk who had saved him in the vision showed up on his college campus as a freshman and came to his office to inquire about taking art courses with him. Carlos reacted with amazement and disbelief. The two felt instantly that they knew each other and a friendship ensued. The realization that he had somehow in his vision been able to look into the future affected Carlos deeply and reinforced his determination to return some day to Iona.

Carlos is of mixed Spanish, Scottish, Irish, German, and German-Jewish extraction. His surname is somehow related to the Spanish Armada and its demise in the Irish/Scottish Hebrides perhaps off the island of Mull next to Iona. He grew up in a small village in western Pennsylvania in a Catholic family. His parents, young adults during the Depression era, were hard workers who retained their ardent Catholic faith. Each parent, raised on a farm, especially valued schooling since they were able to attend and achieve an education through junior high school, as was typical in small town and rural life in the early part of this century. Carlos's sister, ten years older than Carlos, and like her parents a faithful Catholic, married when Carlos was a child. She and her husband had four children and they remained in the small town where she and Carlos were born. When Carlos was sixteen his father, a small town entrepreneur and eventually a laborer on the railroad, died during a working day of a heart attack, having had a history of cigarette smoking. Confronting death has been continuous in Carlos's life, and this event drastically affected him. To begin with, Carlos had to go with the mother and sister to the county coroner's morgue to identify his body, which was a traumatic experience. "I had never been in a morgue before. Seeing my father on a slab, on a table, and my mother screaming, crying, and kissing him even, I just collapsed on the floor and cried." But Carlos's mother went on to manage their country all-general and grocery store until her retirement, then assisted in the preparation of dinners for his only sibling's catering service.

Carlos was encouraged by his family to pursue an education, particularly in art education, and he completed three degrees and the equivalent of a fourth—including an MFA degree, having concentrated in painting and sculpture; the equivalent of a second MA degree in art therapy; and a Ph.D. in comparative arts. He credits his developing his natural artistic ability to his sister's influence and encouragement. Carlos married when he was twenty-six and after he completed his Ph.D., he and his wife, a mathematics teacher, moved to a university community on a mountaintop in the South where they raised their own three children. Although teaching full time, Carlos continued to avidly pursue his own creative productions.

Throughout his life Carlos has suffered from respiratory difficulties related to various allergies. When he was a year old and had respiratory pneumonia, a nurse informed his parents that he was clinically dead. His frantic parents rushed him to a hospital, where a low pulse rate was discovered and he was placed in an oxygen tent. Under hypnosis with Dr. Ward, Carlos recalled the feeling that "the child I had been, died" and "the light creature" he had been previously "took over the dead baby's body . . . Coming into the body was very painful," Carlos says. He experienced intense "resistance to taking on a body . . . I love having a body," Carlos pronounced but then added, "but I did not want to come [to Earth]. The body is prone to so many problems. The body reacts to everything. It is like a jellyfish on the beach; every stimulation, every incoming microbe. The cellular structure is continually shifting and changing. Growing up was hard. Aging is hard. All the things the body goes through. And, it is never still, never truly peaceful."

Under hypnosis, Carlos described the sensation of reentering his infant body around age one. "I felt like I was sliding into it, like you put on socks and shoes and trousers. I would pop into the fingers and pop into the toes and pop into the muscles. It was painful and it hurt. I did not like the feeling of it; I thought it was messy; I thought it was nauseating; I thought it was disturbing. It was a fat, grubby little baby that could not do anything. It had no real presence. It was such a different dimension. It was such a descending."

In a hypnosis session with Dr. Ward, Carlos recalled feeling that despite the pain of becoming embodied again around age one, he somehow "volunteered to come to physical Earth . . . I chose to accept the body," he said. When I asked him why he had "agreed," he spoke of his responsibilities as a teacher and an artist. He is deeply troubled about the failure of human beings to treat "their potentialities very well" and has been concerned all his life by our predatory destruction of the "Earth garden." Using art, he is teaching "'the aesthetic of transcendence' . . . Through art processes I am helping people to be more empathic that they might better understand and identify with those things that are not of themselves, so that they won't destroy them."

The first encounter experience Carlos recalled occurred in the late summer or early fall of 1940, when he was three and a half years old, in association with a display of aurora borealis (northern lights). Such events were quite uncommon in Pennsylvania, and Carlos recalls the apocalyptic reaction of some of his neighbors. The experience was awesome for Carlos too and has remained with him powerfully all his life; he says the spectacle of the colored lights in the sky affects the way he paints as an adult.

As a child, Carlos was so intrigued by this event that he did not want to go to bed. "I was angry because my parents made me go to bed eventually. I thought it was the most significant experience of my entire young life and I was told to get dressed in my pajamas and go to sleep. But my dad worked on the railroad and awakened at four-thirty A.M. in order to be at work by six o'clock. I did not get in bed immediately, however. Instead I went and looked out the window. Then I yelled at my parents, 'Mommy, Daddy, I see an angel.' My dad yelled back, 'That's nice. Now, you get in bed.'" The child, Carlos, could not understand his parents' attitude on "the most important day of my entire life." The angel image was "a yellow light or a yellow haze" that moved toward him and he associated it with the blond-haired, androgynous angel statue he had seen under the family Christmas tree which fascinated him. As a child he used to play with the statue of the angel as if it were an airplane, making it fly around at the base of the tree and land on the limbs or on the top of the manger. The angel was the most important image of his recollections of that season.

Under hypnosis Carlos recalled that out of the "amorphous yellow angel shape" there emerged "one of the skinny little white creatures with the big eyes," a type of alien being he has seen in several hypnotic regressions. "It was a little creature like myself, about my height. It was skinny and it had a large head for its body and it had very large eyes. The eyes were almond shaped, catlike." The eyes were "brilliant blue sometimes, as if they were illuminated from within." The creature was "bald, no hair, no features on the body. I was not sure about the hands. I never felt they were quite hands like my hands although I have seen this creature several times." Instead of fingers it seemed to have "claws or pincers. I am never sure if there were two, three, or four digits, but it seems to be two." Later Carlos felt that he might be confusing the "hands" of the various creatures he is aware of, who he feels might be categorized into four races or species types. He senses that the hands he referred to are most likely those of larger robotic creatures which actually vary in shape and form depending on the functions performed. He feels that the various species are interrelated and co-functional.

Then Carlos recalled how as a child he lifted "in the air in a mirror image of the small creature" and went out through the window and "flew" with the being. "It took me all over the hills around my home and flying was to me, wonderful. I just thought it was great fun. I even once turned over on my back and then returned to the regular position." He flew beside the creature and seemed to be held and propelled by the energy of the light in which they were immersed. "I was aware of distances. I was aware of movement, of color as deep hazes in the sky. I was aware of distant stars. There were lights beyond the stars and planets, way beyond." Sometimes he feels he has intermixed this memory of flight with memories of being lifted and transported through a light tunnel from one site to another, such as to the interior of a craft. Flying is a recurring motif of some of his more vivid dreams.

As he took his "tour of the farming neighborhood," Carlos had the feeling that he had done this before, "before I was a little kid, pre-creature," i.e., before he had become embodied with human countenance. It was an "awesome [a word Carlos uses frequently in our discussion] experience" to go like this "back to the light." Carlos finds it difficult to describe the quality of this experience of metamorphoses and transubstantial bodily-material changes in form and energy, an experience he has had during other encounters. He feels that he literally "dissolves" or "cellularly disassembles" through a painful process of "breaking from material form into the light energy," i.e., he becomes the sky or the light itself, which "permeates everything." The creature accompanies his return to "the energy light place. It took me back to where I had come from—before I came into my body, before I became a body—which was light, an energy light place." The creature itself, he said, "is only a form of light, emerging from out of the light."

Although geographic in one sense, the experience of the light was also "out of space. It was not space/time." I asked him where his consciousness was located in the experience. "I was consciousness; the experience was consciousness, a pure soul experience. Soul is the endlessness of it," he said. "The essence of the experience is of an energy which is pre-form." Carlos also tried to describe the "beauty" of the movement. He was still talking of the experience from age three and a half, but I suspected he may have brought in associations from later abductions. Carlos is adamant that the event was as he described it and that what came up in this hypnotic session was not an intermingling of associations from various experiences. He also says that other later experiences affirmed this imagery of the light from the first hypnotic session. "This transformation was from my second hypnosis session. The return to the light-energy form at age three and a half later was part of a reinstruction to assist me in reaccepting the original intent of accepting human form. The experience of the light reoccurred at least in three hypnotic sessions."

Carlos observed, "You are within the universe's energy. You are that movement of energy and light, but you are aware of the 'spacelessness' of it. The word 'spacelessness' is as close as I can come to an objectivity of any kind . . . I think I've tried to paint that feeling or sensation actually without knowing I was painting it. It is like swimming under water. You can see forms, and you can see distance, and you can see light. But you cannot see your body. Your body is water anyway, so if you can see beyond the scope of the body then that is what it was like." The experience was "joyous, one of the most breathtaking experiences I have had in my life. I think I have painted this light all of my life. I mean literally." Carlos has frequently created large art works—oil paintings and mixed media drawings—of skies, clouds, and vistas. These paintings and drawings have culminated in an exhibition entitled *Lightfall* first presented in November 1992; the entrance piece is of a photograph of a lightfall taken on his second trip to Iona during a period of a loss of consciousness.

When he was five, Carlos again developed respiratory pneumonia with a high fever and near coma that threatened his life. As his doctor thought his infections were related to allergies (and possibly also a result of his continual exposure to cigarette smoke), his parents took him across the state to Philadelphia for tests. In the course of his hospitalization, Carlos "had over two hundred, perhaps as many as four hundred pricks on my skin" to test for possible allergic agents. In the middle of the testing he developed measles which resulted in his being quarantined so that his parents were not allowed to visit. This was a traumatic period for him.

Under hypnosis with Dr. Ward, Carlos recalled what might have been an out of body experience during this illness in which he was visited once again by three or four little creatures with large eyes like the one that had come to his house when he was three and a half. He returned once again to a place of light, energy, or power from which he could look down on his body.

Carlos attributes his recovery from pneumonia to his experience of the application of a healing light energy beamed into his body by the alien creatures. He describes the process as "like laser beams coming into my body through the soles of the feet and the hands, and possibly through the sides of the lower torso, radiating throughout the whole body, expanding and changing color as the light grew to fit the whole body interior, thereby healing it." At the point where the fever broke "the yellow interior core, which was surrounded by an orange mass further surrounded by a hazy pink-mauve-red color array, and then the healing light was edged by very crisp blue and green bands at the inner surface of the skin. The green and blue edge was the breaking point, the cooling so to speak. Then my body reacted again after this long, long illness." At this point, finally, the creatures "brought me back. I went into the body."

In the hypnosis session, as in the experience itself, Carlos felt an intense reluctance to "come back" to his body. "I was crying and sobbing not to come back to this life, to this awareness. But the creatures brought me back." Carlos recalls continuing to cry and scream as he came out of the session, during which he had had considerable difficulty breathing and experienced the sensation of itching and changing body temperatures. "I was crying and then I was angry. My emotions were running all over the place." As he walked down the hallway, leaving Dr. Ward's office to go to the bathroom at the other end of the building, Carlos felt fierce and wild to the point where he was afraid he could kill someone, "like a lion. I felt shamanistic." With long, shoulder-length hair, Carlos looks rather leonine.

Carlos recalls no other abduction-related experiences from his childhood or adolescence. But from his undergraduate years at a state teachers college in western Pennsylvania to the present, he has had a number of them. When he was still a student, while attending a family reunion, he saw "a huge—baseball field in size—round, flying craft, shaped like a saucer, upside down on top of another saucer, separated by a dull metallic-like band, of windows probably, around the center of the outer circumference. It was, other than the window band, highly reflective and a shiny silver. Most of my relatives and I were out in the yard, other than my mother, grandmother, and aunts who were in the kitchen. The spacecraft stayed in one spot without moving for at least twenty minutes, long enough for everybody to look at it, long enough for me to go into the house, drag my mother out on the porch to look at it. Then the ship took off and was out of sight within seconds in a clear, straight, silent streak."

A year later he saw a large fireball one night, "bigger than my car but not as big as the spaceship" a mile from his home. "The fireball flew parallel alongside my car but out in the field and some feet up in the air, and then it shot back through the woods, splintering into four smaller lights, which broke away in four directions as it disappeared." At first Carlos thought it must be a meteor, but it did not behave like one and he was nonplused by its appearance. He does not remember stopping, but arrived home two hours later than he should have. He has recalled a few sketchy details of an abduction experience in relation to this episode.

Carlos's most distressing abduction memories, marked sometimes by nausea and other physical symptoms, are those that have involved his children—he and his wife have two sons, who are now twenty-eight and twenty-six, and a twenty-four-year-old daughter who recently married. Carlos wept in my presence when he recalled the trauma of being unable to protect them in their youth. "I am paralyzed and they take the boys from my arms," he laments, trying to control his sobbing. He says he has seen his daughter on a ship as a little girl "reaching out to touch some of the instruments and I did not want her to do that. I was so scared about her touching them!"

Carlos suspects that his health problems, including his allergies and respiratory illnesses, are related to probes and implants from his abductions. He has had two nasal septum operations because of breathing difficulties, frequently gets sinus and respiratory infections, and continually takes over-the-counter medication to keep his nasal passages clear. When Carlos was in his forties, a tumor was removed from behind his right eye, which he attributed at the time to leaking asbestos in his work area at school but now suspects is related to probing and implantation during his abductions. He believes that his eye surgery impaired his ability to focus and has forced him to need glasses. He also questions the cause of a later hard growth on his forehead outside the area of the interior tumor. Several years ago a mark or bruise on his lower body was diagnosed as cancerous. The cancer went into remission after an operation, but Carlos attributes the remission to the transmutation of light energies that he experiences through gardening and daily painting with watercolors at sunset and also relates it to the energies transmitted to him by the abduction process.

Carlos believes that identifying with other species in his abduction experiences also helps create and develop ecological values. The outdoors has appealed to Carlos since his childhood—"I played in nature a lot. I grew up in the woods, playing the Tarzan myth"—and he has always had a strong connection with animals. Dogs and cats were constant pets in his youthful country home. He had two horses during his junior high and high school years and spent hours riding in the hills of Pennsylvania. "I think I am really good with animals. I feel we communicate." He is surrounded with dogs and cats in his adult life, which follow him on daily walks around the lakes, in the open fields, and through the woods near his home. Carlos supports animal rights and has spent much time working for environmental awareness.

In his hypnosis sessions with Dr. Ward, Carlos recovered some memories of the Iona episode, which entailed two significant encounters, and he has recalled further details in lengthy conversations and two regressions with me. For several reasons, it is difficult to obtain a coherent, temporal narrative of the encounter. Carlos's way of thinking tends to be nonlinear with associations that move about in time. Further, a collapse of space/time order or structure is common in the experience and recall of encounters, but this is particularly so in Carlos's narratives. In the case of an experience as rich in images, sensations, and emotions as the Iona encounter, a precise order of events is virtually impossible to establish. Where such narrative order seems possible, I have stayed with it. Otherwise the basic themes of these experiences, which covered approximately a six-hour period, are set down in a sequence that seems logical or appropriate but is not necessarily temporally correct. I will begin with those events that Carlos remembered consciously.

When Carlos returned to Iona twenty years after his first visit, in April 1990, he had professional reasons to do so—research on the Christian hymns and poetry of St. Columba and the imagery's possible relation to Druidic poetry honoring the earth and the feminine in nature—but his personal wish was "to validate the visionary experiences." Carlos thought that if he could find a particular cave that had come up in a hypnotic regression a year after his first visit to Iona, or if he could find remnants of the stone gate, neither of which was present on any maps of the island, it would verify both his vision and his regression experience. (In the regression, he and the young monk had hidden, crouching low in a small cave on that black, damp night.) Carlos planned to spend only two days at the most on Iona. But car trouble on Mull over the Easter holiday forced him to extend his stay on Iona to about ten days. And thus the encounter that is the centerpiece of this case occurred.

His first day on the island, Carlos climbed on the rocks he had seen in the vision. He felt that he knew intuitively where the cave was, though he had not seen it on his first trip. The water was high. He saw fissures in the rock but could not find the cave. He thought to himself, "I just imagined this," but reported "it troubled me terribly." He took a few photographs and left. A night or two later he returned when the tide was low ("I'd never thought about the tide"), walked around the cliff in one of the rock structures, the one nearest the large stone on the beach he had fallen beside, "and there was the cave." He entered it and "hunkered down" inside the cave as he had in the vision. The impact of this discovery was powerful for Carlos. He "felt great" because this meant to him "that my visionary experiences were real, that they were validated, and that I was dealing with truth."

On Easter Sunday, April 15, after Carlos had been on Iona for three or four days, he started a journey across the mountainous island to St. Columba's Bay, a three-hour walk from the village past the Bay of Seals, in order to obtain green pebbles which he intended give to his children. He started off at noon, in order to make it back before dark, climbing over the plains to a mountain summit in the central western portion of the island. He reached a plateau near the top of the island and walked perhaps thirty to forty steps off the path to urinate against a ledge. When he turned back toward the path he found that he was strangely confused and dazed, even dizzy, and he had trouble walking. Everything looked different and he did not recognize his surroundings, although he realized he was on the island. At first he could not find the path and when he thought he had found it he did not recognize where he was on it. Shortly he realized he was going in the opposite direction he intended—down rather than up. When he turned around to continue his progress up the mountain, he observed that probably two or three hours ("watches don't run on me so I do not ever carry watches") had passed and it was too late to be able to complete the five- or six-hour trip he had planned to St. Columba's Bay and still be back before dark. Deciding he would go the next day, he headed back, somewhat stumbling as he was still in a trance state, and soon arrived at the *maercher,* a kind of field above a beach, that overlooked the Bay of Seals.

From this field he saw over the bay before him a great, long, peach-colored shaft of light descending from the thick clouds to the surface of the water. It seemed to him "miraculous, awesome, eerie, and uncanny." When it struck the water, the shaft made a huge peach-colored circle from which peach-colored mist rose. He saw thousands of sparks inside and outside of the circle "jumping all over the water like the sparklers that kids get on the Fourth of July." (Later during hypnosis he described the beam as a "lightfall," a "tunnel" connecting to a spaceship.) Remembering he had his camera with a close-up lens hanging around his neck, he shot a picture, focusing on the water while walking backwards so as to get as much of the beam, the clouds, and the circle on the water's surface in the picture as possible. Carlos describes this as "the most incredible, natural event I have ever seen."

I have seen his photograph, which includes smaller light shafts radiating downward off the main one to the water that Carlos had not noticed at the time. The picture was taken with slide film. After a print was made, Carlos took it to two colleagues in the physics department at his local university. At first they speculated that the beam could be a solar pillar, but then ruled that out on the basis of its shape, the wide arc, and the haze and spark activity in and above the water where the light struck. Also, Carlos calculated that, at that time, the sun would have been setting farther to the northwest than the beam was located. Carlos himself teaches photography and does not believe that it was a sunbeam, and does not refer to it as a "beam" or "ray of light," but rather, as a "lightfall." The more senior of the two physicist colleagues looked up Iona and reported excitedly to Carlos that St. Columba had seen such a light too.

After taking the picture, a light came over and enveloped Carlos and he "blanked out," falling toward the ground. He is not certain if this was the same light he saw or another that was directed at him, but the color was the same pink-peach. When he "woke up" it was after dark and he was sitting on the beach "about a hundred or more yards away in a complete daze." Although he could not see the beam of light anymore, he did see a cloud "that was glowing and pulsating with an orange light." The sun had set already. He took another picture (which shows small, short light beams coming from the cloud). Then he started back toward the town, walking in the darkness. He took yet another picture, but the light glow was dimmer and the rays were gone. Remarkably, immediately thereafter Carlos had no memory of any of these events. He walked over the mountain to St. Columba's Bay, and, grinning, said to me, "and I got my green pebbles." In fact, he had no recollection of the light beam until four months later when he looked at the slides from the trip for the first time.

The actual encounter, as recalled under hypnosis, began with a shift of consciousness after Carlos urinated and he experienced himself ascending to and in through the bottom of a ship in a laserlike tunnel of light. On the ship he found himself facing "a sweet little creature" who took him along passages on the ship. At the beginning of the encounter the creature seemed to draw him from one enclosure to another by reaching out its arm. Several types of alien beings were involved in his encounter. There were "little light creatures" like the one who accompanied him through the ship and escorted him down passageways. "They scurry around dronelike, very busy, and pay no attention to me" once within the structure, and they performed various functions. Their heads were round and white, with no hair, "like a bald person." He perceived their eyes as "bright, deep luminous blue," rather than black which is more commonly reported by other encontrants; Carlos explained, however, that the color changes, and it is not just a matter of perception. "The color has to do with communication and control."

Carlos reported that the large eyes sometimes look as if they have "goggles" over them, especially when they are seen at night or outdoors, which might be part of or parallel to the eye structure. "There is some confusion to me about whether this is actual 'flesh' or if what I describe [under hypnosis] are in fact goggles or a part of a helmet." A close-fitting helmet appeared in a later hypnotic session; prior to that I asked him to describe the eyes, and "it is possible I was describing either or both. It is like looking through thick glasses, but I see their material, their viscous flesh. Their eyeballs or lenses are transparent so it seems black to people in the dark." Behind the "goggles" Carlos has seen a vertical slitlike pupil, which he describes as catlike, and a quite wide circular iris "with browns and reds moving around," narrowing and widening, contracting and expanding "not unlike our eyeball." The changes of color seem to "go all around the circle of the iris."

To Carlos the creatures do not fit a stereotypical description, and some whole species seem androgynous. In his examination experiences, Carlos has found that there is frequently a larger, "most likely female," entity who seems to control some of the programs (she calmed Carlos when he became afraid in one session). He described her as gray, but said often the haze about her is changing colors, "radiating roseness, mauveness, pinks, and orange." There is "always a mist around her . . . She is thin, resembling the little ones, only she is taller and therefore appears more elongated; she has the same eyes and there is hardly a nose and hardly a mouth."

Carlos described rooms of varying sizes on the ship as having curved ceilings and passages between them. One room he called "a rotunda; the room is large." Another had "a lower half and an upper half" with "a lot of electrical-like ceiling lines, like the veins in a brain. On the side between these halves were window or screen areas which were all around the center of the space. They could walk on that balcony and look out, it is like a two-way mirror; it offers a projection place or a screen as well. It is as if these window/screens are made of a combination of metal/crystal/mirror/glass." If he were to represent this in art, Carlos said he would create a cloudscape or landscape on suspended plastic sheets, one in front of another. He had done works on chiffon with acrylic earlier, after his work in Edinburgh, utilizing similar materials, which he suspects may be related to this perception of the screen which doubles as a window. He continued describing the room, "There is a balcony level at these windows with a railing. From the balcony there are slanted surfaces on the tops of machines which extend from the floor level of the balcony and railing down to the floor beneath. At the base are desks, yellow-beige in color, with instruments on panels at the desks. Small creatures are busily moving around in the lower space or they are seated at the controls at the base of the slanted walls." Carlos recalled walking on the balcony and looking out of the windows that ran horizontally.

During the experience in Iona, as in other abductions, Carlos experienced intense physical pain, fear, and nausea. Most disturbing to Carlos were large, robotlike creatures with large, black eyes that appeared to have reptilian and insectile facial qualities and insectile body characteristics. "I don't have any problems facing the little ones that are so blissful, or the taller ones, but the ugly ones scare and repulse me," he said. The reptile-faced, insect-bodied robots were "brought forth" by the female entity to perform specific functions. "She is like a doctor/philosopher/psychologist of sorts. This is an operation, but it is more than merely a physical examination," he says.

The instrumentation "fucking hurts the nerves," he said. But, he added, "it is not the cutting; that is not the hurt. It is your own fear that you do not know what is happening, because although it has happened before and you sense this, you forget previous times to some respect although memory is awakened during your contact with them, and you see these robotic machines coming at you that look so weird and strange." Nausea, Carlos said, comes when the fear becomes too intense, particularly in relation to the reptilian machine. "I see it as this creature that scares me and I get nauseated, and the nausea is from the fear. It is not from what they are doing . . . The robotic creature, like the female-type entity, has a pink or rose haze around it. I'm afraid of it; it is monstrous. It has an insectian quality in its body with reptilian facial features. It comes directly at you . . . I look away. It is like a larva inside the leather—a hard, dark, scary machine. We see it this way because we are interpreting its surface as flesh which seems to be a leather/metal combination. The robot is an operational functioning mechanism, perhaps a biomechanical creature; it is a mental construct of theirs. They can form it, and then we can perceive it."

Because the looming perception of this creature in our first hypnosis session nauseated Carlos, I suggested he block it out and continue with other aspects of the experience we were pursuing. Once out of the hypnotic trance, however, he stated that he wanted to face his fears in a hypnotic session, and to a large degree, this was why he had sought my assistance. It was important to him to face his fear of these robotic creatures; he did not feel fear with the other species. The nausea and fear had persisted from his very first hypnotic regression, he said, during which he could not speak at all, but previously it had always been related to the examinations or to those scenes involving his children and their separation from him. In our second hypnosis session he was able to confront these perceptions in detail and move through this part of the experience.

In spite of the distress associated with this encounter, Carlos simultaneously experienced this Iona encounter as purifying, enriching, and even ecstatic, a paradox which is consistent with other abductees' experiences as they are described in many researchers' theses.

This aspect of his experience is best conveyed in his account of events surrounding the vision of peach-colored light. After Carlos took the first picture of the lightfall, he walked backwards while looking through the camera lens to get another fuller image that would include the clouds above the bay. Then he recalled that the light, perhaps a beam, came over him and he felt a "tingling" in his body and started to fall down. He fell backwards in what seemed a brilliant flash of light initially, "with my hands up to protect my eyes from the light . . . I was surrounded by this huge circle of peach-pink mist just dancing and moving all around me, and then I was taken up or lifted upward into the ship and I think I might have been nude but I'm not sure. I was another time. I don't know how the nakedness happened when it did." He had difficulty here distinguishing among his memories of various encounters; he knew that nakedness was a condition of his examination at some point in the encounter phenomenon.

As he was going up through this beam of light he saw "the edge of the ship in the clouds" and "went in again through the bottom." After this, within the craft, he saw a group, perhaps between five and nine, of "the little white creatures." They were standing in a "white, shiny, luminous haze . . . I knew they were trying to teach me something. The eyes were the last thing I saw, those blue eyes, and then they completely disappeared into the fog or mist. They went through a whole process of color changes before they totally hazed out into light and this was beautiful." The mention of fog made me wonder if he was really on the ship at that point. "I am," Carlos said. "The haze was in this vessel." Later, Carlos expressed his sense that the empathic nature in perceiving was part of the message; throughout his art teaching career, he has always stressed that it is essential to identify with the subjective nature of the "objects" involved with the creative work.

Early in our second hypnosis session Carlos expressed feeling intense sensations as in reliving the experience with the light he perceived its transformative nature. Initially I confused this reference to light as indicating the beam coming over him when he was "shot back up." However, attempting to clarify these experiences later, Carlos affirmed that there was a difference between the light experience in which "a sort of physical, cellular, molecular change occurred in my body" as he was taken up into the craft and "that ecstatic experience which seemed transformative, also physical but with an intent and within a spiritual dimension. The ecstatic aspect of this experience, although similar in some manifestations, was meditatively provocative." He described "beautiful" tingling feelings that grew into a full orgasm in which his body went into a convulsive spasm lasting nearly half a minute before an interruption occurred in the process while he gasped, moaned, screamed, panted, and even growled. The intensity of his reaction caused the bed on which we were doing the regression to shake violently. After this experience subsided Carlos was "in darkness" but could "see the light." He resisted coming back from this sensation but his consciousness was already actively returning at my suggestion. He continued for a while to feel tingling sensations as we spoke and he described an "alive" feeling or vital peace. "It *is* life!" he affirmed.

It is possible to think, as Carlos does, of any abduction experience in terms of the transmission of information—from the aliens, or whatever they may represent, to the experiencer; from the experiencer to the person selected to report the information (myself in this case); and, finally, from the reporter to his readers or listeners. In each instance the reporter selects and interprets among the various data, stressing some bits of information over others, which itself is a kind of interpretation. In Carlos's case information about the technobiologies involved in the transmutation or metamorphosis of the body-as-matter into forms of energy seems to constitute the core meaning or central importance of his story. This is potentially of great significance for our understanding of such matters as how human beings are taken through walls or windows on the way to the ships or how their bodies can be taken through space. What is most remarkable to me is the extent that Carlos has been "shown," or otherwise let in on, how this process works. Why he has been given this information remains mysterious. It seems fitting that he is an artist who is deeply resonant with light phenomena to begin with. But that does not tell us very much. The April 15, 1990, abduction on Iona was particularly rich in information regarding these processes. And Carlos indicated this is a worthwhile place to pursue our investigations.

There are two periods in the Iona encounter that are noteworthy in this respect. The first occurs as the light beam comes over Carlos at the start of the second part of the abduction when he is being taken up into the craft. The second relates to some sort of crystal instrumentation on the ship. Carlos says that the crystalline machinery has come up in some sessions with Dr. Ward. It seems to have do with televised/projected imagery of a miniature holographic nature, in which particular and multiple personal life scenes are played out while he watches.

Carlos identifies "sparks of light" that are "essential little cores of being-energy" with the lightfall itself. After the beam of light came over him—which he regards as connected to sexual energy; as, for example, the frenzied orgasm he relived during hypnosis—his body seemed to go "in layers . . . expanding and contracting in the mist." He felt a vibratory tingling sensation in his body and then the sense that "my body is dissolved or diffused into its transparency . . . The body just dissolves and goes up. Then I am transparent. I have a sense of the interior transparent shell of the body which is not part of its physicality but yet it is connected. It is the shape and mold and form of the physicality." He continued, "The molecular structure, cellular structure of the body, just goes out into the light . . . It is the transformation from one state of being into another state of being, but you carry the core of a residual shape . . . it is like a ghost image. The image is the memory of the body, and it is clear and it is there and it has form and it shoots (moves with a certain intense direction and strength); I shoot; I go up. I'm head first." I asked where he goes then. "I go into the light, and the light is molten-core, volcanic, liquid fire, but it is light and it is white and my transparency is adaptable to that and there is a silver clarity to my shooting up." The light is "like a viscous membrane that you go through, but it is not a solid thing. It is receptive (the body as to the experience of the energy change), and I don't know; it is like I envision a female orgasm (would be)."

As he ascended and entered the vessel, and went fully into the light within the craft, it became orange, yellow, and white, a kind of "spectrum of color." After this he was in the "rotunda" in the spacecraft, entering at the balcony level. He believed he probably ascended alone this time, but saw ten or twelve of the small, white creatures once he was in the craft. Carlos then found himself in the large, instrument-filled room previously described, with many busy creatures all about him. His presence was not acknowledged by any of them, so he had a sense of anonymity and there was a lack of expectation or anticipation regarding him. He was, however, guided by one of the creatures down a ramp to a lower level where there was a kind of black, marblelike platform base or narrow hallwaylike floor around a central, circular platform upon which were crystal structures. The floor resembled a "dance floor" composed of "dense matter." It was not solid but he could stand on it, feeling that he was standing on open space beneath him in the black, dense matter. The floor in this circular floorway was in the interior of the room but around the outside of the platform. Also on this lower level, he saw little "desks" with buttons like computer technology on the outer edges of the large, high-ceilinged room.

Little creatures were sitting at these and were involved with the machinery. A higher balcony area with windows could also be seen above them; the balcony with its own walkway opened toward the inside but had a railing. The walkway or balcony ramp surrounded the entire circle and had exterior windows.

Carlos had been on a kind of table which was also "a block of crystal" within which or upon which he was situated, depending, he clarifies, on the aspect of the investigation performed on him. The female creature, described previously, was present and "like a spiritual doctor" and brought forth the reptilian-faced, insectile-bodied, or robotlike entities who in turn were to perform "an operation" that was achingly painful and was carried out by an instrument Carlos described thus: "Whatever these crystals are, metal-like more than glasslike, there is light. I can see it [one of the particular crystal instruments used in the examinations]. It is like a squared tube of crystal with the sides lopped off so that at the ends, each tube appears eight-sided, but it is big in the middle and little on the edges, like it is mitered. And then the end of it is shaped like a step-pyramid. It shoots laser light into the body, but it feels like a needle because it hurts, and it resembles a needle."

Carlos defined some differences in the seemingly overlapping expressions of memories that appeared in his hypnosis sessions. "The following situation as described is from another time other than the Iona experience in my later life; I think it is from when I was a child at five years old with pneumonia. In the previous paragraphs where the 'sketchy details' are mentioned, I believe I described what probably are related memories, in particular to the encounter when I was a young man seeing the fireball. One recall under hypnosis seems to provoke more than one memory simultaneously, but some of the processes of transformation differ, apparently for various purposes or intentions. There are subtle differences depending on the function of the metamorphosis or the investigation which I have come to believe is also metamorphic, i.e., that functional changes are introduced. Those who focus on the examinations only provide a simpler interpretation of the experience, a surface explanation that is more easily ascribed. The crystal 'table' was used during examinations, at least once, but probably each time an examination occurs. It is possible that other 'encontrants' refer to a table and even remember a table because it is more pertinent to our regular life experience, i.e., that that particular symbol comes forth and we infer it as a table per se . . . The large crystal structure, however, is located in the center of the lower rotunda area, and is a different mechanism; it is operationally and functionally distinct from the smaller instruments utilized in the exams. The larger crystal structure is usually involved in a 'teaching' situation, but the teaching is inherently metaphorical as opposed to in any way being verbally or definitely explanatory."

Carlos considers the grammatical tense and person of his descriptions a valuable detail of the process of trance description, and urges a reader or researcher to pay close attention to how the memory is interacting with the perception. Sometimes he has to describe the experience he has just, in the immediate past of the hypnotic regression, witnessed or experienced, but his verbalization comes and goes from being that experience to describing that experience—the reflective significance of time is important to him. He feels that the hypnotic process, although clarifying the memory and helping to bring it forth, also further confuses research translations because of subtle (sometimes) structural manifestations of vocal/verbal utterances.

Carlos continued with his clarification and included as well in this description parenthetical details to clarify or fulfill the verbal shorthand of hypnotic trance descriptions, "I went through that body—cracking (apart) an experience regarding the light which is not unlike when I was healed from the fever. First, the light is in the body. The light is a gold core which reaches out through the yellows and oranges . . . the center core is inside an aura. (This takes place) Inside the body. It (the light mechanism) continues (invading and permeating the interior of the body—muscles, tissue, organs, blood, nerves, et cetera) to the edge right next to the skin; it is when the skin dissolves, an atomic cellular dissolving, and then the green and blue precise edges are there (at the dermis and epidermis), and it (the light manifestation, the light mechanism of healing) can form the flesh, and it can take the flesh (i.e., change it)." This process is associated with itching. Carlos tended to rub or scratch his body during his hypnotic recall. Then he felt "pressure, a tightening. The body gets really tight, but at the same time, it is fleshing out. It is like it's being pumped up . . . I can feel it just dimly now, but it is like it gets bigger. The light spreads out; it fleshes full. The white molten light is in the center where the yellow was (and has expanded and now has become). I see the light (spread or array) as yellow in itself, and then it expands outwardly becoming (i.e., having been or having come from) white in the center and yellow at the outer edge of the white haze as it expands, and then your body is broken and you are suddenly free, and you can go one of two places. In this time, I am golden light. I saw the goldness initially outside of myself. I . . . It was internal to me. It is the reverse of what we normally do. It is not me out here observing inside; rather I am seeing my (own) inside (i.e., body interior) from this shell (i.e. body form or structure as a transparent edging of the body shape) that is clear (and) that disappeared at the point of the breakthrough . . . I saw the crystal-clear form of me. It is the ghost image I referred to earlier (in the hypnotic session), but it is clear."

Carlos feels somehow that this process of transmuting his body into light is related to "the process of creating (a) changing (operation) and hiding (or making invisible) the spacecraft." Earlier he had described the process of "when I turn into the light" as "an incredible pain as my body changed, and it feels like the body is swelling and becoming so swollen and hard that it breaks; I was frightened by this process . . . My body just felt as if it puffed out. Then the swelling breaks and you are now the light." Once he had become light the experience was "joyous and blissful." Before becoming physical or embodied once again, however, "as light I was dancing in this big circular space first alone, then with the creatures who were similarly transformed. It was just joyous and I was so happy. It felt miraculous to be this light and to do this moving, dancing, and sharing . . . to share this incarnational joy with the creatures, my companions." This was celebratory in one sense, but Carlos also believes it is an aspect of the teaching process involved, that the activity is metaphorically experiential. Carlos indicated that this mechanism of essential empathic identification with the other is inherent to his or anyone's creative capacity, that his artistry re-creates the metaphorically experiential process as artistic image making. He made distinctions, however, in the depths of the artistry involved, that this depth or comprehension of the creativity as content or meaning is a human component of degrees of understanding.

At this point Carlos experienced himself as gold light and associated the energy that produced the transformation with the "hollow black open 'hallway' floor which surrounds the crystal structures on the platform in the middle of the rotunda. The floor is open and hollow; the blackness comes from its depth; sort of; it is a place of movements, from within it, an energy or movement comes upward to where I 'stand' or float on it and it is like air, but it isn't a cold shaft of air or anything . . . it, like, ionizes me. I mean it is like an electrical fabrication all through the outside of my body, and within the heat inside the generating crystal structure within the circular floor space is the core of light. Whew! I finally got it out!," i.e., expressing the process, although Carlos said he realized he expressed it awkwardly.

After reading the above transcript, Carlos commented, "I believe human beings avoid acknowledging 'material' such as art or 'experiences' that are manifest with meaning, and that for those who do apprehend these possibilities, there is a tendency to become caught up with an interpretative agenda, instead of listening to the depth . . . The experiences are verbally difficult to describe; they are more metaphorically imaged." Also he has, he said, so many images coming through his perceptual/memory/image process that frequently only a word or brief phrase will refer to the activity and then, at that moment of attempting to verbally refer to the imagery, he is, simultaneously, trying to hold on to another set of images experienced, while speaking any kind of even remotely close descriptive word in order to connect with it later. "Sometimes, there are not words there for the experience . . . Speaking in hypnosis is not unlike writing in a dream diary in the middle of the night when a dream provokes one to wake up momentarily. One awakens and the residue but illusive imagery is in the mind, so one jots down a word or two hoping there is enough to empower one in the morning to bring more of the memory to the conscious mind."

During our second hypnosis session Carlos had sensations of anal probing in the examinations which might not have occurred during this abduction and may have been recalled from other times. "They are clarifying that the inside of me is okay, and they are operating on me, i.e., examining me organwise, muscularwise, et cetera . . . If there is anything not right, the process can be healing (sometimes, but not always)." He experienced probing; radiating light "looking at" his heart, ribs, and other parts of his body; "checking my physical well-being"; and "checking to make sure I have the constitution to continue.

"They do a light scan. It's with the rose-pink light, again . . . different colors do different things." The instrument itself seemed to metamorphose into light. "The light went into my body," Carlos continued, "You see, they go in, and it is like an instrument, but it dissolves like I dissolved, and that is part of the process here, the metamorphosis; they can cause the change with their laserlike instruments . . . In my heart," he said, "I feel an extreme heat; I think it is healing something. Clearing arteries or something."

Sometimes, Carlos said, they can enter "through my soles, of my feet." Other times the invasive body checking is in body openings. The anal probe, for example, creates an "itchy" sensation that extends throughout his body during the instrumentation. The itchiness seems related to the absorption of the light energy. He suggested, "It's light matter, but it may take some sort of more solid form than we know. And it does what it needs to do to know my development, in terms of instruments that are different from our own, but we see them as needles . . . well, it is not painful (in this procedure) unless we fear; it is almost erotic in a total body sense." (He is referring to the entire process, not just to the anal sensations. But . . .) "When they're in here (the anus) I am feeling the radiation going out in circular directions from the inside of my anus to wherever it goes—down the legs, down to the feet, up through here [gestures up the torso] . . . it is like the whole rear end/rump is hot and pulled out or expanded with the heat sensation felt throughout the body. It is itchy; it is itchy on the surface skin on the exterior of the body because it is being pushed out similar to when I break through the skin (in some of the other processes and procedures of metaphoric experience). The radiation is from the anal probe but I'm feeling the energy all over. I mean it has that orgasmic quality, like breathing is orgasmic and erotic, but it's not the big one, the cosmic orgasm when I manifest as light-space and energy. You know, it is itchy and there is energy there, and it is erotic because it *is* body."

After such probings or other types of examinations Carlos suggested that sometimes there might be physical manifestations or health symptoms that are not intended by the creatures, that are realized after the abduction; he believes there is an irony involved with healing, the process and instrumentation of which leaves other symptomatic annoyances, such as scars, warts, bruises, lumps, rashes, on the body.

Despite the distress involved Carlos gives himself over to this process. "If the creatures need to probe and push and (have me) ejaculate to understand the continual process of the body, through its metamorphosis on this earthly physical level, then I know that and I am a willing subject." He feels that "when my body is given to the probing mechanisms of the light I am energized; I am re-energized . . . That is not just a healing. If my physical body changes, i.e., experiences cellular changes, the resulting physical sensations are understood. Then the healing that is occurring within me occurs to keep me and enables the teaching to be given."

When he is in the altered state of hypnosis, Carlos seems to be able to perceive the process of transformation of the creatures themselves into light. "Their bodies go from being the little white creatures they are to light. But when they become light, they first become cores of light, like molten light. The appearance (of the core of light) is one of solidity. They change colors and a haze is projected around the (interior core which is centralized; surrounding this core in an immediate environment is a denser, tighter) haze (than its outer peripheries). The eyes are the last to go (as one perceives the process of the creatures disappearing into the light), and then they just kind of disappear or are absorbed into this." He says that they, like himself, and all humans, who he infers to be of a light-energy source, "are light creatures" but "biologically different from us . . . We are or exist through our flesh, and they are or exist through whatever it is they are."

There are times when Carlos feels that he is himself alien in the sense of feeling isolated and also identified as an alien being, existing on "more than one level of consciousness," perceived as different, "a hybridization." He and they are "go-betweens between the knowledge source of being in the universe" and the beings on Earth. His and their transmutations are somehow connected. During the encounter and even separate from them he may experience himself in the head or "helmet" of the perhaps reptilian or other alien creatures. "I feel like I'm looking out through its 'helmet' (or if the helmet is a reptilian head structure, its biotic skull) . . . It's not a helmet really." He persists in using this word for something "they put on." Carlos adds that he believes there is a comparative structure in the biorobotic creatures' head construction and in the helmets worn by the others. He thinks the functions of both are similar.

In his first hypnosis session with Dr. Ward, he felt they might have masks or screens that augment their vision, which may be relevant to the helmet as a mechanism and an aspect of uniform apparel. "The helmet they wear helps them to see various aspects such as disease progressions, forms or manifestations of oxidation, concourses of chemical atomization, temperatures, radium exposure, internal organs, et cetera." When he was given the opportunity to have the helmet on or at least to examine it, "I looked through it, and I saw with the robotic or alien quality. I feel what I observe is recorded. Looking out from its interior, the eyes of the helmet-mask bulge out." The helmet is "the same shape (as the aliens' heads, and the eye bulge parallels their own facial structures), so when we observe these we get real scared because they're weird looking . . . We can see double eyes, dark eyes, observing them with these on from outside the helmet, although we may not be aware that they are wearing these. The real eyes of the creature and the reflective helmet or mask 'eyes' are seen simultaneously and this can be disconcerting."

From inside the robotic "head" Carlos feels he can see temperature and other biological processes. "It is not unlike what we do with computers and electrical generators. When I am a creature" or "in the creature's examination 'structure' (it's not a uniform or costume exactly) I'm studying too."

"Studying what?" I asked.

"Humans," he replied. "I've been waiting fifty years to say this, you know . . . What our little television computer machines do here on Earth is similar to what happens inside these helmets. There are mechanisms on the inner brow. When it sticks out on the lower brow above the eyes, inside are mechanisms, the means to altering what enables the various manners of seeing. It is just like looking into machinery that is a human form . . . it is like looking into the workings of a microchip in a computer or other electrical operations. And there are lots of microchip connectors on both sides of the brow. Inside the helmet is full of these. I'm not sure how they function, but when I'm in them . . . [he hesitates, then adds] I'm not going to tell you that, I don't think . . . right now. But I am able to see the heat of the human. I can see at night. I can see the form, and I can see what I want inside those who are observed. I can turn them into light . . . I mean not turn *them* into light, but I can put light in them and look around (i.e., the mechanism enables its wearer to project such a light), to assist with or enable healing." He becomes in some respect, I infer from this, like the aliens with their large eyes, an examiner or analyst.

"On Earth," Carlos says under hypnosis, "there are times when blocks of time and ordinary consciousness disappear, and I'm (I become like) a creature for awhile, but of which I am basically unaware. (Perhaps it is to augment a learning process.)" Carlos suggests later it is at least a means to identify with another species, perhaps to induce an eventual acceptance of their status and existence. "The information or visionary aspects transfer to accessible imagery." Through hypnosis we can infer these lost brief images of time. Perhaps, Carlos suggests, this is related to "sometime before my life when I went through a stage of being a creature. I took my time considering if I wanted to humanly manifest and then I volunteered to come down to Earth." This capacity to achieve an alien perspective, an act of identification, if not form, is important for Carlos's art and his capacity as a healer. From an alien vantage point he can look "out the window (as if from the balcony on the ship) and survey the land beneath . . . That's where the summit is coming from! My summit work . . ."

At this point in his narrative, Carlos paused reflectively. He then referred to a collection of thirteen large mountain summit-vista mixed media drawings and another set of larger oil paintings he had been creating, and continues to create, for a huge panoramic work (four sets of thirteen works) for which he plans to design and construct a building that will be a place of meditation, prayer, and environmental awareness.

He also simultaneously has been working on seven sets of thirteen drawings each, having to do with the theme Eros and the Mer, which he believes are related to his Iona experiences as well. Carlos exhibited his first set of these works in an installation *Cloud Meditation Chapel* in November 1992, an exhibit moving around to various galleries since then. This exhibit's name is *Lightfall—The Mystical Passage* and features the Iona photographs in the entrance area of the exhibition. The number thirteen in numerology and mythology infers wisdom, he explained—wisdom is the attribute associated with the goddess, Sophia; Carlos points out that wisdom is needed in the world now in respect to the human tendency, through greed, to deconstruct the natural process into human norms. His concern is for wisdom in action about the earth, other species, and the ecological universe within which the earth and its inhabitants are a part. To Carlos, the number thirteen also implies an opening, an irrational number pertinent to the irrationalities (not nonrationalities) of the creative spirit, a means for participatory reception, rather than being a closed or formal number such as twelve. He believes there is astronomical—planetary and sun—significance as well as the number is expressible in religious terms; he reminds us of the number of the apostles or disciples being twelve plus One.

After enumerating these connections between his alien experience and his art, Carlos continued, ". . . I walked around the interior of the ship and looked out at the vistas from up there. As an artist I thought my insight or inspiration about the earth as a form or image of consciousness was from the mountain views, from climbing the mountains. I'm seeing the absolute vista in all 360 degrees as I walk around it, moving from window to window. I know that such views for my human counterparts are a spiritual high; each focus or moment of concentration with such viewing is a moment of inspiration. The spiritual essence takes visual form, and everybody feels good when they see the sunset or sunrise; they are enlivened by the natural world. It can be healing. We ordinarily, so caught up with our activities, just do not take time to deal with the experience of nature like we might. We acknowledge the phenomenal beauties such as clouds or the sunrise, deer in the field, fish in the water. But we often impose our force and will on it, to alter it or to destroy it. But, the change in nature is an energetic force every morning and every night. That's how I was healed of the cancer. God, it's all coming together. It is amazing, the integration; do you know?" I asked him about the cancer. "Well, I every night did watercolors in the changing, shifting light from daylight into darkness through the sunset, and the light transfused me, marking my physical being. Painting relaxed me, and the cancer went away. Working with the flowers and soil in the garden helped as well, similarly."

Whether he is inside a creature's helmet on the ship in Iona, or perceiving the earth from a creature's perspective from the sky, Carlos sees great beauty. "I see the islands rising up. I see the mists rising, and I see the sun going down over there. I see the clouds forming. This is such a paradise." He continued, "I see myself (now, in the capacity of and) as a little creature, but I'm observing . . . Earth is a garden, and these creatures are gardeners in our sense . . . I'm a teacher, and that is the capacity of communication and process. Art is very real, functioning, spiritual—and the earth is really in need of art's significance right now."

Discussion

Neither Carlos nor I can separate cleanly the dimensions of his narrative that are metaphoric and mythic from those that occur in, or are of, our, or any literal, physical world. His case almost begs for the obliteration of this distinction which has been so convenient, if not essential, to the Western perception of reality. Perhaps it is sufficient to begin an interpretation of his experiences by saying that to him they are powerfully real, and leave as perhaps unanswerable now the question of the domain or universe in which they belong. What or who the alien beings, or as Carlos prefers, "creatures" or "light-beings," are remains unknown. At the same time the profound relationship between them and Carlos lies at the core of his transformations. "I am a shaman/artist/teacher," Carlos said under hypnosis, attributing his evolution to the alien encounter experiences. "They are teachers," he continued, but the relationship is in some way reciprocal for "they are also really interested in learning of us." The shaman, he said, "uses techniques to alter the psyche, and what the shaman is doing is playing with the emotional discourse between teacher and community, between shaman and student, between the person who travels and the person who remains or who lives a life here. I teach by emotion and experience." He believes that teaching and the results of the process imply subtle transformation of the spirit. "Teaching, like creating art in truth, is a spiritual activity."

As a teacher of the arts and theater, Carlos seeks to provide his students with a powerful, transcendent aesthetic experience that opens them to the wildness or wilderness and wonder of nature and creation. He uses his painting and writing to bring a deeper awareness of the environment. He also has become politically active in his function as a protector of the earth, and very soon after returning from Iona, helped start a Green Party in Tennessee, working on the platform. He wants to convey in his message "the plenitude of the being of the earth." Again and again, he connects the evolution of this earth consciousness to his relationship with the creatures. "Their function is protection" and "they're keeping track of my changes." He strives to overcome for all those he encounters the sense of separateness from the earth, to reconnect us with it and to create an appreciation of its fragility. The encounter experience "fills my art with . . . with imagery, gives meaning to the landscape of the mind's perception of the earth."

Also central to many aspects of Carlos's case, especially his various transmutations and metamorphoses, is light in all of its energetic displays and manifestations. Again, light is real to Carlos and also metaphoric, "an apt metaphor," he wrote me later after our August 1992 sessions, "for both creative energy and spirit (September 16, 1992)." Light for him is inseparable from spirit and profoundly associated with sexuality on a cosmic level. "My cosmic orgasm while under hypnosis," he wrote, "was physical and ecstatically spiritual—light wrestling with my humanity, the angelic dimension with all of its imagined demonologies wrestling with my humanity."

Carlos is unusually aware of the relationship of his encounter experiences to the various dimensions of his personal transformation and spiritual growth. It is worth examining these processes in some detail.

The traumatic experience of those abduction aspects of an encounter, the helplessness, the probing, and the pain open him to his "light creatureness" and contribute to opening him to "the world of the soul . . . The soul is the world. The soul is me," he says. The pain and trauma seem at times inseparable from the expansion of consciousness that Carlos experiences in relation to his abductions, especially in his inability to protect his children. He also acknowledges that when he awakens in the night and has insomnia, he leaves his bed and his wife in order to try to protect her from the possibility of this intrusion on her sleep and well-being. However, fear and a feeling of protectiveness are only some of many feelings the abductions evoke.

One of the most powerful emotions Carlos experiences is a sense of awe. Toward the end of his first hypnosis session with me he said, "Awe is fear. Awe is mystery. Awe is ecstasy. When I use the word 'awe,' this is the substantial piece of being in a larger (world, universe, movement, energy) . . . where exists all the energy forces that are tension and stress and movement and energy and electrical components and atomic molecular and all of that. That is awe . . . We have the capability of awe, but we don't deal with it." A month later he wrote to me, "To sense that a predator is immanent is part of the fear, but for me there is a fascination as well that undermines the very predatory nature. The fascination is that of an awe, felt deeply behind the mysterious and perhaps ominous presence."

The closeness of death seems to be an intimate part of Carlos's life, and themes of death and rebirth are related to his alien experiences beginning in infancy and early childhood when he was "reborn" into a new body at one year old and saved from death by the healing powers of the alien being at age five. "Our art works [including his own] are full of death," he writes. "My own life is full of death. My life has been to witness death, over and over and over." Carlos's intimate connection with a cosmic source that he, like many abductees, calls "Home," has made him long to die. "I pray for it almost every day," he says, "and I have as long as I can remember." He loves his physicality, yet the experience of his embodiment for Carlos is filled with pain and loss, including having lost a loved one to murder. He fears that "I won't be allowed to die" and will "just keep on getting older and older . . . The human experience is loss," he says, "But there isn't anything material to lose 'cause it is all there. But you see, the personal loss is too difficult, I have lost so many people to death in this life that I've loved . . ."

The elaborate examinations by the beings that Carlos (like so many other experiencers) has undergone have been more than exploratory. He has the sense that they change the "energy structure" of his body. "It is an operation, but the operation is simultaneously an exam of what I am as a muscular-flesh-bone-structure human being. They know there is a component here that is a whole organism. The whole organism is a receptacle, as a sensory modality . . . it is like we are a machine but we are not of course. We are a bio-real substance that has sensory modalities enabling perception and memory which chart a history in us. One of my functions in this is that I am checked, and sometimes healed, and I am connected to my environment which includes all creatures." What is done to him, Carlos feels, is to preserve his body's stability and integrity, "a renewal," to enable him to be a teacher, while also becoming a kind of alien/human hybrid in the process.

Carlos's speech under hypnosis is not, to me as a researcher and hypnotherapist, always altogether coherent when he describes these complex processes. As stated above he ascribes this to an onslaught of images and thoughts coming almost simultaneously. As he begins to describe one, he is "caught by another." He may have articulated only a word or phrase when another word or phrase connecting to another thought or image comes out; and he may grab at particular words in order to describe something but they may not emerge as being entirely descriptive. "It is like writing shorthand, using a word as a connector and as a symbol, as a marker, to relate to later," Carlos says.

For example, I heard the following on our tape of a session, "I'm examined, and the examining . . . and the healings . . . are pushing and probing of the discovery of my changes in the molecular body. Metamorphoses . . . diffusion . . . is the process that I have to undergo mentally/physically . . . in the fleshness . . . of a physical being, in order for the directions we're taking to go through . . . the processes . . . as natural . . . and not dissolve into an unnatural process that is too rational or too nonrational, or too irrational . . ."

For Carlos the irrational component in one's psyche is referential to the creative process and the spiritual dimension of a human being; rationality and nonrationality are terms, that when used together, imply a dualistic or polar mechanism in verbal discourse. He abridges his commentary parenthetically to confer a fuller implication of his speaking. "When I'm cut apart (i.e., cut open, as in a cut made during an examination), or probed, or stuck into," Carlos says, "it is (sometimes) an intensification of the healing processes in me. But it (i.e., my body) has to be examined on the basis of each stage of my being there (i.e., my physicality in its evolution and development), being here (i.e., implying the changes from experiences), in order for the process to maintain its sense of completion and change . . . because the metamorphosis is a continual change, and there will never be a stasis in this realm. Every time there is a contact there is an alteration."

Carlos, in turn, feels that his openness is a method of teaching and communication, that his "openness is both an example (as a means to communication) and (operates in the process of communication) by . . . like (the process of) . . . osmosis. And, in a parallel sense to the contact with the creatures, I'm (also) continually pushing and probing and teaching and changing everybody I (am or come) in contact (with) (through shared experiences, art processes, et cetera) the same as I am affected by each person in each situation." In his art he is "practicing" and "reinforcing the metamorphosis" because he believes that his "art is both a way of exploration and sharing, an external form which manifests from the creative process as it operates through" him.

At their core Carlos's encounters have brought about a profound spiritual opening, bringing him in contact with a divine light or energy, what he calls "Home," which is the source of his personal healing and transformational powers. In our sessions, when he comes close to this light he becomes overwhelmed with emotions of awe and a longing to merge with the energy/light/being. Space and time dissolve, and he experiences himself as pure energy and light or consciousness in an endlessness of eternity, "a pure soul experience . . . I go back to the source because I'm not *just* human. I need to go back to the source in order to continue."

Carlos, like so many abductees, has developed an acute ecological consciousness. He is deeply concerned with the earth and its fate. The question of whether this is an unintended by-product of a process that he, no more than any of us, can fathom, or is an integral part of the alien phenomenon, cannot, of course, be answered. Carlos clearly believes that the aliens, however awkward, or even brutal, their methods, are trying to arrest our destructive behavior. "They're species gatherers. It is not just human species, animal species. If I feel a cruelty in the (use of their machines, their robots), they know as well that *our* cruelties, *our* willingness (i.e., the imposition of our will), *our* limits—are self-destructive and therefore destructive to all sorts of things." The earth and the systems with which it is connected are in danger of "collapsing." The aliens "are like little tiny drones of a vaster complexity which is "in the service of survival." They are "Earth gardeners," he says, "trying really hard to instruct us to find a plenitude and not to be caught in the human impulses towards extinction." They want us to find a "plenitude in the environment, a plenitude of the garden Earth."

Carlos's case opens us to unfathomable mysteries. Yet there are hints of meaning, of patterns that we might begin to grasp. His experiences seem tied to the fate of the earth and the tearing of the cosmic fabric that the destruction of its life-forms is bringing about. During our first hypnosis session he said, "If we explode, then this garden of paradise, even with its predatory nature, this strange, beautiful little ball in the universe . . . if it goes, then there is a loss for everything because it is one means of being that is a comprehension of the unities of things." The earth, he feels is "essential" to this unity for the universe, "I do not know if there are others like it." He feels we are confronting "an apocalyptic final hour" and this must be "met and challenged . . . We experiencers of this spiritual apocalypse constitute a paradigm of initiation," he wrote to me. "We are being initiated, but we initiate."

Among the "lessons" he has learned in his abduction experiences is the need for human beings to expand the scope of their empathy, to identify more widely. "If human beings were empathic and learn to identify with that which is not themselves, then they will be less predatory and destructive." For this to occur it may be necessary for us to gain a sense of where we belong in the universal order. The breadth and depth of Carlos's perspective derives from his alien connections and even his ability to see through alien eyes. "The last image in my hypnotic time with you," he wrote, "garnered by my immediate memory as we talked afterwards, was of my standing, as a small white creature with my own large luminous blue eyes, on the balcony circling the interior in the over-craft. I was looking out the window as I walked around the circular space, and I was seeing this beautiful paradise, Earth, in every direction."

Interestingly, Carlos compares near death experiences, which he has also had, to the alien abduction phenomenon, which he regards as having far more transformative power. Through the alien encounters there is "access to the bliss, and the near death experience is . . . a momentary place in between. It is a soul place, to gather up." Physical and emotional healing, with which the aliens seem intimately concerned, is an important part of this transformation. "You are diseased and then you are healed. With each healing, the emotional growth is established and connected in the human realm and I can go and utilize that towards teaching others."

CHAPTER FIFTEEN

ARTHUR: A VOLUNTARY ABDUCTEE

Arthur called me in January 1993 at the suggestion of a woman with whom he had spoken of a dramatic UFO sighting at age nine which had been experienced by other members of his family as well. She was familiar with my work with abductees and Arthur was curious to explore the incident further. Arthur was thirty-eight when he called me, a highly successful young businessman with beautiful homes on both coasts, deeply committed to the democratization of capitalism, creating a sustainable environment, and protecting the future of the planet. He and other members of his family attribute his remarkable degree of social and ecological awareness and sense of responsibility to the profound and lasting effect of his experience at nine years old.

I have selected his case, the investigation of which has only just begun, to conclude the series because I believe it offers a positive example among the possible human futures that lie before us. To a large degree it is our collective behaviors, as expressed through institutions, that impact the earth's ecology most profoundly. Among these institutions, for-profit business corporations, which impact every part of the globe, are perhaps the most powerful agents of planetary destruction that human beings have created. On the other hand, corporations administered mindfully, with an enlightened awareness of their relationship to the earth's environment, may help to arrest devastation and become one of the most important instruments available to us for restoring and preserving the health of the planet.

If, as Arthur believes, his alien encounters were instrumental in the evolution of his sense of personal and corporate responsibility, he and others like him have much to teach us about what purpose may lie at the core of the abduction phenomenon. Our contact with Arthur, who travels a great deal in connection with his work, has been limited to an initial interview, one hypnosis session, and a few telephone calls. He was able to attend one support group meeting several days after the regression.

Arthur is the fifth of six children of a prominent, conservative, Roman Catholic, East Coast family that includes many successful judges, lawyers, and businessmen. His father was a successful attorney and his mother a real estate mogul. He has a very large extended family which he describes as quite traditional and also "extremely close." He and his siblings were raised as "spoiled rich kids," Arthur says, and there were maids and servants to attend to most of his basic needs. He suggests that some of his values may have come from those who worked in the household "who were just really good people." Arthur has spent much of his life on a huge rural estate that his family has owned for over a hundred years. The incident from when he was nine years old took place near the estate. He recalls no traumatic incidents of any kind in the bucolic setting in which he spent his childhood.

Arthur grew up with a passionate love of nature and spent a great deal of time in the woods and fishing in and around the family estate. For as long as he can remember he has had a special relationship to animals. Whereas his grandfather and great grandfather would shoot and stuff animals, Arthur, like a kind of contemporary St. Francis, would communicate with animals, including porcupines, skunks, woodchucks, rabbits, and birds. To illustrate this he told us a story that took place on his ex-wife's farm. She had many rabbits on the farm and they trained the rabbits to be affectionate by lying down on the ground "with your chin on the floor"; at that height the rabbits would not be afraid of a human being. Then they would be "incredibly quiet," and the rabbits would grow curious, come closer, and "let you scratch their heads."

Rabbits, Arthur said, have a muscle behind their ears which they cannot reach to clean or massage but which gets tired. "So if you just put this finger between the ear and this one on the outside, and you go like this [demonstrates in the air the scratching/rubbing movement] eventually they just get so trusting they put their chin on the ground and it's a state of zuz, a way to describe this." In this "euphoria," Arthur said, the rabbits will lick your nose, and if you continue to "zuz" their heads they "just get totally zuzzed and it creates that bond."

Arthur told this story in the first half hour of our time with him as a kind of analogy to illustrate how to communicate with alien beings. "The way you communicate with these beings is telepathically, but the only way that you can achieve that telepathy is by eliminating fear. That fear will block it. You won't be able to communicate with them until you can get rid of the fear" and other "negative emotion . . . If you feel negatively, if it is fear or anger or destructiveness, or anything like that, they don't communicate that way . . . They simply want to be communicated with by not being feared, and that is the kind of reason they have a difficulty with humans."

The beings, Arthur believes, are trying to induce "that kind of euphoria." Perhaps "the light does it," but one way or another "they can induce that." They "are speaking a different language," but "if you react with fear, negatively, they just can't. But if you don't react with fear, you can communicate with them and the communication is in itself incredibly inspiring. I don't know how to describe it, but I know, at least my family thinks that it has affected my life in the way that I treat society and my commitments to different causes and stuff." Arthur can also "zuz" fish, he said, "I know how to zuz a trout out of a stream." Rabbits and other animals think "in as profound a way as we think" in their own way, he said. "You know it is beautiful in its own sense because it is so trusting, and there is a tremendous amount of love and trust and affection. It is just basic. There is a life energy there." During his childhood Arthur tended, like many abductees, to get significant throat and sinus infections. He also read a great deal, and even in the third, fourth, and fifth grades remembers liking to get "into debates" with his teachers "about religion and things like that."

Arthur has created many successful companies whose total worth he estimates to be in the tens of millions of dollars. Each one, he says, has funded specific philanthropies. For example, Arthur and his ex-wife put most of the profits from a snack food company they sold for fifteen million dollars into buying a piece of land for a migratory park and put the rest of the money into B-Green environmental awareness and similar programs. Currently he is working with several hundred U.S. corporations to license them to use special technologies under the condition they give their royalties to particular foundations.

"Anybody can set up a company and make millions of dollars," Arthur said. The challenge is "to figure out how to employ that to a higher goal of affecting other people by having your company be a model, its products being models." The results then are much more rewarding, he said. "People are motivated by more than just money," Arthur observed. His ex-wife puts environmental messages on her packaging of various products and they receive thousands of appreciative letters from consumers. She tries to give handwritten answers to each one. But although this may be worthwhile, "it is selfish," he says, "because you get more out of it than what you are putting into it."

An article in a New England newspaper in March 1993 described in detail the creative business activities of Arthur and his ex-wife. The article told of imaginative recycling schemes, including a recycling program for the homeless, a new conversational packaging technology Arthur had conceived, and "little essays" about the environment on the packages. The article contained a quote from Arthur's mother-in-law who calls him "indefatigable" and "the sort of individual you wind up begging to take your money and put it into his pocket." In our most recent contact with him in August 1993 (he called to "check in" and ask when the next support group meeting would be) Arthur told Pam that he had licensed his new packaging technology to several major food store chains. He said he has "lots of projects that are in good shape," including neighborhood gardens, a project for "more democratic capitalism," and a "women's discovery foundation" he is funding out of the profits from selling the new packaging technology.

Arthur believes that homelessness and environmental destruction are critical issues. To understand what it was like to be homeless, he lived on the street for a while as a homeless person. Ozone depletion and out of control population growth are to him two "primary" threats to the earth's ecology. Out of his own pocket Arthur has been funding "trim tab" pieces of legislation that might constructively impact the environment "on a worldwide basis." He also makes "a tremendous commitment of time dealing with politicians whom I am not crazy about" on these issues. He has worked tirelessly to put environmental and human rights considerations and standards into important international trade treaties and agreements as well.

A year ago, Arthur said, he knew little about the federal government or the world trade system, "but something made this a priority, and I have committed an incredible amount of time and money to just help influence everything from congressional elections to getting groups that are usually contentious or adversarial at best, like the unions and [a leading industrial organization], to get together. . . . I don't even know them. These are total strangers. And I don't even know why. I know that [this] will make sense, and I know it is a necessary thing to have to do it, but it is, like, okay, this is just like what I am doing."

Arthur says that he is "totally in love" with his ex-wife, Alice, still, and she remains "my best friend." They separated without contention because she wanted to live on a farm and "I just can't live on a farm . . . I think that the way you love somebody," he said, "is to do everything you can to help them fulfill their potential, and to be as happy and healthy as they can, and when they are happier and healthier that is just going to generate more love back." One of Arthur's former girlfriends and Alice, he said, are deeply close to each other and concerned for each other's feelings. "That is kind of the way for most of the people who I am close to in my life, male and female and just friends."

From the time Arthur was a small boy and continuing into his teenage years, he would sometimes lose periods of time and come home late. He knew the thousands of acres of the family estate very well. Nevertheless, he would sometimes find that the hour was later than it should be and discover himself facing in a different direction than he expected. On one occasion, when he was perhaps six or seven, he saw what he took to be the lights of the manor house through the trees, but realized that he was a quarter of a mile away, the lights were "two white lights like headlights, but they were farther spaced and they were like halfway up or midway up the tree." He was confused as to his direction ("somehow I got completely turned around") and "it was much later because it was dark." In his early teens Arthur recalled at least two episodes of missing time. On one occasion he was following a deer trail in the forest and ended up "hours later heading back to where I had just come from." When he remembered seeing the alien beings during the episode when he was nine they seemed to him like "old play friends."

The central UFO-related experience of Arthur's life occurred in the summer of 1967 when he was nine. Although we explored this episode in detail under hypnosis, Arthur recalled a great deal consciously in our first meeting. He, his mother, his older sister, and one or possibly two of his brothers were returning home toward the family estate after seeing a movie in a nearby town. He figures the movie must have been a comedy "because my parents never did take us to anything except comedies." His mother was driving, Arthur and his sister were in the back seat, and his younger brother and perhaps one older brother were in the front seat. As they were driving along on a dark country road, his mother said, "that plane is flying awfully low." Then she said in a worried voice, "that is not a plane," and made everybody get down in the car. "I saw her hand come down and kind of push us down . . . we were wired on candy," he said, and thought, "this was going to be pretty exciting for me at least."

Arthur and his sister were in the well in front of the backseat, and he saw an object about "a hundred to a hundred and fifty feet" above the car. "It had lots of different lights, but I couldn't tell you what colors any specific light was, it was just different colored lights." The light "was coming in my mother's window and my window," and Arthur found it peculiar that the light came down to the ground but seemed to create no shadows. "All the air was lit up," but Arthur knew the object he had seen "was the light source . . . It was the most incredible light. If there is such a thing as pure white. Totally pure, and it was everywhere." The feeling in the light was "like swimming and being in water and having the water being phosphorus." He saw different lights on the object, "like Christmas tree lights, but there weren't bulbs." He recalls specifically seeing red, white, and a "pinkish violet" color.

Arthur's mother and younger brother were intensely afraid, while his sister was mystified, but he felt mainly excitement. "There was absolutely no fear. In fact it was the exact opposite. It was complete euphoria, that is the only way to describe it, better than the best sex or the best anything you could have, better than fly-fishing for a week." His sense of the encounter was difficult to describe, "almost completely benevolent." The perspective from which Arthur recalls seeing his family seemed odd to him, as if the car were a convertible and he could look down on them into a car without a roof. His mother must have pulled over because the car was stopped at the side of the road.

As he crouched down in the well Arthur thought to himself "this is really cool." Then he experienced some sort of thought communication from the object. These thoughts "were being offered is the best way to describe it." He was told "you are going to get more," and "the way you can get more is by not being afraid and by being receptive—a million different things at the same time. It is so difficult to describe . . . There was almost a sadness that it couldn't be learned, but a gladness that it was being learned." The more receptive he could be, and the less he would be afraid, "the greater the understanding, the more profound it felt." He could not recall then seeing the beings, but thought of them as "little playful angels," light beings that had "features, but not really definite features."

At this moment Arthur began to feel slightly anxious for the first time in the session, "just a gut feeling." His consciousness seemed to split away somewhat and he felt "like when I am communicating this I am not here and it is like someplace else." Yet for Arthur it was "not a threatening kind of fear. It's that it's awesome. I can't understand it . . . Big is such a dumb word, but it is so big . . . The subjects" he felt the beings were "trying to communicate" were "so kind of new and big and just incredibly beautiful, but just massive and not what I am used to. I have a little mind like this, and they are kind of dumping a lot of water and it is pouring over the sides, so you are like whoa, you know what I mean?"

Arthur felt that his anxiety and the mild dissociative experience he had just had were serving to "slow down the velocity" of the information that was coming to him. "They [the beings] know they can overwhelm us," he said, "and the anxiety is the block or the thing that is in between . . . If they blow us away or scare us to death then they can't communicate actually."

Arthur is not certain how long the car was stopped. It took about twenty minutes to drive to the manor house from the movie, which he estimates probably let out at eight-thirty or nine. His grandmother was waiting for them and would probably have been upset had they come back more than two hours late, so he estimates that incident involved no more than one or one and a half hours of elapsed time ("normally we would have stopped for ice cream"). Arthur does not recall just how the trip home was resumed. No one spoke in the car, and his mother did not ever tell her mother or her husband about the incident. He found this unusual, because "in my family if you buy the wrong shoe size we talk about it, so that it is incredibly unusual in our family because we talk about everything."

It seems that no one who was present that night talked of the incident for twenty-five years until a huge family reunion in the summer of 1989 in which hundreds of members from the various sides of Arthur's family gathered to commemorate the hundredth anniversary of the family estate. During this event Arthur's sister, Karen, came up to him and asked if he remembered anything about "that night driving home from the movies." At first he could not remember anything at all, and she asked then "Does Mom remember any of this?" He could not answer that either, so Karen "described her perspective of what had happened that night . . . She told me a bit, and then it was like a trigger, like it set off bits and then my bits set off her bits." According to Arthur, Karen told him that "an incredible light filled the car" and their mother was very scared. Other details conformed to Arthur's account given above. Karen, he says, "remembers angels. She has a thing for angels."

Pam spoke with Karen on the telephone four weeks after our initial meeting with Arthur. Her account is similar to his, except that she remembers the light as "bluish white." As Arthur had said, its brightness made it possible to see the inside of the car "as plain as day." Karen's reaction was one of amazement, and she is open to the possibility of alien abduction but does not recall further details herself. Unlike Arthur she does not find it remarkable that their mother did not talk of the incident. If she "doesn't understand something, she just doesn't talk about it." Karen attributes the forgetting of the incident more to "neglect" than fear.

Once the memories of this incident had come back to him, Arthur became interested in asking his mother and brothers what they remembered. At a dinner out, in the spring of 1990, when his two oldest brothers, their wives, Karen, his girlfriend, his mother, and his younger brother were present, Arthur asked his mother about the 1964 incident. She said she saw a cigar-shaped vehicle, "like a dirigible," with "lots of different lights" that was going at the same speed as the car. She was "scared to death," though "mostly for us—my mother is pretty much that way." She remembered pulling over and trying to get the children down below the seats. Arthur's younger brother's memory was similar to his mother's. He too was very scared, remembered the light and "being stuffed down" in front of his seat. The two other brothers did not recall the incident and may not have been present.

Arthur had met Donna, the girlfriend who was present at the family dinner, soon after the summer family reunion. She told him of an incident from her childhood when she saw an "angel" on her windowsill and that she continues to communicate with angels. She told him that all her life she had dreamed of meeting someone like him. After this, he told her of his experiences and she encouraged him to explore them more fully and encouraged him to contact me. Donna also believes that Arthur's involvement "in all these different projects and such and causes" is related to his UFO experience. Donna "and my sister have been the two people to say that I am the one who has to find this out because they think it has affected my life the most, and, you know, I am curious if I would have been a different person without it." Arthur has the sense, related to information he received during his UFO experience at age nine, that there is "something that I am supposed to communicate, but I don't know what the message is . . . The nine-year-old experience is the trigger or the key to something else . . . something that is important for other people. And it is simple and elemental, and it is like in me but I don't know how to communicate it."

The regression was scheduled for March 25 to fit Arthur's travel schedule. After our last meeting Arthur had talked with Ted, the brother who had not been at the dinner in the spring of 1990. Although Ted could not recall whether he was in the car at the time of the 1964 incident, he was "one hundred percent convinced that there was something significant there." Ted is inclined to think that he was there, "simply because he's such a traditionally conservative Republican person and wouldn't normally believe in UFOs or anything, but he's absolutely convinced they exist."

Arthur told a strange story of photographs that his great grandfather, who was one of the early photographers in this country, had taken of "elves and little people," with chins "pointier" than humans', set in rock formations. Arthur suggested that they were probably little clay models that his great-grandfather, who was a creative inventor, placed in these formations and then photographed. But what is less clear is the source of his ancestor's inspiration for this activity. Perhaps, Arthur suggested, his own imagination had made some sort of connection with the little people he had seen in his grandfather's photographs. In any event, the photographs left him with the impression that "I met little people." Ted is also "just so convinced that these people exist. For him to agree to that is like getting Richard Nixon to do it." Before beginning the regression Arthur reiterated his conviction that "there's some message I'm supposed to get."

At the beginning of the hypnosis session I set the stage by taking Arthur back to the night of the incident when he was nine and asked him to report any feelings, images, or bodily sensations that occurred. The movie, he now recalled, was *The Mouse that Roared,* with Peter Sellers, "and I think we ate a lot of junk, like popcorn and stuff." He also remembered once more who was in the car and what his mother was wearing. She said something like "there's a plane flying awfully low," but soon in "a nervous voice" she said "that's not a plane." The large craft had lights that "went across like a *T* almost, but they fanned out on the back." Arthur saw "red and green and yellow." His mother put her hand behind her seat and told the children "to get down on the ground," but "I remember thinking that was so cool, whatever it was." The craft came from behind the car to the left, "out of nowhere," and "just plopped down right on top of us," perhaps fifty or a hundred feet above, and seemed to fill "the whole sky."

With conscious recall Arthur had not been able to make out details of an actual vehicle, but now he remembered something like ribs, a hatch, and seams, and the vehicle seemed to be made of a "really shiny, light metal, like metal when it's superheated." It appeared as if "the thing itself was lit, was the source" and, in addition, had little points of light. Again, Arthur said, this was "the coolest thing I've ever seen . . . I was ready to go," he said, and "I wasn't going to stay behind the seat, for sure." The light from the craft flooded the car and the air around it, and it seemed as if "the car was in water." The light was everywhere, and there were no shadows.

Arthur could see his mother, brother, and sister, who "were like crumpled down" in front of their seats. Although he could see his younger brother, Peter, below the front seat, Arthur did not see another boy in the car, which made him doubt there was anyone else in it. Next he remembered the car being off the road on the ground, but he still could not recall a moment when his mother pulled off. He could see nothing then except the light, and had the sense of the car being below him and the "feeling of being lifted." I sensed in Arthur a tightening or tension, and his critical mind beginning to work harder. I encouraged him to stay as much as possible with his sensations and raw images without judgment or trying to figure things out. There is a moment in the session I cannot pinpoint, when he seemed moved virtually to tears as he realized the car was no longer on the road and he was in unfamiliar surroundings.

Arthur experienced "many different feelings," including being "scared," but "something was telling me not to be afraid the whole time." As "incredible" as it seemed to him, Arthur had the sensation that he was being lifted up above the car "and I couldn't talk to anybody," for they were "huddled" safely in the car and he knew "they couldn't hear me." Then and now he felt "just confusion . . . It was like all the cells in my body were moving up, but my body was staying where it was, and it was separating." This feeling of not being "even part of my body anymore" seemed to be the source of Arthur's fear. He could see his mother and brother in the car, but he could no longer see his sister in it. But his perspective was as if "on top of the car" and looking down through its roof. Still confused, he believes that somehow he went "through the roof" of the car, but "I can't remember how I got through there."

Then a "message" came to Arthur, "like they're concerned" and "they don't want to hurt me . . . They can tell I'm afraid," and this communication "definitely helps" to calm him. Then (the source of the communication is not clear at this point) "they're trying to tell me there's a thread, like a spider thread, and that it's between us, but it's really fragile . . . 'Don't be afraid, or it will break the thread,'" he was told. I asked if this were a literal, physical thread or a metaphor for some sort of connection. Arthur could actually see a "light like a thread, a spider's thread that's lit" in the sky, surrounded by black. Then he saw more than six "little lighted beings clustered all together" who were telling him not to be afraid and not to break the thread with his fear.

I asked Arthur to describe the beings. They looked "like an embryo," he said, "just little gentle things." They were "luminescent," and semitransparent, with large heads and small bodies, "little skinny arms and little fingers. Maybe not five fingers." In addition, he saw "little legs," smooth faces with "a little mouth, little noses" and "no hair . . . like babies, like embryos," he said again. The eyes were all black and rounder than ours. The beings seemed to be "so close to each other they're touching each other," Arthur noticed. "They're discrete," but "their arms are always touching each other . . . It's like a bunch of rabbits. They huddle like rabbits."

I asked Arthur what was holding him in the air and he said it was "the thread," which he estimated to be perhaps an eighth of an inch in diameter, "like a kite string, maybe . . . It connects me, but I don't know how it connects to me." The beings seemed "pretty funny looking," yet "just cute" in "the way they behave . . . They're like bunnies," he said. "They don't want to harm anybody. Bunnies don't want to harm anybody. They want to play." Although "I could break it if I wanted to," the beings pulled Arthur upwards on the thread, lifting him "into the room" as if on "a rope or something." The thread was like an arc, or "half an arch." He went up it along a curve with darkness underneath. "I couldn't have walked up because it was too steep, and I wasn't walking. I was going up on an angle."

Arthur found himself "in space . . . I could see stars, and it was like this really smooth kind of beveled thing [this was the first explicit statement that he was being drawn toward the craft]. Looked like a round edge, like if you took a shallow bowl and put it upside down, really shallow bowl." It had the appearance of buffed stainless steel. The thread or string seemed to be bathed in a light that was everywhere, and as the beings were telling him "not to be afraid" Arthur was "just going" along it, "standing erect," pulled as if by an unseen force. The string seemed to go into the craft, "like a phone line or something." Arthur could now see the beings inside the craft, which I found confusing as he seemed to be able to see them while he was ascending on the thread. "They were never down below," he said. "I could see them through the string," but they were "in a bubble of light, and it is probably in the steel thing." This may have made sense from his point of view, but only introduced other problems for me, which I chose to let pass.

Arthur could not recall exactly how he got into the spacecraft—"there wasn't any door or anything." Once inside he found himself in a huge room with a wavy, irregular floor, surrounded by a ring of pink lights. The beings were chattering happily and were glad that he was not afraid. "They're like, you know, old play friends . . . Their main thing is that they don't want to hurt anybody." I was a bit shocked when he said that the room was as big as Boston's Fenway Park "upside down." The beings went on "chattering," communicating without voices or sound, giggling and touching each other, and seemed to want to play. They touched Arthur's face "very gently. It's like you touch a rabbit . . . They touch like my shoulders and my face is what I remember, with their hands," as if out of "curiosity," and they move "all around." There was also mind play, "like they throw fun things at your mind and you throw fun things back." I asked for an example. "Well, like blobs of color at you," Arthur said, "and just really pleasant feelings. Just a million different kinds of pleasant feelings . . . It's the way you communicate with bunnies," he repeated.

All of this playfulness, and the concern that Arthur be relaxed and unafraid, seemed to have been a preparation for "something serious" that was to follow. Darker beings wanted him to know about a "field of life" that he had lost, out in the country where it is green and there are "leaves and flowers and grass and birds and fish and everything." But there was a "big blob of darkness that's gonna fall over it. They want me to experience what it feels like." This blob is like "a massive water flood, and it's gonna go over the entire planet and just kill everything." I asked why they want him to know that. "Because I think they want to help," he said. "They don't want it to happen—it's so contradictory." I suggested he allow himself to go deeper into the contradiction. "They realize we're not bright. We're kind of stupid," he said. The destruction can be avoided, but the only way they can "do something about it" is "by communicating." At the same time "they're afraid of making us afraid," for "if they make us afraid then we'll be ineffective."

The blob, Arthur said, was like a great "water balloon," black and huge, that will cover the whole planet and suffocate it. I asked how this idea was "envisioned" by him. "They want to tell me something serious," he said and "they tell you by you feeling it." As he saw the blob he felt the fear and the smothering and the sense that "everything's gonna die." At the end of what seemed like a kind of staging or demonstration, the beings removed the blob, having communicated how serious the problem was. I asked Arthur how such a disturbing message might have affected him as a nine-year-old boy. He answered rather indirectly that "the most beautiful thing is between people who can communicate with each other like you can with these little guys." The blob was caused by our failure to promote life, to rid ourselves of fear, and to communicate with one another and with nature. The image had made him feel afraid and sad when it was happening, especially as the beings "make you feel so much life, and they show you what it [life] is not."

Arthur returned to the thread that had brought him to the ship, which became for him now a kind of metaphor for communicating and connecting in a loving way, "like we do with rabbits . . . Some people have to go beyond what they think they can do," he said. The only thing that could kill the "life-energy" was fear and the blob was its result. I asked him to say more about fear. "It blocks everything," he said. If we are afraid we cannot communicate with one another, with the beings, or with animals. "We can't even let life into us at all." I was not clear about the source of fear and what we were supposed to do about it. To "stop the blob," Arthur said, it would be necessary "to just challenge little fears at a time. That eventually will get bigger" and "people will get stronger." Take the risk of empathizing, saying hello, being "nice to everybody," as, for example, with a "tollbooth person or the waiter," he suggested. "The reward," he asserted, would be "much better than the fear and the risk . . . We're in a losing battle, but we can turn it around," he said.

I had the sense that we had gone as far we could along these lines and asked Arthur if the beings gave him any other information in the ship. He remembered then another, smaller room with other, "darker people," like "runts" or "little monks" with robes and hoods. These beings were "businesslike" and "not really playful." They felt his neck and behind his ear and pushed a "soft rubber, cold rubber thing" into "the back of my neck." Several of these beings were behind him and he saw "a couple on the right side of me farther away." These entities seemed more "ambivalent" Arthur said, "more complicated . . . They're doing stuff. They're like moving around, and they're pantomiming. But they move pretty smoothly, and they're more serious . . . It's not like the bunnies" where it was more playful and "you can take it in." They too conveyed to him a sense that human beings act destructively and they do not have confidence in our intelligence.

The communication from the darker beings was "more intense . . . It's like a billion, chillion, infinite amount of things going when they hit your brain. It just is like incredible overload, like you can't experience everything at the same time." The message, however, was unequivocal, "Stop fucking around with life, with the planet . . . You guys are total idiots." I asked how this was conveyed to him. Was it via the instrument in his neck? It was coming in "behind my right ear, and it hurts, but it doesn't hurt like a stabbing pain. It hurts like a dull, like somebody's pushing the edge of a baseball bat, or the tip of a baseball bat against the back of my ear—not my ear, my head." It was like a "really big blast of information."

I asked Arthur how this information affected him as a nine-year-old, if it got through. His response reflected his struggle to assimilate the impact. "It was there, I know, but it, I can, I know what they mean . . . Yeah. I guess it did. I can definitely, it went from like playful and nice to like, I don't know, like you have to be serious. You gotta be, this is important. This is serious. But the feeling is like, like a shock." I asked him to say more about the shock feeling. It "doesn't hurt," he said. "Just your entire brain, every cell lights up, or you see you're in this like infinite plasma of incredible light particles. It's not like a lightning bolt; it's like a, but it's a feeling that's communicated."

"What feeling?" I asked.

Like "a roller coaster where you're at the top and they drop you off and you're free-falling. You know, it's like whoa . . . this is definitely no kidding around."

"They want to make sure you get it," I suggested.

"Yeah," he said. "It's like they were grabbing your collar and just saying, slapping you around a little bit but not hurting you, just to say you gotta do this, and you have to talk to everybody else. You have to get it across to other people. You have to teach teachers."

This was the end of the "lesson." Other than the pressure of the instrument in the back of his neck he recalled no other intrusive procedures. He compared the dark people with the lighter ones. The dark ones had "more intelligence or information," and the "little people" were "simpler, or just more gentle." But the dark ones "respected or admired the little people" who were even "higher up in a way. They were revered more . . ."

I asked Arthur if he could describe the faces of the darker beings. "Like deep furrows, brows, eyes but dark eyes, deep-set eyes, kind of old," he said. They wore hoods and dark robes and were "squat" in appearance. The communications of the dark beings, he added, were "just too overwhelming," requiring "a long time for us to sort out" what they were conveying. So they needed the little people to "communicate little things at a time."

We were coming to the end of the session and I asked Arthur if he remembered anything about the return. He could not, and recalled only that his mother was still "hunched down" on the front seat as he came back to the car. "She was like frozen." His brother and sister were now there as well. Arthur does not remember his mother starting the car up, but "I remember us not talking. We didn't talk, which was, for my family, really weird. We're pretty gabby. None of us talked the whole way home." He could not remember at that time what had happened to him. They may have arrived home an hour or an hour and a half late.

Arthur remembered then that he had had an experience during the summer just past on the family estate. Without knowing quite why, he had arranged his entire schedule so he could spend two months there. "I had to go there for another blast." One day while he was fishing there was, in fact, "another blast," and he believes he may have lost as much as five hours. He slept in a "far room" of the manor house that night and his dreams, which he cannot recall, convinced him that he had had another powerful experience but we would have to postpone its exploration to another time. Looking back at what we had done in the session Arthur said the impact of his childhood experience left him with the feeling that he was "part of a puzzle" and had to do what he could "to stop . . ."

Arthur asked to be introduced to other people who were part of the puzzle "to see if we can create, if the thread" could create a "synergy." If "several of us get together it's like a hundred," he said. He agreed to come to the support group four days later. Arthur spoke further of the problem of dosage. The way the beings communicate, he said, "we can only absorb so much." When "we get to the threshold of fear" we "resist absorbing it all." I asked Arthur how he held what he had uncovered today. "It's completely confusing," he said. "Intuitively I believe it," he said, but "I don't intellectually." He had struggled during the regression with "the temptation to flip into fear . . . The fear is never gone. It's always there," he said, but (referring to my presence) "it's like having somebody hold your hand through an operation . . . You have to overcome" the fear and "not let it take you."

The "blob," he recalled, "was to make you feel" how "overwhelming it can, it's gonna be if we don't—just how important it is." The task now was "to get those blasts across to everybody else," he suggested to me.

Pam spoke with Arthur on the telephone on March 31, two days after he attended the support group. He was struck by how much anxiety there was in the group and how little of the sort of empathy with the aliens that he feels was expressed. He attributed this in part to the secure and happy childhood he had had. Arthur felt comfortable with the outcome of the regression and even told a few friends about it, but he did not tell his sister the details of it so she would not be "contaminated," i.e., have her credibility as an independent witness influenced by what he reported. He noted that her philosophy of life is similar to his and wondered if this were because she too has had abduction experiences. Since the regression he has felt more convinced than ever of the message he received from the alien beings and more determined than ever to do what he can to help.

Discussion

The for-profit business corporation is one of the most powerful institutions for getting things done that human beings have invented. The impact of corporations is felt throughout the planet, affecting virtually every aspect of its life. Corporations can be agents of continuing ecological destruction or a potential source of benefit for human beings and the life of the earth. Arthur is a business leader of extraordinary vision. His life has become a model of corporate responsibility (Everett, Mack, and Oresick 1993). On his own and in partnership with others he has committed his time, energy, and financial resources to preserving the earth's environment. He has done so through activities which range from local recycling to political efforts at a national and international level aimed at making sure that the earth's environment will be protected and sustained.

What seems remarkable is that Arthur and members of his family appear to believe that his social and environmental concerns and vision are intimately related to his childhood UFO abduction experience. He wonders "if I would have been a different person without it." Certainly the information that came across to him so powerfully during the abduction, some he had recalled consciously and uncovered more fully during the hypnosis session, was completely consistent with the unusual life he has led.

What was communicated to him by the alien beings was information about the danger facing the earth's ecology, the need for open, loving communication, and the necessity of ridding ourselves of fear in order to most effectively address these concerns. This message was received so powerfully that it seemed to reach every cell in Arthur's body. A dark "blob" that smothered the life of the earth was shown to him as an image of its potential fate. "Stop screwing around with life, with the planet," he was told. "Do what you can to stop the destruction and promote the life of the earth." In addition, his mission would be to be a teacher and "to teach teachers" to communicate the information that seems to have been literally "blasted" into his soul during his abduction experience at age nine and perhaps in others. Surely that is how he has lived his life.

Powerful metaphors run through Arthur's case. The thread that brought him in an arc from the car to the spacecraft is also a symbol of loving connection between living beings. Arthur makes an analogy repeatedly between the way he knows how to communicate with animals—his "zuzzing" rabbits, for example—and the open, loving way that it is necessary to relate to the alien beings. Indeed, he even likens the "little people" he has encountered since childhood to bunnies. The way Arthur has been able all his life to connect with the spirits of both wild and domestic animals seems in some way intimately linked with his fearlessness, as compared to the resistance of other members of his family who were present during the 1963 incident, in accepting the presence and message of the alien beings.

There are other interesting aspects to Arthur's case. His experience at age nine involved two distinct types of beings: light, somewhat translucent, "little people," who were playful, loving, and altogether friendly; and darker, more serious entities that seemed somewhat like the familiar grays. These two types seemed to work together in a kind of good cops/bad cops manner, the little people helping Arthur to overcome fear and to develop a joyous, open consciousness, while the darker beings delivered the powerful, disturbing message. This division seems to be found in other cases. Typically children are prepared through manifestly light, playful, and loving connection with friendly beings for later, more serious, or even threatening communications and projects that are carried out by sterner, more determined entities.

Arthur's case is also interesting from a mental health perspective. His childhood seems to have been uniquely secure and free of trauma or troubled relationships. There is a conspicuous absence of psychopathology in his history or contemporary mental state. Indeed, he seems remarkably stable and balanced, while at the same time highly creative and innovative. It would even appear that it was Arthur's sense of inner security, a kind of adventurous attitude toward life already in evidence in his fearless excitement when the UFO came down over the car ("this is really cool"), that made him especially suited for the assignment that he was given by the alien beings.

Arthur's story, like many abduction cases, raises puzzling questions about the relationship of thought to the physical world, cause and effect, and the vicissitudes of memory. There is something almost organic about the metaphors that run through his case. The dark blob he is shown by the beings and the thread or string that brings him to the UFO seem to exist in a kind of gray area between thought and the physical world. Like waves and particles in quantum mechanics, they seem to be thoughts in one context and something physically real in another. They are not simply one *or* the other, thought *versus* something physical, but rather are *both* depending upon the context. The blob seems to be something that really destroys the earth, more "real" than a metaphor, but it is also an image or symbol, staged by the beings in Arthur's mind, to bring about a particular effect. The thread brings Arthur physically to the ship, but it is also a powerful symbol of relationship and connection.

Even the association between the bunnies on Alice's farm and the perception of the light beings as "bunnies," or the relationship between how to "zuz" a bunny or open to an alien, seem to be organic linkages, closer or more intrinsic somehow than mere analogies. What this may bring us to is the relationship between thought and the physical world, two domains that have been kept radically separate in the Western worldview. But in Arthur's case, thought and physical reality seem inseparable, as if one could give rise to the other in a generative fashion that we do not understand. Perhaps consciousness itself represents a kind of creative source, a ground of being from which both thought and the physical world ultimately derive.

I also wonder about what is cause and what is effect in Arthur's abduction experience. Did he *have* the experience because his psyche was open to such realities, or did the abduction encounters themselves, going back perhaps to earliest childhood and possibly before (we have no past life memories in Arthur's case), bring about the flexible and visionary state of his mind? It may be that the very category of cause and effect is too linear and is not useful in such instances. More appropriate would be a way of thinking that looked upon Arthur and his experiences as interconnected in a way that could only be looked upon as a whole.

Finally, Arthur's case raises questions about forgetting and the triggering of memory similar to what we saw in Ed's experience (see chapter 3). Why didn't he or anyone present during the 1963 incident seem to remember it or speak of it afterwards? And what forces, twenty-five years later, led his sister to open up the subject and thus trigger Arthur's memory and the sequence of events that brought us together? To what extent do the forces of forgetting and remembering reside in Arthur's psyche; or are they imposed from outside by the aliens or whatever source determines *their* activity? And, like the separation of cause and effect just discussed, does this dichotomy even apply? These are but a few of the many questions and mysteries into which cases like Arthur's seem to lead us.

CHAPTER SIXTEEN

ALIEN INTERVENTION AND HUMAN EVOLUTION

Tibetans tell a story of an old frog who had lived all his life in a dank well. One day a frog from the sea paid him a visit.

"Where do you come from?" asked the frog in the well.

"From the great ocean," he replied.

"How big is your ocean?"

"It's gigantic."

"You mean about a quarter the size of my well here?"

"Bigger."

"Bigger? You mean half as big?"

"No, even bigger."

"Is it . . . as big as this well?"

"There's no comparison."

"That's impossible! I've got to see this for myself."

They set off together. When the frog from the well saw the ocean, it was such a shock that his head just exploded into pieces.

Sogyal Rinpoche, *The Tibetan Book of Living and Dying*

When we look back our lives seem to possess a coherence, and even a progression, of which we were hardly aware as events unfolded. When the UFO abduction phenomenon first came to my attention, my curiosity was aroused, and I had a sense that something unusual was happening within me. But I had little notion of the extent to which my explorations would open my consciousness to vast mysteries and uncertainties. And I did not foresee how fundamentally challenged the view of the world in which I had been raised would be.

Each of the thirteen people whose cases are described in this book—indeed, each of the seventy-six abductees with whom I have worked—tells a unique story. The individual differences can probably

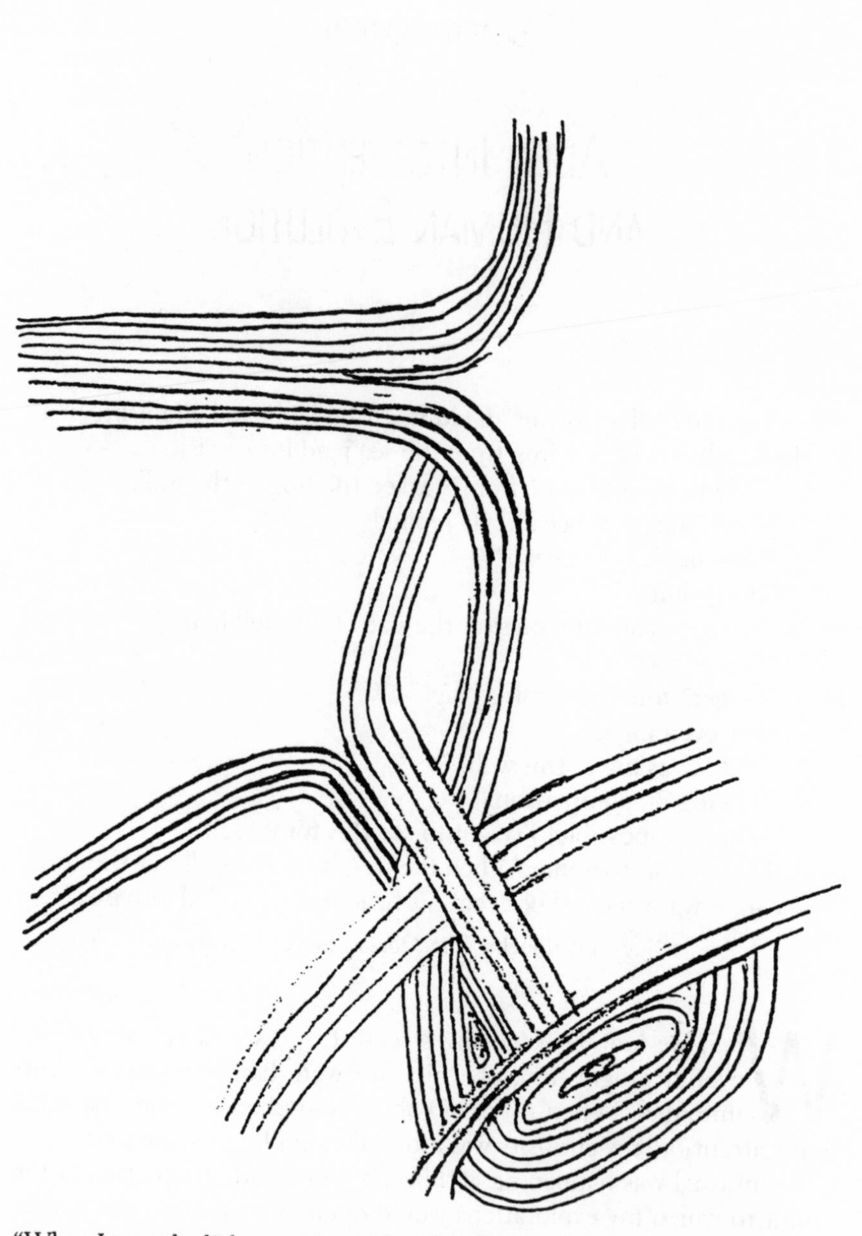

"When I turn the lights out at night, this has been the show for the past week. I can see it with my eyes closed as well as open."
—Anne (see page 405)

be accounted for on the basis of the diverse personalities of the experiencers and the varieties of circumstances within the abductions themselves. But what I have found to be so extraordinary from the beginning of my study has been the readily identifiable patterns that emerge when the case narratives are examined carefully.

It may be argued that it is my own mind that has created this coherence, and that I have shaped and interpreted the data in line with a structure that I already have in mind. In response to this criticism, I can only state that the abduction phenomenon was, in the beginning, as unbelievable to me as it is to any skeptic and that I have tried to be aware of any inclination to form new beliefs and convictions that might take the place of the previous ones that have been so radically called into question. One of the reasons I chose to include so many cases in such great detail is my desire to weave a tapestry of sufficient richness to give readers the opportunity to make their own judgments about the abduction phenomenon. I cannot say that the cases selected have been "typical," because I do not know what a typical case would be, or even that there is such a thing as a typical case. I do believe, however, that the cases I have discussed are illustrative of the range of phenomena that characterize abduction experiences.

There are aspects of UFO abductions that do not obey the physical laws of the universe as we have known them. Some of the phenomena *might* be understood at some future time through great advances in physics. But others, such as the capacity abductees like Paul discover to move their consciousness through space and time, require another ontological paradigm. I do not expect that the material presented in this book will have much impact on the minds of those who believe that the laws of physics as encompassed by the Newtonian/Einsteinian system are the full definition of reality. I hope, however, that the data contained here is of sufficient power and solidity to enable those who are open to expanding their view of possible realities to consider that the world might contain forces and intelligences of which we have hardly allowed ourselves to dream.

WAYS OF KNOWING: METHODOLOGY

In physics, psychology, and other fields the data we obtain is a function of the way we have gone about the task of gaining information. In my own discipline of depth psychology, the discoveries of Freud and his followers about the contents and structure of the human unconscious derived from the analysis of dreams, quite overlooked in the rationalist neuropsychology of his time, the application of hypnosis in medical psychology and the method of free association. Then as now, the development of new ways of knowing requires, however, something more than the application of different technologies or tools. Rather, an expanded epistemology, especially in psychology, may demand the legitimization (or *relegitimization*) of neglected aspects of ourselves as instruments of knowing.

Although psychoanalysis has been responsible for great advances in our knowledge of human experience and the depths and structure of the psyche, it has retained as a way of knowing a good deal of the dualistic, subject/object split that characterizes Western empirical science, including psychology. The patient or client is generally regarded as someone with a problem, separate from the therapist/investigator; the patient is to be helped or studied. Indeed, Freud turned away from hypnosis as a way of working with his patients in part because of the subjective bias it seemed to introduce into the therapeutic process (Mack 1993).

Alternative therapies and investigations of human consciousness appear to share an expanded use of the psyche as part of the exploratory process. The feelings and spirit of the facilitator, in alternative therapeutic situations, as well as his or her rational mind and observational skills, are a vital aspect of the therapeutic or investigative method. This expanded use of the self relies on empathy and is, in essence, intersubjective. Within this framework hypnosis, shamanic journeys, meditation, Grof breathwork, vision quests, and other modalities, which are called in the West "nonordinary" states of consciousness, become natural investigative allies. For they involve, by definition, the opening of the psyche to the deeper realms that lie behind the rational or observing mind. In the end, of course, the rational intellect is essential for understanding and integrating the data obtained through the fuller use of consciousness as an investigative tool. But along the way we—as investigators or readers—must open ourselves through this expanded way of knowing to whatever the patient or subject may report to be his or her experience.

It is perhaps understandable, therefore, that I would have, in effect, "stumbled" upon the UFO abduction phenomenon and that I was then more or less prepared to take on its investigation in the context of my training in the Grof holotropic breathwork method (see chapter 1), which relies upon the use of the whole self—mind, body, and soul—in the exploration of unconscious human experience. In working with abductees, the investigative method I have developed is basically a combination of the old and the new. In addition to standard interviews I use hypnosis, which was Freud's original "nonordinary" state for investigating the unconscious, modified by the use of the breath for centering and deepening the process. In my work with abductees I am fully involved, experiencing and reliving with them the world that they are calling forth from their unconscious. My whole psyche or being is engaged; naturally the rational or observing self is always present, shaping, limiting, and protecting the process.

The empirical methods of Western science rely primarily on the physical senses and rational intellect for gaining knowledge, and downplay feeling and intuition, and were developed in part to avoid the subjectivity, contamination, and sheer *messiness* of human emotion. Yet the cost of this restricted way of knowing may be that we now only learn about the physical world with only limited use of our faculties. In order to learn about the worlds "beyond the veil," as abductees put it, we may need a different kind of consciousness. This means that the process of gaining information about abductions is, to a large degree, "co-creative"—understanding comes to those who will accept it, and what I help bring forth from experiencers is something I am helping them to discover within themselves. But this co-creative aspect does not mean, as my critics sometimes have said, that I impose beliefs of my own about the phenomenon upon the experiencers, or even that I believe literally everything an abductee says.

As much as possible my questions in the sessions derive from what has just been said, or my intuition, based on experiences in therapy not only with abductees, but hundreds of patients, about where I feel the inner experience of the abductee is going. I avoid leading questions, and abductees, in my experience, are quite difficult to lead (they all seem to feel, as Sheila said, "I know what I saw"). Yet, at the same time, I cannot avoid the fact that a co-creative intuitive process such as this may yield information that is in some sense the product of the intermingling or flowing together of the consciousness of the two (or more) people in the room. Something may be brought forth that was not there before in exactly the same form. Stated differently, the information gained in the sessions is not simply a remembered "item," lifted out of the experiencer's consciousness like a stone from a kidney. It may represent instead a developed or evolved perception, enriched by the connection that the experiencer and the investigator have made.

From a Western perspective this might be called "distortion"; from a transpersonal point of view the experiencer and I may be participating in an evolution of consciousness. When we are dealing with a phenomenon like alien abductions, which manifests in the physical world but may derive from some other reality, the question of whether hypnosis (or any other nonordinary modality that can help us access realities outside of or beyond the physical world) discloses accurately what literally or factually "happened" may be inappropriate (see chapter 1 for a fuller discussion of this matter). A more useful question would be whether the investigative method can yield information that is *consistent* among experiencers, carries emotional *conviction*, and appears to *enlarge* our knowledge of phenomena that are significant for the lives of the experiencers and the larger culture.

BASIC FINDINGS: A SUMMARY OF WHAT THESE AND OTHER ABDUCTION CASES TELL US

The experiences of the people whose cases are reported in this book (and of many other abductees) may be divided into three categories: physical or physical-like events; the reception of information; and spiritual or transformational phenomena.

Physical or Physical-like Events

I will stress once again that we do not know the source from which the UFOs or the alien beings come (whether or not, for example, they originate in the physical universe as modern astrophysics has described it). But they manifest in the physical world and bring about definable consequences in that domain. The great majority of abductees have had vivid and powerfully meaningful sightings of UFOs at some time in their lives, although they may not see the outside of a spacecraft during a particular abduction experience. In virtually every case there are one or more concrete physical findings that accompany or follow the abduction experience, such as UFO sightings in the community, burned earth where UFOs have landed, and independent corroboration that the abductee's whereabouts are unknown at the time of the abduction event. Unexplained or missing pregnancies, a variety of minor physical lesions, odd nosebleeds, and the recovery of tiny objects from the bodies of experiencers are also widely seen.

Often these findings are subtle and difficult to prove by the methods of empirical science. They must, therefore, be seen as secondary evidence in support of what the abductees have reported. The sheer consistency and number of these accompanying physical findings make them too important to dismiss. The greatest mass of data, however, comes not from the physical findings but from the reports of the experiencers themselves. Although varied in some respects, these are so densely consistent as to defy conventional psychiatric explanations.

Abductees may be found to have had virtually lifelong encounters with alien beings and UFOs, although in certain cases, like Ed and Arthur, a single abduction event has been of particular importance. The abduction experiences begin with a shift in consciousness on the part of the abductee, which may be signaled by a hum or other odd sound, by the appearance of a light for which no usual source can be found, by the sense of a presence or even the sight of one or more alien beings (as described in chapter 2) or by a strong vibratory sensation in the body (as in Sheila's "electrical dreams"). This change in consciousness may be subtle, but abductees are always sure that they are not dreaming or imagining. Rather they experience that they have moved into another reality, but one that is, nevertheless, altogether real. This is a waking reality, but a *different* one. As one abductee described this shift to me, it is as if the alien beings break through a kind of screen, revealing a new reality to the experiencer.

After this the experiencer is taken by some force, often a beam of light or some other energy at the disposal of the alien beings, out of or away from the house or other place he or she has been, through walls, doors, or closed windows if necessary. Experiencers may see their home and the earth itself recede before them as they are transported into a spacecraft, which is commonly described as metallic and saucer or cigar-shaped and which is revealed to be the source of the light that they saw initially. Once inside the craft the abductees see varying numbers of alien beings as described in chapter two, who are engaged in a rather businesslike way in preparing to administer various procedures. The inside of the craft is generally rather cold, emotionally and physically, sometimes with a musty smell, with computer-like consoles along the walls. The walls tend to be white and curved, although black floors are sometimes described.

The procedures have a medical or surgical-like quality, but the instruments used do not resemble those with which we are familiar. Gazing, staring and mind scanning, and other telepathic communications of various sorts on the part of the aliens often occur initially. Abductees tend to feel that the content of their minds are thoroughly revealed to the aliens. After this there are various procedures administered under the control of a slightly taller and older-appearing alien, spoken of by abductees as the doctor or leader. Although this figure may be resented by the experiencers, he (the leader is usually, but not always, perceived as male) is often familiar to them from abductions going back to childhood and they may feel a strong bond, even intense love toward him, which they sometimes feel is reciprocated.

The procedures include taking of small tissue samples, probing of the head (which is usually felt to be related to taking information from the brain, monitoring the state of being of the experiencer, and inserting or removing of implants), and the insertion or application of odd instruments into or onto other parts of the body, including especially the abdomen, anus, and reproductive organs. Sperm samples are forcibly taken from men, and women experience the removal of ova; fertilized eggs, which may have been genetically altered, are implanted, and later there is the eventual removal of the pregnancy. In subsequent abductions, experiencers are shown hybrid offspring and may even be asked to hold or nurture them. As children, they may have been asked to play with these odd creatures, who may appear listless by our standards. All of this is quite terrifying to the experiencers, although the level of terror is modified by some form of anesthetizing energy that the aliens administer with their hands or rod-like instruments. Other sexual and reproductive activities may be brought about by the aliens, who seem, as will be discussed further later, especially interested in the sexual and emotional aspects of our lives.

When the experiences are recalled consciously, or during hypnotic regressions, the abductees go through an emotional *reliving* of great intensity and power. Otherwise quite controlled individuals may writhe, perspire, and scream with fear and rage, or cry with appropriate sadness, as they remember their abduction experiences. This emotional expression appears altogether authentic to those who are unfamiliar with the abduction phenomenon and witness it for the first time. For myself, being with abductees who are going through these experiences requires every bit of holding energy and caring presence that I can muster. There has not, as yet, come to my attention in any case an alternative explanation for the basic elements of the abduction experiences that my clients are reporting in such overwhelming and vivid detail.

To make our understanding even more difficult, it appears that the penetration of the abduction phenomenon into the physical world is not an all-or-nothing matter. In some cases a person is known to have been missing, can recall, with or without hypnosis, an abduction experience, and has returned with bodily lesions for which there seems to be no other explanation. But in other situations, "complete" abduction does not appear to occur. The individual may have an out-of-body experience while others see that he or she has not left the house. Abductees will also report a great deal of "activity" that may or may not presage an abduction but which can occur often, even on a nightly basis. This includes a sense of the alien presence, disturbing vibratory and other bodily sensations, odd noises in the house, and many kinds of light phenomena that appear in their minds or in the surrounding environment. All of these elements are clearly, from the experiencers' point of view, related to the abduction phenomenon.

The Reception of Information

The transmission of information from the alien beings to the experiencers appears to be a fundamental aspect of the abduction phenomenon. Indeed, for some abductees, including Arthur and, to a certain degree, Paul and Eva, there is relatively little trauma; the experiences seem to be primarily informational or transformational in character. Information during abductions appears to be transmitted in two forms—by direct, mind-to-mind conveyance, or through depiction of phenomena or events on television-like screens. As in Catherine's case, an entire conference room setting was displayed for her benefit in order, apparently, to create the right conditions for communicating with her about the state of the earth's living systems.

The information that abductees receive is concerned primarily with the fate of the earth in the wake of human destructiveness. Scenes are shown of the planet wasted by nuclear war and especially of the earth's environment devastated by pollution and toxic clouds. Sara and Arthur, for example, were shown great black clouds or "blobs" suffocating the earth's living systems, the effect presumably of environmental catastrophe. A number of abductees have been shown apocalyptic images of the earth itself literally cracked open or broken up, followed by elaborate triage scenes in which some people will die, others will survive in some way on Earth, and still others will be transported to some other place where human life will continue in a new way.

Although these prophetic visions may be viewed metaphorically by outsiders, for the abductees themselves they are experienced with conviction as literal and concrete. The visions cause great sadness. In Ed's case even the spirit beings themselves have been afflicted by human folly and destructiveness. Scott is typical of experiencers who have received information from the aliens about past destruction of their planet or environment. Now they are trying to prevent a recurrence.

Although the alien beings seem to be intervening to alter our consciousness in such a way that our aggression would be reduced, they seem genuinely puzzled regarding the degree of our apparently mindless or gratuitous destructiveness. As they communicated to Paul, they do not understand why we would set about to destroy a realm of such transcendent beauty and they seem in some awkward way determined to stop us from doing so. Environment, as Sara has said, means much more than nature or our physical ecology. It refers to the entire context of life itself. It is, in her words, "the creative, life-affirming" place of unconditional love. "You *are* your environment," she said. The aliens seem to be concerned with our "environment" in this total sense.

Transformational and Spiritual Phenomena

The aliens themselves seem able to change or disguise their form, and, as noted, may appear initially to the abductees as various kinds of animals, or even as ordinary human beings, as in Peter's case. But their shape-shifting abilities extend to their vehicles and to the environments they present to the abductees, which include, in this sample, a string of motorcycles (Dave), a forest and conference room (Catherine), images of Jesus in white robes (Jerry), and a soaring cathedral-like structure with stained glass windows (Sheila). One young woman, not written about in this book, recalled at age seven seeing a fifteen-foot kangaroo in a park, which turned out to be a small spacecraft. I heard recently of a case where a number of children were transported into the sky in a small craft that appeared to them initially as a booth at a carnival in which aliens disguised as humans asked if they wanted to go on a journey.

Other experiences relate to the expansion of consciousness and its separation from the body in present time, as in Paul's ability to "travel" to the time of the crash of a space vehicle before he was born, or even to the period when dinosaurs were on Earth. The frequency with which past life experiences are recalled during the hypnotic regressions relates also to the idea of expanded identity, i.e., in some sense the human spirit or soul is not limited to this lifetime but may have extended over hundreds and even, as in Catherine's case, thousands of years. Past life recall becomes particularly powerful when it makes it possible to see, as in Joe's and Dave's cases, a continuity of personal growth over more than one lifetime.

Many of the abductees written about in this book have experienced a dual identity as both a human and alien being. As an alien they see the world from the alien perspective, and, as in Joe's case, carry out the alien side of the reproductive enterprise. Peter, on the other hand, was troubled to discover memories of his mating as a human being with an alien female. The alien self is sometimes perceived as the lost or abandoned soul of the human self, linked once to a common source. The felt task then becomes one of integrating the split-off alien self or soul and becoming whole once again.

Many abduction experiences are unequivocally spiritual, which usually involves some sort of powerful encounter with, or immersion in, divine light. This phenomenon is pervasive in Carlos's case and is present in many I have studied. The alien beings, although resented for their intrusive activities, may also be seen as intermediaries, closer than we are to God or the source of being. Sometimes, as in Carlos's case, they may even be seen as angels or analogous to God. A number of abductees with whom I have worked experience at certain points an opening up to the source of being in the cosmos, which they often call Home, and from which they feel they have been brutally cut off in the course of becoming embodied as a human being. They may weep ecstatically when during our sessions they experience an opening or return to Home. They may, as in Sara's case, rather resent having to remain on Earth in embodied form, even as they realize that on Earth they have some sort of mission to assist in bringing about a change in human consciousness.

Related apparently to this opening to the divine source is the experience that some abductees have of great cycles of birth, life, and death, repeating over long stretches of time. This may become particularly apparent when past life experiences are relived and the abductee is permitted to recall the actual experiences of death and rebirth. A related phenomenon, which I will describe in more detail later, might be called the reification of an archetype or metaphor. Tubes, passageways, threads, et cetera may be literally seen, or passed through or along physically, but at the same time they symbolically represent important transitions from one state of being to another.

THE IMPACT OF ABDUCTION EXPERIENCES

Most of the individuals discussed in this book have suffered the multidimensional trauma that is associated with abductions—the helplessness as terrifying intrusive procedures are inflicted on their bodies, the isolation from family and friends, the inevitable shock to their worldviews, and, especially, the realization that the experiences may recur at any time to them or their children. "We're not relieved of our continuing unrelenting other world melodrama," wrote Jerry, whose three children appear to have been visited.

Yet the abduction phenomenon is not simply traumatic. Experiencers may be left with fears, nightmares, and other sequelae of severe stress together with small bodily lesions, sinus headaches, gastrointestinal symptoms, mild neuropathies, and psychosexual dysfunctions that appear to be related to their abduction encounters. At the same time, however, in my caseload there is evidence that the alien encounters have been responsible for healing conditions ranging from pneumonias and leukemia to limbs paralyzed due to muscle atrophy from poliomyelitis. Furthermore, many abductees seem to gain powers themselves as healers. Although abductees may continue to resent the abduction experiences and fear their recurrence, at the same time many in one way or another come to feel that they are participating in a life-creating or life-changing process that has deep importance and value.

In addition, many abductees, including the cases discussed in this book, appear to undergo profound personal growth and transformation. Each appears to come out of his or her experiences concerned about the fate of the earth and the continuation of human and other life-forms. Virtually *all* the abductees with whom I have worked closely have demonstrated a commitment to changing their relationship to the earth, of living more gently on it or in greater harmony with the other creatures that live here. Each seems to be devoted to transforming his or her relationships with other people, to expressing love more openly, and transcending aggressive impulses. Some abductees, like Eva, Peter, Carlos, and Arthur, wish to use their evolving perspective to influence others and have become teachers of a new way of living. In addition, abductees seem, especially once they confront and integrate their experiences, to be especially intuitive; they sometimes demonstrate strong psychic abilities, including clairvoyance or the ability to perceive at a distance. Further research is needed to document these capabilities.

Virtually every abduction researcher must at some point confront the question of whether these expressions of personal growth are a consequence of trauma—a stressful confrontation with unknown forces—or are an intrinsic aspect of the phenomenon itself. After all, it may be argued, victims of war, rape, abuse, and other traumata may grow emotionally and spiritually simply because their experiences lead them to discover new and deeper resources within themselves.

Comparisons to the "Stockholm syndrome," in which victims of abusive behavior come to identify with the purposes of the abuser, do not take into account the subtleties of these encounters. Dr. Judith Herman, a specialist in the effects of chronic trauma on battered women, political prisoners, and victims of torture (Herman 1992), cites an inability to recall details of the events as one of the reactions to coercive control. Herman notes that when a victim is threatened with violence or pain, is denied control over bodily functions, experiences unpredictability in the enforcement of rules, and receives intermittent rewards, the victim's sense of personal power is destroyed and the individual comes to feel helpless. Victim's of such trauma identify strongly with their captors, "voluntarily" choosing to remain with them in the absence of physical restraint.

It is true that abductees may experience terror and some pain with the examinations performed on them. The aliens' actions are also in many ways unpredictable, sometimes frightening and sometimes rewarding. However, in contrast to the narrow and self-serving purposes of human abusers and political kidnappers, the beings reveal a shared purpose, and offer the possibility of openings to an inclusive, more expansive worldview that is powerfully internalized by many abductees.

My own impression, gained from what abductees have told me, is that consciousness expansion and personal transformation is a basic aspect of the abduction phenomenon. I have come to this conclusion from noting in case after case the extent to which the information communicated by alien beings to experiencers is fundamentally *about* the need for a change in human consciousness and our relationship to the earth and one another. Even the helplessness and loss or surrender of control which are, at least initially, forced upon the abductees by the aliens—one of the most traumatic aspects of the experiences—seem to be in some way "designed" to bring about a kind of ego death from which spiritual growth and the expansion of consciousness may follow.

But my focus upon growth and transformation might reflect a bias of mine. The people who choose to come to see me may know of my interest in such aspects of human psychology, and may be aware that I consider my work with abductees to be a co-creative process. In some cases—Arthur, for example—the commitment to environmental sustainability and human transformation antedated contact with me.

A word must be said about the strain that the abductees' experiences places upon spouses and other intimates. Such experiences may dominate their lives, and as Peter and Jerry did, abductees may require a great deal of support from wives, husbands, and others, who may feel at the same time that their own needs are not being met when the experiencer's are so preoccupied with what has happened to them.

Furthermore, if those close to an experiencer are not themselves experiencers, they may find the whole situation difficult to believe, as their sense of reality is being assaulted. They may wish to deny the whole matter despite strong evidence that something important is happening to their loved ones.

Or parents, who may themselves be abductees, can become alienated from their children when they cannot bear what is happening to their offspring; in some instances it is even more painful for a parent—I have seen several cases where the child experiences an alien being as their "true" parent. In some instances, as in Eva's case, strain in a marital relationship may arise when the abductee feels that he or she has in some sense outgrown the spouse, who may be unable or unwilling to follow the abductee into the expanded realms he or she describes. More work is needed in supporting family members of abductees who need to understand and participate in their loved one's new experiences.

SOME IMPLICATIONS OF THE ABDUCTION PHENOMENON

As the power of her abduction experiences sank in Sheila said, "It simply does not make sense in the world as I know it to be." Faced with this fact Sheila, like each of us, has a choice. We can continue to try to make the phenomenon fit the world as we have known it, jamming it into a kind of procrustean bed of consensus reality. Or we can acknowledge that the world might be other than we have known it. Then we are free to see where our thinking leads us. I cannot discourage those who try to discover conventional explanations for the abduction phenomenon. I would only point out that as a clinician, I have spent countless hours trying to find alternate explanations that would not require the major shift in my worldview that I have had to face. But as I discussed in chapter two, and as I have tried to make clear in the case narratives, no familiar theory or explanation has come even close to accounting for the basic features of the abduction phenomenon. In short, it is what it is, though the ultimate source of these experiences remains a mystery.

Clinical Implications

Several of the people described in this book were seen by psychologists, psychiatrists, and other physicians who were not familiar with the UFO abduction phenomenon. Scott was subjected through much of his childhood and adolescence to medical tests for epilepsy, was given anti-epileptic medication over a period of years, and was even hospitalized to evaluate and treat his strange abduction-related symptoms. Sheila was treated in unavailing psychotherapy for many years and evaluated extensively by mental health professionals who remained doubtful about the reality of alien abductions and UFOs. This resulted in considerable, and in my opinion unnecessary, stress for her. Paul spent years in psychotherapy with a psychologist who could not accept the reality of his abduction experiences. Jerry and her first husband went to several counselors who offered useless conventional interpretations of her abduction-related conflicts about intimacy and sexuality. There are specific challenges for mental health professionals who treat abductees.

There can, of course, be genuine value in trying to establish whether some other condition might account for, or exist in addition to, abduction-related symptoms. I have been working recently with a young woman who is struggling to come to terms with both alien abductions and experiences of sexual abuse. Careful interviewing, however, has shown that one set of experiences does not explain the other, and that this woman is quite capable, when given the opportunity, of distinguishing the different effects of each. But evaluations of possible abduction cases should be pursued by physicians and other clinicians who are at least familiar with and open to the reality of the phenomenon, even when they do not "believe" in it. There is sufficient information available now in books, popular periodicals, and other media, if not in professional journals, to suggest strongly that something that defies conventional explanations is happening to many people. As Sheila said, there is no excuse "for an ignorant stance."

The abduction phenomenon also raises interesting questions about the nature of memory and the control of consciousness. As discussed in chapter one, prevalence or incidence polls of UFO abductions are rendered almost meaningless by the fact that long forgotten abduction memories may be triggered by an event—Ed's walk along the coast in Maine, or Arthur's conversation with his sister at a family reunion—the timing of which is itself unpredictable. What are the forces that keep the memory out of consciousness during the years—more than twenty-five in these two cases—during which there is seemingly no recall of the events in question? The abductees themselves feel that something more than simple repression is at work—that some repressing force is imposed by the alien beings themselves. Ed, as I reported, remembers being told that he will remember "when you need to know." Sometimes this not remembering appears to protect the abductees from a distress that they could not handle, especially in the case of children. But we have little understanding of how this repressing force works, or, for that matter, why an altered state of consciousness, facilitated in a caring, protective setting, is so effective in recovering abduction memories.

These odd vicissitudes of memory appear to be part of a larger phenomenon. The aliens, or some other agent, appear able to control the minds and perceptions of abductees. It is common during abductions that spouses and others who are in the room when the abductee is taken seem to be "switched off," i.e., rendered in some way unconscious during the abduction. This may be quite aggravating to the abductee calling for assistance in his or her helpless distress. This alteration of consciousness may be related to the differential perception of UFOs themselves on the part of witnesses. There are reported occasions in which some of the people in a particular setting may not see a UFO that is quite clear to others (Crawford 1993). Research into these matters may, indeed, lead us through new doors of perception.

The alien abduction phenomenon appears to have something to teach us about the redemptive and transformational role of emotion in human life. The terror, rage, grief, and, on a few occasions, joy expressed during my sessions with experiencers, are among the most powerful I have ever witnessed. For me and others who have attended the sessions, as well as for the abductees themselves, it is this intensity of recovered emotion that lends inescapable authenticity to the phenomenon. *Something,* everyone who goes through these sessions agrees, has *happened* to these people, whether or not it is possible to identify the source of what has occurred.

Furthermore, this intense emotion, especially as it is felt and discharged through bodily movements and powerful vocalization, appears to have transformative power, especially when the facilitator can be fully with the experiencer through the course of its most powerful reevocation. The bodily responses seem quite literally to drive the experience into new realms of psychical awareness. When this occurs an expansion of consciousness or broadening of knowledge becomes possible. Peter, for example, found that it was the very intensity of his bodily experience when reliving his abductions that permitted—perhaps forced—him to acknowledge the reality of the alien beings themselves, which was a basic step in his spiritual journey. It is possible that experiencing terror, or "pushing through" it, is an intrinsic or necessary aspect of breaking the psychological boundaries that limit our perception of reality. When Catherine found that she was able to experience within herself—and express fully to the alien beings—her terror and rage, a more reciprocal, meaningful, and creative relationship with them became possible. Indeed, the alien abduction phenomenon seems to open abductees and those who work with them to deeper realms of human emotion, whether or not that is a specific "purpose" of the phenomenon.

The abduction phenomenon also seems to offer new perspectives on human destructiveness. The aliens, as illustrated, for example, in Peter's and Paul's cases, seem genuinely puzzled about the extent of our aggressiveness toward one another and especially our apparent willingness to destroy the planet's life. As Paul said when speaking from an alien point of view, "We don't understand why you choose destruction." Indeed, the extent of this destructiveness, as reflected to us from the alien vantage point, demonstrates the inadequacy of our biological and psychological theories of aggression and the need for a fresh look at this aspect of our individual and collective selves. "An organism that gets to be at such a degree of destruction should flip back and learn upon itself," Paul said. As one who has long studied the psychological consequences of the nuclear threat, I can only agree.

Implications for Physics, Technology, and Biology

The first reports of UFOs raised questions for contemporary science, which has dealt with the issue largely by ignoring or denying the whole matter. How did the spacecraft get here? What are its propulsion systems? How did they project light and heat of extraordinary intensity over huge distances, or accelerate and change direction in apparent defiance of the laws of gravity? These are questions difficult to address within the parameters of modern physics. The abduction phenomenon has only added new ingredients to old technological puzzles. How, for example, do the aliens pass people through walls? Carlos describes the sensation of having his cells vibrate and dissolve as he is transported by a light beam, leaving behind a kind of ghostly shape. But exactly what this transformational process consists of we have virtually no knowledge.

What is the mechanism whereby cuts and other lesions are apparently healed so promptly? One man told me of a gash several inches deep that appeared on his leg following an abduction. Yet this cut virtually disappeared in twenty-four hours. What is the process whereby abductees are tracked, so they can be found whenever and wherever the beings wish? Are implants involved in this? It has been suggested that the aliens are many thousands of years ahead of human beings in their mastery of various technologies. Perhaps so. In any event we cannot begin to answer any of these questions within the framework of modern science. As is frequently remarked upon in this field, multidisciplinary studies combining physics with comparative religion and spirituality, are needed to further consider how the interdimensional bridging properties of the abduction phenomenon might work.

Philosophical Implications

When I say that the UFO abduction phenomenon seems to require a shift in our worldview, this may be thought of in two senses. On the one hand the aliens and the UFOs themselves seem, as discussed above, to act in ways that defy accepted laws of physics and principles of biology, requiring that we at least extend our knowledge of the material world before we can understand how UFOs function. But these problems are more or less in the domain of science as it has evolved over the centuries. After all, the capabilities of contemporary space shuttles and electronic communications systems would appear altogether magical if confronted by someone from the Middle Ages. Yet their creation are real accomplishments of science and technology in the Western tradition. But there are more fundamental questions raised by the abduction phenomenon which seem to lie outside the ontological framework of modern science and appear to be unapproachable by its methods. Foremost among these is the problem of defining in what reality the abductions occur.

Quite a few abductees have spoken to me of their sense that at least some of their experiences are not occurring within the physical space/time dimensions of the universe as we comprehend it. They speak of aliens breaking through from another dimension, through a "slit" or "crack" in some sort of barrier, entering our world from "beyond the veil." Abductees, some of whom have little education to prepare them to explain about such abstractions or odd dislocations, will speak of the collapse of space/time that occurs during their experiences. They experience the aliens, indeed their abductions themselves, as happening in another reality, although one that is as powerfully actual to them as—or more so than—the familiar physical world.

Sara, who is one of the most educated individuals in my sample, spoke perhaps for many experiencers when she described one of her abductions as "dimensional merging . . . You can't really evaluate it in the language and physical descriptive terms of this dimension," she said, "because it really wasn't happening here. It was half happening here and half happening somewheres else." Catherine, in a regression in January 1993, some weeks after the last one I reported on, spoke of a "place" she remembered between times of embodiment on Earth. In that "place" bodies were not solid, appearing only in a kind of energy outline. "This was from a long, long, long, long time ago," she explained. "This was before any of us had lives here. This place is in a totally different universe. It's not in our Earth space/time dimension."

A young woman abductee I will call Anne tried to explain to me during a regression her sense of converging time frames that occurs in her abductions. Anne felt as if she were functioning simultaneously in different times. "All times can come to one place," she said. "This is real. It's not philosophical," she insisted. "I can really go to another time frame and [my experiences] can pull me from other time frames to here."

Our use of familiar words like "happening," "occurred," and "real" will themselves have to be thought of differently, less literally perhaps. When an event seems to have occurred in another dimension, as described, for example, by Catherine and Anne, and we know about it principally through powerfully felt experience—when consciousness is the principle available epistemological tool—how do we decide what is "real" or "true"?

The word "dream," or the idea of dreams, provides a good example of how a familiar term has to be looked at more carefully, even redefined. When abductees call their experiences "dreams," which they often do, close questioning can elicit that this may be a euphemism to cover what they are sure cannot be that, namely an event from which there was no awakening that occurred in another dimension. On several occasions I have seen a look of distress, even tears, on the face of an abductee who is realizing that an experience that he or she had chosen, more comfortably, to consider a dream had occurred in some sort of fully "awake" (another word that might need to be redefined) or conscious state, however different this might be. The problem is complicated further by the fact that dreams are an important way that we normally process and integrate experience during the night. Therefore, it is not surprising that, since abductions are themselves powerful and disturbing experiences, they may frequently give rise to true nightmares or dreams that re-create in modified form the abduction experience, even during the same night that the abduction occurred.

Again, it will require careful questioning to establish whether what is being reported is (1) an abduction, distortedly reported as a dream;

(2) a dream that relives an abduction experience; or (3) simply a dream that contains UFO-related material but does not necessarily reflect an abduction experience. The second scenario may be distinguished from the first by the fact that a moment of waking, of *realizing* that a dream has occurred may be identified; or there may be elements of experience that do not occur within the familiar abduction scenario. Finally, we know that in the abduction the event proceeds inexorably to the end, and that the person is unable to end it or force him- or herself to "wake up." Ordinary dreams that contain UFO elements may be separated from dreams that recapitulate abductions in that they tend to be less vivid or intense, and the dreamer does not have the sense, as in the abduction-related dreams, that an actual experience is concealed behind the dream representations.

Allow for the time being that there is little knowledge about the domain from which the alien beings derive—perhaps not even language or concepts to describe it. Yet acknowledge too that *something* is going on that cannot be dismissed out of hand. Then we are living in a vastly different universe from that which I, for one, was taught about at home and in school. This universe, the one that Western science has analyzed and categorized so successfully, consists of matter and energy, arranged exquisitely perhaps, but, devoid, as far as we can tell, of intelligence or intelligences that can be discovered by its methods.

In this worldview the various spirit entities, God or gods and other mythic beings that peoples throughout the world, including in our own culture, experience as altogether real have no objective reality. They are the subject matter of psychology and psychopathology, anthropology, religious study, and science fiction, the projection outward of the perceptions and images of the brain. If we make these entities real, it is through metaphor and symbol, as poets do so well. To acknowledge that the universe (or universes) contains other beings that have been able to enter our world and affect us as powerfully as the alien entities seem able to do would require an expansion of our notions of reality that all too radically undermine the Western scientific and philosophical ideology which Tulane philosopher Michael Zimmerman calls "naturalistic humanism" (Zimmerman 1993).

The alien beings that appear to come to us from the sky in strange spacecraft present a particularly confusing challenge to such a naturalistic or objectivist ideology. For they seem to partake of properties belonging to both the spirit *and* the material worlds, bridging, as if effortlessly, the division between these realms which has become increasingly sacred and unbreachable since science and religion went their separate ways in the seventeenth century (Toulmin 1990). On the one hand these beings seem able to be seen by the abductees, who feel their bodies moved and find small lesions inflicted upon them. On the other hand the beings seem to come, like intermediaries from God or the Devil, from a nonembodied source, and they are able to open the consciousness of abductees to realms of being that do not exist in the physical world as we know it. Before concluding I will speculate about why the barrier between the spirit and material worlds has become so entrenched in the West. For it is this false dichotomy that makes our confrontation with beings who do not respect this gulf so shocking.

Spiritual Implications

I am often asked how experiences that are so traumatic, and even appear cruel at times, can also be spiritually transformative. To me there is no inconsistency here, unless one reserves spirituality for realms of the sublime that are free of pain and struggle. Sometimes our most useful spiritual learning and growth comes at the hands of rough teachers who have little respect for our conceits, psychological defenses, or established points of view. Zen Buddhist teaching is notorious for its shock treatment methods. One might even go further and argue that genuine spiritual growth is inevitably disturbing, as the boundaries of consciousness are breached and we are opened to new domains of existence.

The alien beings that abductees speak about seem to many of them to come from another domain that is felt to be closer to the source of being or primary creation. They have been described, however homely their appearance, as intermediaries or emissaries from God, even as angels, by Carlos for one. The acknowledgment of their existence, after the initial ontological shock, is sometimes (see Peter's and Catherine's cases, for example) the first step in the opening of consciousness to a universe that is no longer simply material. Abductees come to appreciate that the universe is filled with intelligences and is itself intelligent. They develop a sense of awe before a mysterious cosmos that becomes sacred and ensouled. The sense of separation from all the rest of creation breaks down and the experience of oneness becomes an essential aspect of the evolution of the abductees' consciousness, as Joe spoke of in relation to his own and his baby son's development. Joe felt the choice he was facing was, as he put it, between oneness and insanity. The aliens themselves may come to be seen as a split-off part of the abductee's soul or Self; his or her own reensoulment requires the integration of this separated dimension.

Many abductees with whom I have worked, including several not written about in this book, experience a kind of ecstasy which, as in Carlos's case, can reach orgasmic proportions as they feel themselves open through their experiences (or the reliving of them) to a divine source or creative center of being in the cosmos. This source is, to the abductees, inexpressibly luminous and filled with color, and they may weep when they find themselves in its presence, for separation from it was painful beyond words. When they do open to the source once again, they may call the experience a "return" and protest once again having to fulfill even a newly agreed upon purpose as a human being.

As their experiences are brought into full consciousness, abductees seem to feel increasingly a sense of oneness with all beings and all of creation. This is often expressed through a special love of nature and a deep connection with animals and animal spirits. Sometimes there is a strong identification with one type of animal. Deer are, for example, "totem" creatures for Dave; Carlos feels a special kinship with lions; and Arthur has a remarkable capacity to connect with a variety of animals. The aliens themselves, as we have seen, may appear at various times to the abductees in animal form. The connection that they have with animal spirits, a kind of shamanic dimension, remains to be explored.

Finally, many if not most of the abductees with whom I have worked intensively come to feel that their enhanced spiritual awareness must be translated into some sort of teaching or higher purpose. Even as they are saddened, and even become hopeless about the ecology of the planet and the fate of the earth's life-forms, they feel that their experiences are, ultimately, about preserving life and that they must do something toward this end. Ed, Joe, Jerry, Eva, Peter, Sara, Arthur, and others each feel that they have a particular mission or responsibility to teach a different sort of consciousness concerning the human place on Earth. Some, like Peter, Ed, and Eva, have even changed their jobs or are seeking to shift the direction of their work in order better to fulfill their life's new purpose. Peter envisions a future "Golden Age" of learning and opportunity, which he hopes to help bring about.

Implications for Human Identity

Each abductee experiences in some sense an expansion of his or her sense of self, of identity in the world. Paul wondered how we had come to mistake the "shell" of our being for the "whole" of it; and Eva, in her recall of the Soul Bird story of her childhood, recognized there were compartments of self which she was opening up and merging in order to move from fragmentation toward wholeness. The change that abductees experience is fundamentally that they may no longer feel themselves to be separate from other beings. They shed their identification with a narrow social role and gain a sense of oneness with all creation, a kind of universal connectedness.

This opening to a wider identity appears to be a direct result of the abduction experiences, if not a central aspect of the whole process. The change appears to be the result of two related elements. The abduction experiences themselves shatter the illusion of our control, and demonstrate forcefully that we are helpless in the face of forces and beings whose purpose we do not understand. Each abductee discovers that he or she is but one intelligent being in a universe populated with various other entities that are not "supposed to" exist. Human beings are not lords of the earth, they realize, but children of the cosmos who must find their way to live in harmony with all manner of creatures on the earth and elsewhere. This is a terrifying lesson in humility that opens the psyche to a wider perception of the universe, including the beings and entities that inhabit it.

At the same time, as I noted, abductees become open to the presence of a divine source, which fills their being and gives a sense of connection with a universal consciousness from which they have come and to which they will return. Past life experiences, which extend the sense of self over time in both embodied and unembodied form, create a further expansion of the feeling of what it is to be a human being. Finally, the peculiar sense that many abductees gain during the regressions that they have a dual human/alien identity reinforces all of the above processes. For the alien self is felt to be a kind of missing part, a soul link to the universal source or consciousness, the *anima mundi* from which they have been cut off.

Political, Economic, and Religious Implications

The Western scientific/materialist worldview has been hugely successful in its explorations of the physical world, revealing many of its secrets and using this knowledge to serve human purposes. We have overcome the harshness of winter, reduced suffering through advances in medicine, and learned to communicate electronically with those who are far away. At the same time we have applied our knowledge to creating weapons of destruction that can now easily destroy life as we know it. Our use of modern technology to tear resources from the earth is bringing the biosphere to the brink of collapse. We are a species out of harmony with nature, gone berserk in the indulgence of its desires at the expense of other living beings and the earth that has given us life.

The task of reversing this trend is momentous. Even as we recognize the peril we have created, the vested interests that stand in the way of discovering a balance in our relationship with nature are formidable. Huge corporate, scientific, educational, and military institutions consume many billions of dollars of material goods and maintain, as if mindlessly, a paralyzing stasis that is difficult to reverse. For international business the world seems at times to be nothing more than a giant market to be divided up among the cleverest entrepreneurs.

But there are psychospiritual vested interests that resist change and that are perhaps even more powerful than these material ones. These interests are reflected in the attachment to the notion that the physical laws we know describe all that is, and that if other beings reside in the cosmos they will behave more or less like us. The U.S. government-funded SETI (Search for Extraterrestrial Intelligence) program, which operates on the assumption that extraterrestrial intelligence could be found by sending radio waves out into the universe, illustrates this bias. The possibility that advanced intelligences might not choose to communicate with us through such a tiny or limited technological aperture, seeking perhaps some fuller opening of our consciousness, seems not to have occurred to its inventors. As philosopher Terence McKenna has suggested, "To search expectantly for a radio signal from an extraterrestrial source is probably as culture-bound a presumption as to search the galaxy for a good Italian restaurant" (McKenna 1991).

It is not altogether clear to me why we become so attached to our ways of seeing the world. Perhaps a comprehensive scientific paradigm, like any ideology, gives a sense of mastery and power. Mystery and the sense of not knowing are antithetical to the need to maintain control and seem, at times, to inspire such terror that we fear that we might blow apart, like the frog in the Tibetan story when confronted with a universe too vast to comprehend. This might explain why it is the intellectual and political elite in our culture that seems most deeply wedded to perpetuating the materialist view of reality. The UFO abduction phenomenon, which strikes at the heart of the Western paradigm and reveals us to be utterly *without* control, is more readily accepted at the grassroots level than by the culturally sophisticated or most intellectually advanced among us. For it is, to a large degree, the scientific and governmental elite and the selected media that it controls that determine what we are to believe is real, for these monoliths are the principle beneficiaries of the dominant ideology.

This "politics of ontology" (Mack 1992) is then the primary arena in which the reality and significance of the UFO abduction phenomenon must be confronted. Before its potential meaning for our individual and collective lives can be realized it has to be taken seriously and moved out of the sensationalizing tabloids into the mainstream of the society so that the sophisticated media is free to give up their supercilious tone. For our own government and other governments around the world the abduction phenomenon presents a special problem. It is, after all, the *business* of government to protect its people, and for officials to acknowledge that strange beings from radar-defying craft can, in seeming defiance of the laws of gravity and space/time itself, invade our homes and abduct our people creates particular problems. This may explain why government policy in relation to UFOs has been, from the beginning, so confusing, a kind of garbled mixture of denial and cover-up that only fuels conspiracy theories.

There are other political implications of the abduction phenomenon. Politics, local, national, and international, is, after all, a game of power. We seek power to dominate, control, or influence a sphere of action. But the abduction phenomenon by its demonstration that control is impossible, even absurd, and its capacity to reveal our wider identity in the universe invites us to discover the meaning of our "power" in a deeper, spiritual sense. Ethnonational conflict, which derives ultimately from the fact that we define ourselves exclusively in parochial regional terms (what Erik Erikson called "pseudospeciation") is the source of prodigious suffering and represents a vast threat to human survival. The more global, even cosmically, interconnected identity that is implicit in the UFO abduction phenomenon, might, at least, offer a distraction from our interminable struggles for ownership and dominance of the earth. At best it could draw us out of ourselves into potentially infinite cosmic adventures. But all this depends on taking the phenomenon and its implications seriously.

The economic implications of the abduction phenomenon are inseparable from the political ones. The loss of a sense of the sacred, the devaluation of intelligence and consciousness in nature beyond ourselves, has permitted the stronger among us to exploit the earth's resources without regard to future generations. Growth without restraint has become an end in itself, as the reports of economic "indicators" endlessly intone, ignoring the inevitable collapse that cannot be far off if the multiplication of the human population continues unchecked and the pillaging of the earth does not stop. Furthermore, if the acquisitive impulse (euphemistically called "market forces") is not controlled, inequities in the distribution of food and other goods that do remain may deepen, giving rise to potential chaos and war without limits. The UFO abduction phenomenon does not speak directly to this issue. It does not, *cannot*, "save" us. But, as will be discussed shortly, it seems to be intricately connected with the nature of human greed, the roots of our destructiveness, and the future consequences of our collective behavior. For the abductees, and the rest of us if we pay attention, the encounters are profoundly enlightening in the fullest sense.

The UFO abduction phenomenon presents a particular problem for some organized religions. From the beginnings of history groups of human beings, recognizing the power and potential perils of spirit forces "out there," have taken upon themselves the task of guiding us through the "ultimate matters" (Zock 1990) of life. Religious leaders instruct us in the nature of God, and determine for us what spirit beings or other entities may exist in the cosmos. The Catholic Church in the Middle Ages, for example, in its zeal to impose a particular sort of monotheism based on the Trinity, quite ruthlessly suppressed the nature-worshipping polytheism of much of Europe.

There can be little place, especially within the Judeo-Christian tradition, for a variety of small but powerful homely beings who administer an odd mixture of trauma and transcendence without apparent regard for any established religious hierarchy or doctrine. It is one thing to acknowledge that "spirit" resides in the universe and "we are not alone." It is quite another for "spirit" to show up in such odd and threatening form, created partially in our own image. At best, this would seem puzzling and difficult to integrate. At worst, to the polarizing perception of Christian dualism these dark-eyed beings must seem to be the playmates of the Devil (Downing 1990). Eastern religious traditions, such as Tibetan Buddhism, which have always recognized a vast range of spirit entities in the cosmos, seem to have less difficulty accepting the actuality of the UFO abduction phenomenon than do the more dualistic monotheisms, which offer powerful resistance to acceptance.

ALIEN INTERVENTION AND HUMAN EVOLUTION

WHAT MIGHT THIS BE ABOUT?

It is difficult to ignore the fact that the UFO abduction phenomenon is taking place in the context of a planetary crisis of major proportions. Human power and greed, made invincible by technologies that are ravaging the earth's environment, are bringing the planet's biosystems to the edge of collapse. Positive efforts to arrest this process are evident at every level, but the overall destructive trend continues. Abductions seem to be concerned primarily with two related projects: changing human consciousness to prevent the destruction of the earth's life, and a joining of two species for the creation of a new evolutionary form.

CHANGING CONSCIOUSNESS IN RELATION TO THE EARTH'S ECOLOGY

Nothing in my work on UFO abductions has surprised me as much as the discovery that what is happening to the earth has not gone unnoticed elsewhere in the universe. That the earth itself, and its potential destruction, could have an effect beyond itself or its own environment was altogether outside the worldview in which I was raised. But it would appear from the information that abductees receive that the earth has value or importance in a larger, interrelated cosmic system that mirrors the interconnectedness of life on earth. The alien abduction phenomenon represents, then, some sort of corrective initiative.

Anne, the abductee to whom I referred earlier and whose case is not included in this book, learned from her experiences that "the whole universe is self-correcting, because if one part of the universe can be . . . like a feedback machine, the whole thing has to be self-correcting like a feedback machine." She likened the universe to a tapestry. "It's all connected. If you take one part of the tapestry, and you put a hole in it or you rend it, you wreck the parts that are next to it. If you take a thread out, the threads that are next to it all get bumped and jostled about so you've got to correct it . . . If you make a mess in one part of the universe," she continued, "you jostle the next part over, and the part that's able to move in or to adjust will do so."

Virtually every abductee receives information about the destruction of the earth's ecosystem and feels compelled to do something about it. But, as we have seen in the cases presented here, this is not received merely cognitively, like a lecture. Abductees experience powerful images of vast destruction, with the collapse of governmental and economic infrastructures and the total pollution and desertification of the planet. This knowledge is felt profoundly in their bodies, and I have been greatly moved as they sob on the couch and experience heartache so intense that they can barely bring themselves to speak of it. It is the kind of knowledge that must be translated into action. Writer and futurist Jean Huston, at the Congress of the World Parliament of Religions in Chicago in September 1993, commented that all myths begin with a form of betrayal. Perhaps the human betrayal of the earth itself is giving rise to a new myth of interspecies relationship and creation.

The Hybrid "Program"

The pioneering work of Budd Hopkins and David Jacobs has shown what is amply corroborated in my cases, namely that the abduction phenomenon is in some central way involved in a breeding program that results in the creation of alien/human hybrid offspring. Comparing the experiences of various abductees it appears that during the abductions sperm are forcibly taken from men and ova from women, after which this germ plasm is brought together and altered. This process is called "genetic" by abductees and investigators, but we have no actual evidence of this. The altered conceptus is reinserted in the uterus during a subsequent abduction, allowed to gestate for some weeks, and then removed. After this the hybrid fetuses are "incubated" in tanks or cylinders (as shown in Catherine's case and drawing) until the fetuses are old enough to live outside of them in some sort of room on the ships. Periodically the abductee mothers and fathers are brought to see the hybrid offspring and encouraged to hold and love them, which is one of the most disturbing aspects of the whole process. For the abductees are naturally filled with conflict at the prospect of forming a deep bond with an odd offspring that they can only see rarely at the pleasure of the alien beings.

Despite their resentment of the forced and traumatic nature of this process, virtually all the abductees with whom I have worked come to accept their participation in this program. I believe that "identification with the aggressor," as is sometimes suggested, is a much too simple explanation of the abductees' attitude toward the hybrid process. Both men and women come to feel despite their anger that they are taking part—even that they have chosen to participate—in a process that is life creating and life-giving. Furthermore, for most abductees the hybridization has occurred simultaneously with an enlightenment imparted by the alien beings that has brought home forcibly to them the failure of the human experiment in its present form. Abduction experiencers come to feel deeply that the death of human beings and countless other species will occur on a vast scale if we continue on our present course and that some sort of new life-form must evolve if the human biological and spiritual essence is to be preserved. They do not generally question why the maintenance of human life must take such an odd form.

I ask abductees repeatedly to explain to me why, if these hybrid beings are to represent the species that will repopulate our planet after the prophesied environmental holocaust, they appear so listless and frail on the ships, hardly vital stock to perpetuate the human or any other race. Only Peter has offered an answer by insisting that to him the hybrids do not appear listless, but have a vitality of their own. Jerry in a recent abduction has seen quite beautiful, angel-like, young adult hybrids with porcelain skin; their function was to show her on a screen the earth's inevitable future if the present course is not changed. For me the future role that the hybrid offspring are being prepared to play represents one of the most puzzling aspects of the entire abduction phenomenon.

Those investigators who perceive the UFO abduction phenomenon from an adversarial perspective tend to interpret its meaning one-sidedly. The aliens are using us, the argument goes, for their own purposes, replenishing their genetic stock at our expense after some sort of holocaust on their own planet. If they make us feel that there is something worthwhile about the whole process, this is the result of deception. I would not say that the aliens never resort to deception to hide their purposes, but the above argument is, in my view, too narrow or linear an interpretation.

My own impression is that we may be witnessing something far more complex, namely an awkward joining of two species, engineered by an intelligence we are unable to fathom, for a purpose that serves both of our goals with difficulties for each. I base this view on the evidence presented by the abductees themselves.

First, many abductees, for example, Scott and Peter, become aware in the course of our work that the alien/human union serves the reciprocal needs of each species, completing aspects of the identity of each that are missing or have been lost in the course of evolution. Roughly speaking, the aliens have remained less densely embodied and closer to the creative source in the universe from which human beings have been cut off. Through their interaction with the abductees they bring them (and all of us potentially) closer to our spiritual cosmic roots, return us to the divine light or "Home," a "place" (really a state of being) where secrets, jealousy, greed, and destructiveness have no purpose. The aliens, on the other hand, long to experience the intense emotionality that comes with our full embodiment. They are fascinated with our sensuality, our warmth, our capacity for eroticism, and deep parental affection, and they seem to respond to openhearted love. They act at times like love-starved children. They delight in watching humans in all sorts of acts of love, which they may even stage as they stand around watching and chattering as the abductees perform them.

Second, the human/alien relationship itself evolves into a powerful bond. Despite their resentment and terrorization, the abductees may feel deep love toward the alien beings, especially toward the leader figures, which they experience as reciprocated, despite the cold and businesslike way the abductions themselves are conducted. The aliens may be perceived as true family, having protected the experiencers from human depredations, disease, and loss. The leader may be seen as a familiar, loving, and wise figure, known by the experiencer since childhood and, ultimately, forgiven for the change from playfulness to a more serious or grim purpose that occurs in the abductions when puberty is reached and the hybrid-creating process begins.

Anne describes an amusing incident that occurred at the end of one of her abductions in which an alien being seemed unwittingly to betray his affection for her. She had been returned to her bed at home and was, presumably, asleep. But she awoke to see one of the beings "looking at me with a loving look . . . just looking at my face, looking at my features . . . looking at my eyes, looking at them with such emotion, such love . . . When he found out I was awake," she said, "he freaked out . . . His eyes [crinkled], they [got] smaller. I think his mouth did open, but it was like a silent kind of, he didn't really make any noise when he screamed." She saw in his face "a hint of fine, long bone structure." When she looked into his eyes the expression was one of "Oh, my God, you're awake and I'm in a shitload of trouble," and "then he went whoosh out the window . . . floating horizontal, like lengthwise . . . I think he was like a doctor-in-training," Anne observed, and "he wasn't supposed to take it personally . . . an alien would be in trouble for waking somebody up because they don't want us to know that they're around."

The connection that human beings experience through looking into the eyes of the aliens seems to be a central feature of the acknowledgment of the existence of the beings and the establishment of the bond itself. Abductees have repeatedly described to me a loving, totally engulfing feeling they experience when they look into these huge, black, all-knowing eyes. This contact is "fifty times more powerful" than any human-to-human connection, one woman said to me. For Peter and other men the eye connection serves to restore a lost "brotherhood," sundered when we were both—humans and aliens—separated from a common primordial source. The experience can be one of profound, total, even blissful merging. Some abductees have experienced actual alien-human sexual connection or partnership—Joe, for example, in his alien identity and Peter as a human connecting powerfully with an alien partner with whom he is parenting hybrid offspring.

It needs to be stressed that we do not know if any of the above phenomena exist literally on the purely material plane of reality, despite the apparent physical manifestations, such as perceived pregnancies and hybrid babies. The aliens stress the evolutionary aspect of the species-joining process, the repopulation of the earth subsequent to a total environmental collapse. But all this may be in some mysterious way a kind of play of consciousness, embodied in some sense, yet at the same time separate from our physical bodies. It could all be "educational," a kind of mythic drama, intended by a transcendent intelligence to move our being to a higher level. Or the merger of the alien and human species might be more literally real, its very awkwardness the result of the prodigious difficulty of bringing together a densely embodied race such as ours with more nearly spiritual entities like the alien beings.

ANOTHER INTELLIGENCE AT WORK: WHY DOES THIS SEEM SO MOMENTOUS?

At the end of one of her regressions Anne said, "Something else is interested in us that we don't want to know about. This is happening. It's not just a happy little dream where you can feel like you're important. This is really a responsibility, and things that you don't want to see happen are going to happen." Yet within our culture, at least for those who determine for us what we are to accept as real, the very existence of this other intelligence, this "something else" that is "interested in us," is difficult to accept. Why should this be so, since every culture from the beginning of recorded time and throughout most of the world, even in our own time, has accepted the existence of other intelligences in the universe?

I have discussed already some of the reasons why the reality of the abduction phenomenon has been so difficult for our culture to accept—the material and philosophical vested interests, for example, attached to the Western worldview. But I believe there is a core belief in our culture that is violated by the alien abduction phenomenon, namely the total separation of the spirit and the physical world. We have made that gulf inviolate, relegating to religion the spirit (subjective) world and assigning to science the material (objective) domain. We simply do not know what to do with a phenomenon that crosses that sacred barrier. It shocks the foundations of our belief structure. Our minds have no place to put such a thing.

The Dalai Lama once pointed out that the devastation of the planet's ecology was destroying not only the habitat of plants and animals but the realms in which the spirits reside as well. Perhaps this has left them no choice but to manifest in our world, to appear to us in the only "language" that remains to us, the language of the physical world. In the context of the planet's crisis they have had no choice, but must find some way, however difficult, to come to us. Perhaps our consciousness has become so atrophied that we are simply unable, on our own, to be open to the spirit world. Could it be that the break of our carefully crafted psychological boundaries, the dramatic reopening to a world from which we have distanced ourselves, this shocking reanimating of the senses whereby we might know the spirit world, has made the alien abduction phenomenon seem so hard to believe? Perhaps we have created the spiritual conditions that have made this necessary.

FURTHER IMPLICATIONS FOR HUMAN CONSCIOUSNESS

With the opening of consciousness to new domains of being, abductees encounter patterns and a design of life that brings them a profound sense of interconnectedness in the universe. For Dave this took the form of meaningful coincidences, similar to what Jung called synchronicities, which he seemed to discover everywhere he went, especially after he came to terms with his abduction experiences. Thoughts and ideas may appear more organically connected with the physical world during abduction experiences than ordinarily seems apparent in everyday life. Metaphor becomes tangible or reified. Peter's "fourth step" through the wall of a Nantucket house during his August 1992 abduction was both a literal physical action *and* at the same time powerfully symbolic of his decision to accept his passage from one plane of reality to another. For Arthur the illuminated thread or string which, quite literally, was the source of energy that carried him to the spaceship when he was nine came to stand quite concretely for the ties that have connected him with others during his life.

A forty-one-year-old health care executive spoke to me of large tubes through which he passed during one of his abductions "into the next plane where there was this light . . . It was like a birth because there was fluid" in the tubes. Going through the tube was both literally and metaphorically a "rebirth," a passage "through the veil to the other side." Catherine in an abduction subsequent to the ones described in my chapter about her also spoke of a tube or tunnel through which she and others passed from a spirit plane outside space/time back to the embodied physical state on Earth.

Abduction experiences also open the consciousness of abductees as I work with them to cycles of birth and death that are reminiscent of the Tibetan transitional realities or *bardos* (Sogyal Rinpoche 1992). This is most clearly illustrated in the past life experiences that are emerging increasingly in our sessions as I have become willing to listen to them. These reports suggest that individual consciousness may have its own line of development, separate from the body. In our sessions abductees such as Dave, Joe, Catherine, and Eva have told with great feeling of periods of embodiment, followed by deaths quite vividly reexperienced. They speak of a return to or diffusion into some sort of primal or universal creative consciousness or source, and then a later rebirth through a woman's womb into a new embodiment on Earth.

The transition at the time of bodily death seems to include, as Joe recounted, a feeling of being literally lighter and thinner. "It's good to be back," he said. "This is much more real." Eva reported "going up, expanding, joy." She went "into a white, gold light" and saw a dove released from a cage. "That's my soul," she said. In certain cases the alien beings, who seem to reside or belong in this fluid dimension (although they periodically become embodied to some degree on Earth), seem to have been with the abductees through more than one lifetime. As discussed in considerable detail in Dave's and Joe's cases, it is possible to trace a development or evolution of consciousness in the sequence of lives that they have experienced over time.

It is not necessary to postulate that a past life identity belongs literally to the individual abductee in the same way that our bodies are only ours while we inhabit them. As biologist Rupert Sheldrake has suggested, it is possible that there exists a kind of eternal collective memory on which we all may draw. One might, Sheldrake suggests, "tune in to particular people in the past who are now dead, and, through morphic resonance, pick up memories of past lives." This does not prove, he says, that "you were that person" (Sheldrake 1992). This idea is consistent with the observation that the psyche or memory of the abductees seems to be able to travel, especially during the opening of consciousness that occurs in our sessions, to wherever or whenever the evolutionary requirement of the moment seems to take it.

THE PARADIGM SHIFT

Needless to say none of this makes much sense within the modern worldview brought to us by Western science, whose "governing assumption," in philosopher Richard Tarnas's words, is that "any meaning the human mind perceives in the universe does not exist intrinsically in the universe but is projected onto it by the human mind" (Tarnas, in progress). To Tarnas "this complete voiding of the cosmos, this absolute privileging of the human" is perhaps the "ultimate anthropocentric projection, the most subtle yet prodigious form of human self-aggrandizement" and represents an intellectual "hubris of cosmic proportions."

The experiences recounted by the abductees with whom I have worked during the past four years constitute, I think, a rich body of evidence to support the idea that the cosmos, far from being devoid of meaning and intelligence is, to quote Tarnas again, "informed by some kind of universal intelligence," an intelligence "of scarcely conceivable power, complexity, and aesthetic subtlety yet one to which the human intelligence is akin, and in which it can participate." I am reminded, particularly, of Carlos here. Other evidence, including the thoroughly documented near-death experiences and extraordinarily intricate and symbolic crop formations appearing all over the world, provide additional indications, if we will allow ourselves to realize their implications, of various expressions of intelligence in a universe that is reaching toward us.

As I come to the end of this story I cannot help wondering what it might take to bring about the shift in consciousness, the change of paradigm that is implicit in what the abductees have undergone. It would appear that what is required is a kind of cultural ego death, more profoundly shattering (a word that many abductees use when they acknowledge the actuality of their experiences) than the Copernican revolution which demonstrated that the earth, and therefore humankind, did not reside at the center of the cosmos. UFO abductions and related phenomena suggest first that humans are not the preeminent intelligent beings in a universe more or less empty of conscious life. But abductees' experiences also indicate that we are participating in a cosmos that contains intelligent beings that are far more advanced than we are in certain respects and have the power to render us helpless for purposes we are only just beginning to fathom.

Each abductee appears to me like a pioneer on a hero's journey. For as they undergo their own ego-destroying terror, and allow us to know about their experiences, their consciousness opens to the existence of unknown dimensions of the cosmos and the human psyche, which themselves appear increasingly to be profoundly interwoven. My own work with them, perhaps, has enabled them to acknowledge their experiences, and come to terms with the importance of the gifts they have to offer.

I am often asked why, if UFOs and abductions are real, the spaceships do not show up in more obvious form. "Why don't they land on the White House lawn?" is the reigning cliché. The most popular answer to this question among those who take the phenomenon seriously is that the aliens do not *dare* to manifest themselves more directly. Government leaders would panic, might attack them, and surely would not know how to avoid scaring the rest of us.

I believe that there is a better answer to this question, one that is more consistent with the information contained in this book. The intelligence that appears to be at work here simply does not operate that way. It is subtler, and its method is to invite, to remind, to permeate our culture from the bottom up as well as the top down, and to open our consciousness in a way that avoids a conclusion, that is different from the ways we traditionally require. It is an intelligence that provides enough evidence that something profoundly important is at work, but it does not offer the kinds of proof that would satisfy an exclusively empirical, rationalistic way of knowing. It is for us to embrace the reality of the phenomenon and to take a step toward appreciating that we live in a universe different from the one in which we have been taught to believe.

There is considerable debate among investigators of the abduction phenomenon about whether, given the harsh and often terrorizing methods the aliens employ, the intelligence at work might be evil or mean us harm. Obviously in considering this question one enters a realm of interpretation that goes beyond the available evidence. We have only sketchy knowledge, for example, of what the hybrid "program," which seems to lie at the center of the abduction phenomenon, is really about. Yet my overall impression is that the abduction process is not evil, and that the intelligences at work do not wish us ill. Rather, I have the sense—might I say faith—that the abduction phenomenon is, at its core, about the preservation of life on Earth at a time when the planet's life is profoundly threatened.

What is the vision of the possible human future that the abductees have brought back to us from their journeys? UFO abductions have to do, I think, with the evolution of consciousness and the collapse of a worldview that has placed humankind at a kind of epicenter of intelligence in a cosmos perceived as largely lifeless and meaningless. As we, like the abductees, permit ourselves to surrender the illusion of control and mastery of our world, we might discover our place as one species among many whose special gifts include unusual capacities for caring, rational thought, and self-awareness. As we suspend the notion of our preeminent and dominating intelligence, we might open to a universe filled with life-forms different from ourselves to whom we might be connected in ways we do not yet comprehend.

The connecting principle, the force that expands our consciousness beyond ourselves, appears to be love. In the discovery of a fundamental, loving interconnectedness, we might overcome the sense of fragmentation and evolve toward wholeness as individuals, family members, and planetary citizens. From this perspective, the earth would no longer be simply a marketplace, its lands and resources divided among competing human groups. The earth would become the jewel in the crown of our being, the place where we experience once again our connection with a cosmic Source from which we have become too separate. As our psyches open, we could abandon the dualistic thinking that has divided mind from matter and the physical from the spiritual world.

The alien beings have come to the abductees from a source that remains unknown to us. We still do not fully grasp their purposes or their methods. It seems clear however, that "they" have had to come to "us," appearing in material form so that we might know them. Some have speculated that the alien beings have mastered time travel and come to us from the future. Sometimes they even communicate that this might be so. We do not know. But the guiding or regenerative myth of the abduction phenomenon offers a new story for a world that has survived many holocausts and may yet be deterred from a final cataclysm. The abduction phenomenon, it seems clear, is about what is *yet to come*. It presents, quite literally, visions of alternative futures, but it leaves the choice to us.

REFERENCES

Barron, J. 1991. Mysterious Fireballs Across the Sky Leaves a Trail of Questions. *New York Times*, 7 March: A20.

Basterfield, K. In press. Abductions and Paranormal Phenomena. In *Proceedings of the 1992 Abduction Study Conference at MIT*.

Basterfield, K., and Bartholomew, R. E. 1988. Abductions: The Fantasy-Prone Hypothesis. *International UFO Reporter* 13(3):9–11.

Bloecher, T., Clamar, A., and Hopkins, B. 1985. Summary Report on the Psychological Testing of Nine Individuals Reporting UFO Abduction Experiences. In *Final Report on the Psychological Testing of UFO "Abductees."* Mt. Rainier, Md.: Fund for UFO Research.

Bullard, T. E. 1987. *UFO Abductions: The Measure of a Mystery*, 2 vols. Mt. Rainier, Md.: Fund for UFO Research.

Carpenter, J. 1993. Multiple Participant Abductions. Paper presented at the Seattle UFO Research Conference, 17–18 July.

Chandler, D. 1991. Stargazers Thrill to See Meteor Dart Across Sky. *Boston Globe*, 7 March.

Chiang, H. 1993. UFO Sightings and Research in Modern China. In *MUFON 1993 International UFO Symposium Proceedings*, edited by W. Andrus, Jr., and I. Scott, 41–58. Seguin, Tex.: Mutual UFO Network.

Clark, J. 1990. *UFOs in the 1980s*. Detroit, Mich.: Apogee Books.

Clark, J. 1991. Airships: Parts I, II. *International UFO Reporter* 16(1):4–23; 16(2):20–21, 23, 24.

Clark, J., and Coleman, L. 1975. *The Unidentified: Notes Toward Solving the UFO Mystery*. New York: Warner Books.

Crawford, F. 1993. Controversial Correlations and the Question of Consciousness. Paper presented at Ozark UFO Conference, Eureka Springs, Ark., 2–4 April.

The Dalai Lama. 1992. Conversations at Dharamsala Meetings, 15–17 April, in Dharamsala, India.

Dean, G. In press. Comparisons of Abduction Accounts with Ritual Maltreatment. In *Proceedings of the 1992 Abduction Study Conference at MIT*.

Downing, B. H., Ph.D. 1990. E.T. Contact: The Religious Dimension. In *MUFON 1990 International UFO Symposium Proceedings*, edited by W. Andrus, Jr., 45–60. Seguin, Tex.: Mutual UFO Network.

Druffel, A. 1991. "Missing Fetus" Case Solved. *MUFON UFO Journal* 283:8–12.

Druffel, A. In press. Resistance Techniques Against UFO Abduction. In *Proceedings of the 1992 Abduction Study Conference at MIT*.

Druffel, A., and Rogo, D. S. 1980. *The Tujunga Canyon Contacts*. Englewood Cliffs, N.J.: Prentice Hall.

Eliade, M. 1957. *Myths, Dreams and Mysteries*. New York: Harper and Row.

REFERENCES

Eliade, M. 1965. *Mephistopheles and the Androgyne.* New York: Sheed and Ward.

Erikson, E. H. 1969. *Gandhi's Truth.* New York: W. W. Norton and Company.

Everett, M., Mack, J. E., and Oresick, R. 1993. Toward Greening in the Executive Suite. In *Environmental Strategies for Industry: International Perspectives on Research Needs and Policy,* edited by K. Fischer and J. Schot, 63–78. Washington, D.C.: Island Press.

Fowler, R. E. 1993. *The Allagash Abductions: Undeniable Evidence of Alien Intervention.* Tigard, Ore.: Wildflower Press.

Fowler, R. E. 1979. *The Andreasson Affair.* Englewood Cliffs, N.J.: Prentice Hall.

Frankel, F. E. 1993. Adult Reconstruction of Childhood Events in the Multiple Personality Literature. *American Journal of Psychiatry* 15(6):954–58.

Fuller, J. G. 1966. *The Interrupted Journey.* New York: Dial Press.

Ganaway, G. K. 1989. Historical Versus Narrative Truth: Clarifying the Role of Exogenous Trauma in the Etiology of MPD and Its Variants. *Dissociation* II(4):205–20.

Goodenough, W. H. 1986. Sky World and This World: The Place of Kachaw in Micronesian Cosmology. *American Anthropologist* 88(3):551–68.

Grof, S. 1985. *Beyond the Brain: Birth, Death, and Transcendence in Psychotherapy.* Albany: State University of New York Press.

Grof, S. 1988. *Adventures in Self-Discovery.* Albany: State University of New York Press.

Grof, S. 1992. *The Holotropic Mind: The Three Levels of Human Consciousness and How They Shape Our Lives.* San Francisco: HarperCollins.

Haisell, D. 1978. *The Missing Seven Hours.* Markham, Ont.: PaperJacks.

Herman, J. 1992. *Trauma and Recovery.* New York: Basic Books

Hind, C. 1993. Abductions in Africa: Worldwide Similarities. In *MUFON 1993 International UFO Symposium Proceedings,* edited by W. Andrus, Jr., and I. Scott, 16–25. Seguin, Tex.: Mutual UFO Network.

Hopkins, B. 1981. *Missing Time: A Documented Study of UFO Abductions.* New York: Richard Marek Publishers.

Hopkins, B. 1987. *Intruders: The Incredible Visitations at Copley Woods.* New York: Random House.

Hopkins, B. 1990. A Special Report for Members of IF: The Ongoing Problem of Deception in UFO Abduction Cases. New York: Intruders Foundation.

Hopkins, B. 1992. The Linda Cortile Abduction Case. Parts I, II. *MUFON UFO Journal* 293 (September):12–16; 296 (December):5–9.

Hopkins, B. In press. A Doubly Witnessed Abduction. In *Proceedings of the 1992 Abduction Study Conference at MIT.*

Hopkins, B., Jacobs, D., and Westrum, R. 1991. *Unusual Personal Experiences: An Analysis of the Data from Three National Surveys.* Las Vegas,: Bigelow Holding Corporation.

Howe, L. M. 1989. *An Alien Harvest: Further Evidence Linking Animal Mutilations and Human Abductions to Alien Life Forms.* Littleton, Colo.: Linda Moulton Howe Productions.

Jacobs, D. 1992. *Secret Life: Firsthand Accounts of UFO Abductions.* New York: Simon and Schuster.

Jacobson, E. In press. Dissociative Phenomena as a Context for Research and Psychotherapy of Abduction Experiences. In *Proceedings of the 1992 Abduction Study Conference at MIT.*

Kuhn, T. 1962. *The Structure of Scientific Revolutions.* 2d ed., enl. Chicago: University of Chicago Press.

Laibow, R. L. 1989. Dual Victims: The Abused and the Abducted. *International UFO Reporter* 14(3):4–9.

Lorenzen, C., and Lorenzen, J. 1976. *Encounters with UFO Occupants*. New York: Berkley Medallion Books.

Lorenzen, C., and Lorenzen, J. 1977. *Abducted!: Confrontations with Beings from Outer Space*. New York: Berkley Medallion Books.

Mack, J. E. 1992. Helping Abductees. *International UFO Reporter* 17(4):10–15, 20.

Mack, J. E. 1992. The Politics of Ontology. In *Center Review, A Publication of the Center for Psychology and Social Change* 6(2).

Mack, J. E. 1993. Nonordinary States of Consciousness and the Accessing of Feelings. *Human Feelings: Explorations in Affect Development and Meaning*, edited by Steven L. Ablon, Daniel Brown, Edward J. Khantzian, and John E. Mack, 357–71. Hillsdale, N.J.: The Analytic Press.

Mack, J. E. In press. Why the Abduction Phenomenon Cannot be Explained Psychiatrically. In *Proceedings of the 1992 Abduction Study Conference at MIT*.

McKenna, T. 1991. *The Archaic Revival*. San Francisco: HarperCollins.

Martin, J. J. 1993. The Astounding UFO Experience in Puerto Rico. Paper presented at the MUFON 1993 International UFO Symposium, 2–4 July, in Richmond, Va.

Meadows, D.H., Meadows, D.L., and Randers, J. 1992. *Beyond The Limits: Confronting Global Collapse, Envisioning a Sustainable Future*. Post Mills, Vt.: Chelsea Green Publishing Company.

Miller, J. In press. Medical Procedural Differences: Alien vs. Human. In *Proceedings of the 1992 Abduction Study Conference at MIT*.

Miller, J., and Neal, R. In press. Lack of Proof for Missing Fetus Syndrome. In *Proceedings of the 1992 Abduction Study Conference at MIT*.

Moura, G. In press. The Abduction Phenomenon in Brazil. In *Proceedings of the 1992 Abduction Study Conference at MIT*.

Neal, R. 1992. Medical Explanations, Not "Alien." Part II. *UFO* 7(1):16–20.

Noyes, R. 1990. "Abduction, the Terror that Comes." In *The UFO Report*, edited by Timothy Good, 80–101. London: Sidgwick and Jackson.

Parnell, J. O., 1986. *Personality Characteristics on the MMPI, 16PF and ACL of Persons Who Claim UFO Experiences*. Ph.D. dissertation, University of Wyoming, University Microfilms International, order no. DA 8623104.

Parnell, J. O., and Sprinkle, R. L. 1990. Personality Characteristics of Persons Who Claim UFO Experiences. *Journal of UFO Studies* 2:45–58.

Pazzaglini, M. 1992. Paper presented at Dharamsala Meetings, 15–17 April, in Dharamsala, India.

Pritchard, D. E. 1992. Physical Analysis of Purported Alien Artifacts. In *Experienced Anomalous Trauma: Physical, Psychological and Cultural Dimension*, edited by R. L. Laibow, B. N. Sollad, and J. P. Wilson. New York: Brunner/Mazel.

Ring, K. 1992. *The Omega Project: Near Death Experiences, UFO Encounters, and Mind at Large*. New York: William Morrow and Company.

Ring, K., and Rosing, C. J. 1990. The Omega Project: A Psychological Survey of Persons Reporting Abductions and Other UFO Encounters. *Journal of UFO Studies* n.s. 2:59–98.

Rodeghier, M., Goodpaster, J., and Blatterbauer, S. 1991. Psychosocial Characteristics of Abductees: Results from the CUFOS Abduction Project. *Journal of UFO Studies* n.s. 3:59-90.

Rogo, D. S., ed. 1980. *UFO Abductions: True Cases of Alien Kidnappings*. New York: Signet.

Rojcewicz, P. 1991. Fairies, UFOs and Problems of Knowledge. In *The Good People: New Fairylore Essays*, edited by Peter Navaez, 479–514. New York: Garland Publishing.

Rojcewicz, P. 1992. Paper presented at Dharamsala Meetings, 15–17 April, in Dharamsala, India.

Sheldrake, R. 1992. Morphic Resonance and Collective Memory. Paper presented at the International Transpersonal Association Conference, Prague, Czechoslovakia, June.

Slater, E. 1985. Conclusions on Nine Psychologicals. In *Final Report on the Psychological Testing of UFO "Abductees."* Mt. Rainier, Md.: Fund for UFO Research.

Sogyal Rinpoche. 1992. *The Tibetan Book of Living and Dying*. San Francisco: HarperCollins.

Spanos, N. P., Cross, P. A., Dickson, K., and DuBreuil, S. C., 1993. Close Encounters: An Examination of UFO Experiences. *Journal of Abnormal Psychology* 102(4):, 624–32.

Spiegel, D., and Cardena, E. 1991. Disintegrated Experience: The Dissociative Disorders Revisited. *Journal of Abnormal Psychology* 100(3):366–78.

Stone-Carmen, J. In press. A Descriptive Study of People Reporting Abduction by Unidentified Flying Objects (UFOs). In *Proceedings of the 1992 Abduction Study Conference at MIT*.

Story, R. D., ed. 1980. *Encyclopedia of UFOs*. Garden City, N.Y.: Doubleday Books.

Tarnas, R. *Cosmos and Psyche: Intimations of a New World View*, work in progress to be published by Random House.

Tormé, T. 1993. *Fire in the Sky*. Paramount Pictures.

Toulmin, S. E. 1990. *Cosmopolis: The Hidden Agenda of Modernity*. Chicago: University of Chicago Press.

Vallee, J. 1969. *Passport to Magonia*. Chicago: Henry Regnery Company.

Vallee, J. 1988. *Dimensions: A Casebook of Alien Contact*. New York: Ballantine Books.

Walsh, R. and Vaughan, F., eds. 1993. *Paths Beyond Ego: The Transpersonal Vision*. Los Angeles: Jeremy P. Tarcher/Perigee Books.

Walton, T. 1978. *The Walton Experience*. New York: Berkley Books.

Wright, L. 1993. Remembering Satan. Parts I, II. *The New Yorker* 17 May:60–81; 24 May:54–76.

Zimmerman, M. 1993. Why Establishment Elites Resist the Very Idea of UFOs and Reported Alien Abductions. Paper presented at Gulf Breeze UFO Conference, Gulf Breeze, Fla., 22–24 October.

Zock, H. 1990. *A Psychology of Ultimate Concern*. Amsterdam-Atlanta, Ga: Rodopi.

INDEX

The Program for Extraordinary Experience Research, PEER

The Program for Extraordinary Experience Research, PEER, is a nonprofit research and education group dedicated to a deeper understanding of experienced encounters with intelligent nonhuman beings, the phenomenon commonly called "alien abduction." PEER was founded in 1993 by Dr. John E. Mack as a project under the auspices of the Center for Psychology and Social Change.

PEER works with experiencers and the therapeutic community to foster an understanding of these experiences, develop a network of support and education, consider the implications of the phenomenon, and disseminate information to the general public. PEER is interested in exploring and documenting experiences worldwide. We would be happy to hear from individuals who would like to provide us with information or inquire about our program. All information sent to us will be kept confidential. Please do not send us original documents or anything that you wish to have returned.

If you'd like to write to us, please include the form below (or a copy of it) with your letter.

PEER
P.O. Box 382427
Cambridge, MA 02238
USA

Name ______________________________

Address ____________________________

____

____

Telephone (optional) ___________________

Please indicate your area(s) of interest.

I am interested in:

—— providing information about experiences for possible inclusion in your work (Please do not send more than two pages.)
—— helping experiencers in a therapeutic context
—— sharing cross-cultural information about the phenomenon
—— making a donation to the program

Other area(s) of interest ____

____

Would you like to make your name and address available to receive mail?

—— from PEER
—— from other individuals or groups whose work relates to the phenomenon